Urban Wildlife Management

Second Edition

Urban Wildlife Management

Second Edition

Clark E. Adams • Kieran J. Lindsey

CRC Press
Taylor & Francis Group
Boca Raton   London   New York

CRC Press is an imprint of the
Taylor & Francis Group, an **informa** business

CRC Press
Taylor & Francis Group
6000 Broken Sound Parkway NW, Suite 300
Boca Raton, FL 33487-2742

© 2010 by Taylor and Francis Group, LLC
CRC Press is an imprint of Taylor & Francis Group, an Informa business

No claim to original U.S. Government works

Printed in the United States of America on acid-free paper
10 9 8 7 6 5 4 3 2 1

International Standard Book Number: 978-1-4398-0460-5 (Hardback)

Library of Congress Cataloging-in-Publication Data

Adams, Clark E. (Clark Edward), 1942-
   Urban wildlife management / Clark E. Adams and Kieran J. Lindsey. -- 2nd ed.
      p. cm.
   Includes bibliographical references and index.
   ISBN 978-1-4398-0460-5
   1. Urban wildlife management. 2. Urban ecology (Biology) I. Lindsey, Kieran J. (Kieran Jane) II. Title.

QH541.5.C6A33 2010
639.909173'2--dc22   2009020584

Visit the Taylor & Francis Web site at
http://www.taylorandfrancis.com

and the CRC Press Web site at
http://www.crcpress.com

Dedication

We dedicate this book to all those individuals, organizations, and agencies on the front lines, addressing urban wildlife management problems. They represent the unsung heroes of wildlife management who receive little recognition or peer acceptance for their attempts to confront a growing wildlife management phenomenon. They are the futurists, involved in the cutting-edge aspects of human-wildlife interactions in urban environments. In other words, we dedicate this book to all those who realize that urban wildlife management goes far beyond controlling raccoons in garbage cans.

Contents

Preface

We began the adventure of writing this book in 1999 because we were unable to find one publication that addressed all the pertinent issues related to urban wildlife management; rather, information was scattered throughout various books, journal articles, conference proceedings, government documents, websites, data sets, and within the anecdotal tales of our colleagues. Now, with the end of the second journey in sight, we confess that even after going through the process once before, we were *still* rather naïve about the breadth and depth of information sources available on this subject. Neither of us expected to spend two years preparing the first edition of *Urban Wildlife Management* only to turn around and spend another two years preparing the second edition of our book. Invariably, though, in attempting to track down a single piece of information we would stumble across a dozen others, each of which led to still others—all of which made our seemingly simple task of synthesizing the available information into a conceptual rather than in-depth presentation a lengthy and formidable one.

Nevertheless, unearthing, accumulating, and organizing the wealth of information has turned out to be an extremely enjoyable and intellectually stimulating task. The path has led us to well over 500 professional manuscripts and countless articles and stories in the popular media. We examined secondary data sets that contained a wealth of information relevant to the story we wanted to tell in this second edition. We convinced some of our colleagues to share their expertise in urban wildlife research and management in several sections of this book. Even some of our students contributed to the cause. Unlike the initial publication, the second edition contains the results of original research conducted by the authors and/or their students. We continue to discover new information, but at some point one must stop reading and start writing!

It is becoming evident that the challenges and opportunities related to urban wildlife are beginning to take front stage in the wildlife profession. As a result, this book is a much-needed tool for teaching and learning. In fourteen chapters, we examine a range of issues that explain human interactions with wildlife in urbanized environments. We begin with a discussion of the past, present, and future directions of wildlife management in the United States—what we have come to see as the changing landscape of wildlife management. Selected chapters relevant to understanding the presence or absence of wildlife species in urban communities include ecosystem structure and function, urban soils, urban waters, and principles of population dynamics in the context of the impacts of urbanization. Urban habitats and hazards are discussed in terms of two chapters on urban green and gray spaces. The sociopolitical issues of particular importance in urban wildlife management are covered in chapters on the human dimensions of wildlife management, stakeholders, and legal considerations. Special management considerations include three chapters on the ecology and management of selected species that are endangered, introduced, or feral; zoonotic disease management (a new chapter); and overabundant animals, including resident Canada geese (*Branta canadensis*) and urban white-tailed deer (*Odocoileus virginianus*).

The second edition of *Urban Wildlife Management* is a continuation of "the rest of the story," including those issues that are usually left out of manuscripts that focus primarily on how to alleviate the problems associated with nuisance urban wildlife (e.g., M. Conover. 2002. *Resolving Human–Wildlife Conflicts: The Science of Wildlife Damage Management;* J. Hadidian. 2007. *Wild Neighbors: The Humane Approach to Living with Wildlife*). For example, the book provides a basic framework of information that will give the reader an understanding of factors that promote or prevent the presence of wildlife in urban communities.

We were delighted and somewhat surprised to realize that the first edition of *Urban Wildlife Management* was being used as a text or reference document by seventeen colleges or universities. Others who used the first edition included urban wildlife biologists at state departments of natural resources, state and federal agencies, and urban planners and managers in urban areas. Private citizens who have a personal interest in urban wildlife management also purchased this book.

Six peer reviews of the first edition in the *Journal of Wildlife Management, Condor, Human Wildlife Conflicts*, book reviews on Amazon.com, and newspaper editorials were overwhelmingly positive. In addition, the first edition was selected as the 2007 Outstanding Book by the Texas Chapter of The Wildlife Society, and CHOICE Magazine's annual "Outstanding Academic Title List."

The development of both editions of this book would not have been possible without the assistance of other wildlife management professionals and students. Jessica Alderson provided the information for and helped write Perspective Essay 14.1. Even though Sara Ash had to leave us as a coauthor, she provided information and editorial assistance for Chapters 7 and 12. Debra Cowman helped us assimilate the relevant literature on environmental toxicants on wildlife population dynamics in Chapter 6. John M. Davis contributed a lot of information in the first edition that was carried over into the second edition; his excellent contributions include the "biotic communities" section of Chapter 3; urbanization on soil structure and function in Chapter 4; the impacts of stream channelization in Chapter 5; a perspective essay on people's fears of wild places in Chapter 9 and another on weed ordinances in Chapter 11. John also provided many of the photos used in this book. Thanks to Rob Denkhaus and Suzanne Tuttle for their case study on feral hogs (*Sus scrofa*) in Chapter 12. Fran Gelwick and Michael Masser reviewed and edited sections on urban streams and impoundments in Chapter 5. Deb Teachout, DVM, offered support for the development of Chapter 13. Marian Higgins provided the introduction to nature centers in Chapter 7. Ardath Lawson wrote the perspective essay on cemeteries in Chapter 7. Roel Lopez contributed the key deer (*Odocoileus virginianus clavium*) case study in Chapter 12. Robert Meyers conducted an analysis of the Wildlife Services Management Information Systems data set that led to the national and regional overview of the species of most concern and economic impacts of animal damage in Chapter 2. The Quality Deer Management Association (Bogart, GA) was the source of information concerning white-tailed deer densities in each state, as provided in Figure 14.10. Emily Rollison and Sara Ramirez developed a spreadsheet based on a review of over 900 nature centers in the United States that was used to develop a new section for Urban Green Spaces in Chapter 7. Linda Tschirhardt wrote the first draft of Chapter 11 and provided a photo for Chapter 12. We are grateful to each of these individuals for their talents and contributions.

Thanks also to our spouses, Judy Adams and Chet Weiss, for their patience while we spent long hours in front of our respective computers, and when we remained mentally immersed in the book even when we were physically present at home.

Clark E. Adams and Kieran J. Lindsey

The Authors

Kieran Lindsey, Clark Adams

Clark E. Adams is a professor in the Department of Wildlife and Fisheries Sciences at Texas A&M University in College Station, Texas. He has a B.S. in biology and education from Concordia Teachers College, Seward, Nebraska; an M.S. in biology and education from the University of Oregon; and a Ph.D. in zoology from the University of Nebraska – Lincoln. Clark chaired the Conservation Education Committee for The Wildlife Society (TWS), edited the newsletter for the Human Dimensions of Wildlife Study Group, is now a member of the Urban Wildlife Management Working Group, and has chaired many committees for the Texas Chapter of TWS. He is past president of the Texas Chapter of TWS and TWS Southwest Section. Since 1981 he has directed the Human Dimensions in Wildlife Management research laboratory. He and his students have conducted and published many national, regional, and statewide studies on the public's activities, attitudes, expectations, and knowledge concerning wildlife. He developed the degree option in urban wildlife and fisheries management for the Department of Wildlife and Fisheries Sciences, and developed and teaches the senior-level urban wildlife management course. He is coauthor on another book titled *Texas Rattlesnake Roundups*.

Kieran J. Lindsey is the director of the Natural Resources Distance Learning Consortium (NRDLC) at the College of Natural Resources, Virginia Polytechnic Institute and State University (Virginia Tech), in Blacksburg, Virginia. Kieran received her Ph.D. in wildlife biology from Texas A&M University in 2007. She was first introduced to the concept of urban wildlife in 1997 when she became the director at a nonprofit wildlife center in Houston, Texas, and she took to it like Canada geese to golf courses. The center offered a variety of services to the Houston metroplex and surrounding areas, including a wildlife hotline that handled over 20,000 queries annually on every imaginable urban wildlife topic. Kieran expanded the organization's outreach efforts with a regular column on urban wildlife issues for the *Houston Chronicle* newspaper. In 1998 she received a regional Emmy® award from the National Academy of Television Arts and Science as the producer of a feature about the center that aired on Houston Public Television. Following a move to Albuquerque, New Mexico, in 1999 she started a private consulting business, and also served as executive producer, writer, and host of *Wild Things Radio* at KUNM-FM 89.9. In addition to her duties as director of the NRDLC, she teaches graduate distance learning courses on urban wildlife management and human dimensions of natural resource management for Virginia Tech, and serves as the editor of the *Journal of Wildlife Rehabilitation*.

Introduction: A New Wildlife Management Paradigm

In the future, we're all going to be urban biologists.

—Timothy Quinn, Washington Department of Fish and Wildlife (2003)

KEY CONCEPTS

1. There are several lines of evidence that document the need for urban wildlife management.
2. There are similarities and differences between wildlife management in urban and rural habitats.
3. The wildlife profession is not yet prepared to meet the wildlife management challenges of urban environments.

A SNAPSHOT OF THE URBAN WILDLIFE MANAGEMENT LANDSCAPE

Americans may like to think of themselves as a primarily rural nation, but nearly 80 percent of those who dwell in the 48 lower states live in areas classified by the U.S. Census Bureau as *urban*: "a large central place and adjacent densely settled census blocks that together have a total population of at least 50,000." Also known as the "built environment," the urban wildlife management landscape includes places where most of the land is devoted to all things man-made and/or maintained: buildings of all shapes and sizes, manicured lawns and landscaped office parks, cemeteries and vacant lots, strip malls and high schools and warehouse districts. A substantial portion of this land is covered with impervious surfaces, in the form of both structures and pavement. The plant life is mostly native to other parts of the world or highly hybridized, and thus in need of a lot of caretaking, in the form of sprinkler systems, herbicides, pesticides, and fertilizers. Urban in this context includes both the cities and the suburbs. To be absolutely accurate, the focus of this book is on wildlife that lives in human dominated landscapes ... but "urban wildlife" is quick, catchy, and has become the accepted terminology.

Urbanization and the encroachment of humanity into former wild habitats will continue into the foreseeable future. This population shift from rural to urban has changed, and will continue to change, the landscape and the agenda concerning wildlife management. In fact, within the next ten to twenty years, it is entirely possible that urban wildlife will become the dominant focus of wildlife professionals.

Americans may be more urbanized, but they have not lost their curiosity about wildlife and the natural world, both of which are seen as helping to improve quality of life. If you doubt this is the case, consider the fact that in 2006 there were more Americans involved in wildlife watching (71.1 million) than in hunting and fishing combined (33.9 million; U.S. Census Bureau 2006). Even the home-building industry has taken note; promoting close proximity to green space and wildlife has become a common marketing strategy for developers.

Concurrently, there is a growing concern about human-wildlife encounters, especially those perceived to endanger the health and safety of humans and their companion animals. Outbreaks of zoonotic diseases such as West Nile virus, hantavirus, and chronic wasting disease make headlines in both local and national newspapers. The number of nuisance wildlife complaints continues to rise, as does the number of private wildlife control businesses. Clearly, life in the urban wilds is not a return to Eden. Which begs the question—who's tending this garden?

THE NEED FOR A COMPREHENSIVE TREATMENT
OF URBAN WILDLIFE MANAGEMENT

Change often must begin at the grassroots level. The National Institute for Urban Wildlife was the first formal organization of individuals who recognized and wanted to address urban wild-life management issues. In order to start a dialogue the Institute hosted three national symposia on urban wildlife [Chevy Chase, Maryland (1986); Cedar Rapids, Iowa (1990); and Bellevue, Washington (1994)], and two proceedings were published (L.W. Adams and Leedy 1987, 1991). The fourth symposium was held in Tucson, Arizona in 1999 (Shaw, Harris, and Vandruff 2004). Subsequently, the Arbor Day Foundation, in cooperation with the Urban Wildlife Working Group of The Wildlife Society (TWS), took on the task of organizing and hosting biannual national urban wildlife management conferences, beginning in 2001 (Lied Conference Center, Nebraska City, NE). The most recent (2007) was held in Portland, Oregon. A variety of topics were presented at these conferences, ranging from managing wildlife in urban environments, human-wildlife conflicts, public education on urban wildlife, stakeholder recognition, and a host of others. Additionally, while the manuscript for this book was being prepared, the Urban Wildlife Working Group was planning an International Symposium on Urban Wildlife and the Environment, to be held in Amherst, Massachusetts, in June 2009. No records or proceedings have resulted from these meetings thus far.

However, during the 2003 conference, one attendee who represented a city government asked if there was any publication that summarized the issues and complexities of urban wildlife management under one cover. At that time, there was no such document, but the first edition of *Urban Wildlife Management* was published at the end of 2005. Since then, we've received invaluable feedback from colleagues and students, along with requests to include topics that had to be omitted initially due to space limitations.

Ideally, a comprehensive book on urban wildlife management should have been available at least 20 years ago. Lowell Adam's seminal work, *Urban Wildlife Habitats: A Landscape Perspective* (1994), which focused on urban and suburban wildlife habitat, was a first and important step in the right direction. Other authors have addressed individual aspects of urban wildlife management, such as urban ecology and sustainability (Whiston-Spirn 1985; Platt, Rowntree, and Muick 1994), human dimensions (Decker, Brown, and Siemer 2001; Manfredo et al. 2008), human-wildlife conflicts (Hadidian 2007; Conover 2002), urban wildlife law (Rees 2003), urban planning (Tyldesley 1994), and even urban species identification (Landry 1994; Shipp 2000). The first edition of *Urban Wildlife Management* filled a void, but the field continues to grow and evolve. We've done our best to see that the current edition reflects these changes without giving short shrift to the basic ecological principles that are the underpinning of urban wildlife management.

This is not a book on how to address specific urban wildlife conflict issues—other authors (e.g., Conover 2002; Hadidian 2007) have addressed these issues admirably. Rather than providing a prescription for short-term, reactive methods that address symptoms, the information included here will provide professionals in wildlife management and related fields with the information required to set and achieve long-term, proactive management goals that focus on the root cause of urban wildlife management challenges.

As before, this edition can be used as a textbook for both undergraduate and graduate courses on urban wildlife management, urban ecology, or even urban planning. Often, if a textbook is not available, much-needed college courses will not be taught. Urban wildlife management is not a traditional component in university curricula for wildlife biologists, but more courses are available now than ever before. The first edition of this book was adopted by seventeen different universities in the United States. In some cases the book was adopted for use in existing urban wildlife classes, but in other cases new classes were designed with *Urban Wildlife Management* in mind. We're hopeful the second edition will foster a continuation of this trend.

There are two key questions a teacher has to answer when preparing a class: (1) "What am I going to teach?" and (2) "How am I going to teach it?" Applying these decisions to an undergraduate university class on urban wildlife management without a textbook can be a formidable task. There is a growing body of literature in scientific journals and the popular media (both print and electronic) about urban wildlife. The public, and to some degree even wildlife professionals, are unaware of the information on urban wildlife presented in the scientific literature, while the primary focus of popular media is entertainment rather than education. A curriculum for training urban wildlife biologists emerged as we examined the full range of urban wildlife issues in the context of human history and society, natural history, ecology, politics, law, and economics. *Urban Wildlife Management* captures information that was strewn throughout journal articles, conference proceedings, government documents, websites, other books, secondary data sets, and the personal experiences of colleagues. *Urban Wildlife Management* gathers the essential information together under one cover, providing a synthesis document for academic, community, and professional development. Case studies are included to illustrate the concepts. Our literary approach was to continue to tell a story based on a review of hundreds of references, but there is no way to include all of the pertinent literature. The information we provide in each chapter is meant to be an overview of the subjects discussed, not an exhaustive treatment. Many of our chapters have been, or could be, the subject of an entire book. We hope our readers will take the opportunity to expand their understanding of the concepts introduced here.

Urban Wildlife Management has found an audience outside of the university classroom as well. There are many individuals working for the government, nonprofit organizations, and for-profit businesses whose responsibility or job description is to address some aspect of urban wildlife management. Developers, for example, may fail to take into account the surrounding wildlife community while at the same time hoping to attract buyers by incorporating community green space … which can lead to some of the management challenges addressed in this book. An understanding of the cause-and-effect outcomes could significantly change the "business as usual" development process, leading to a planning approach that allows for increased interaction between humans and wildlife while avoiding potential conflicts. Other fields that may benefit from a greater understanding of urban wildlife management issues include public health, urban planning, parks and recreation, sanitation, tourism, transportation, and animal control (domestic/feral and wild species).

As we began to develop this book, initially and for the second edition, our goals were fairly straightforward: (1) compile a body of information that stimulates the reader's curiosity about the urban world in which most of them live and present it in a way that would be accessible for most readers; (2) expand the reader's knowledge about how natural and urban ecosystems work; (3) challenge readers to examine the role their personal actions may have in causing at least some of the urban wildlife issues covered in this book; and (4) give readers an opportunity to apply their new knowledge and understanding through personal actions that promote sustainable approaches to urban wildlife management.

UNDERSTANDING AND MEETING THE FUTURE
CHALLENGES OF WILDLIFE MANAGEMENT

The wildlife profession has been slow to respond to this shift in public interest and need. In 1999, the Urban Wildlife Working Group of TWS conducted a national survey of state wildlife management agencies and land grant universities that offered a degree in wildlife science (C.E. Adams 2003). The survey was designed to determine how well the above-mentioned entities were prepared to address urban wildlife management issues. The results were disturbing in many respects, but the general conclusion was that the infrastructure for urban wildlife management is missing in state departments of natural resources and land-grant universities.

The role of the wildlife manager has changed significantly in the twenty-first century. So has the role of those responsible for preparing the next generation of wildlife professionals—but many wildlife management faculty have yet to respond to the change in public interest and demand, and are either completely oblivious, dismissive, or grossly misinformed as to what the discipline entails. A telling example of the latter was observed during a meeting at one of the premier university departments of wildlife sciences in the United States. The topic of discussion was whether or not to add an urban wildlife management option to the department's undergraduate curricula. One faculty member, a nationally recognized scholar in ecology and conservation biology, commented that he could see no reason for developing a curriculum about raccoons (*Procyon lotor*) in garbage cans, and consequently voted against its inclusion. Often when we disclose that our research interests include urban wildlife, we are asked how we like studying pigeons and rats (neither of which, of course, are actually wildlife, since urban pigeons (*Columba livia*) are feral animals and Norway rats (*Rattus norvegicus*) are an introduced species in the United States, but we digress ...). University faculties must begin to recognize the need to develop new instructional paradigms that meet the challenges of urban wildlife management or they may find they have become largely irrelevant. Human residents of urban and suburban habitats do not stay in their own home ranges—they venture forth into previously undeveloped lands and create urban wildlife through feeding and habituation, so even traditional wildlife professionals are likely to find themselves faced with the issues described in this book.

Students, too, must become aware of this paradigm change. When we visit with students interested in wildlife professions, we often see a reflection of ourselves at their age and level of educational development. Many are drawn to working with wildlife because they enjoy nature and are looking for professional pursuits that offer a more adventurous life, and one that doesn't require much contact with "the public." They picture themselves collecting and analyzing data in the middle of a remote forest, far removed from the aggravation of the human race (other than, perhaps, an equally adventuresome *National Geographic* photographer!). They want to "fight the good fight," gathering information that could someday save a species or an ecosystem from extinction. These budding wildlifers *can* make a difference, but if they want to do so while being employed they may have to change the scenery of their daydreams to residential developments instead of rainforests (Figures I.1 and I.2).

In our formative years and early in our careers we were drawn to the study of wild things for the sake of knowledge, and also because, in our opinion, they were so much more interesting than people. Since those early days we have come to understand there is no place on earth that is not wildlife habitat, and we can make a much more significant impact on the world if we direct more of our attention to understanding people and their relationships with wildlife, and educating them about the wildlife that surrounds them.

We recognize the complex nature of urban wildlife issues prevents the "quick fix." Therefore, to be effective an urban wildlife biologist needs academic training and pre-professional experience in both the traditional curricula and courses that add both the built landscape and people into the mix. In our opinion, the naturalist approach to understanding urban wildlife issues needs to be resurrected in academia. Too much time is spent memorizing the names of stuffed and preserved specimens from museum collections, and too little time is spent on understanding why the animal lives where it does and its relationship to both the habitat and other species found there! Related to this is the need for students to have more time in the field experiencing the situations they will encounter in the real world.

In addition to basic core courses in zoology, botany, taxonomy, genetics, and chemistry, aspiring urban wildlife biologists need exposure to courses in ecology, conservation and management of wildlife, urban forestry, urban land use planning, environmental education, public speaking, and conflict resolution. A thorough understanding of wildlife laws and the legal ramifications of urban wildlife management at the community, city, county, state, and federal levels also is crucial. This type of academic preparation will help students understand how urban communities function and how wildlife

Figure I.1   The traditional image of a wildlife biologist is that of an individual collecting and analyzing data in the middle of a remote forest, possibly even saving a species from extinction. (Courtesy John and Karen Hollingsworth/USFWS)

management issues arise. In fact, extensive knowledge and experience will be needed in all of the areas mentioned above if potential job candidates are to have any hope of rising to the challenges they face after graduation (see Sidebar I.1, Job Description for an Urban Wildlife Biologist).

We continue to be convinced that the future scope and purpose of wildlife managers will be shifting from traditional/rural to urban issues. This shift will occur even though we currently have at least two generations of deer, turkey, and quail managers in the field or waiting in the wings (no pun intended). As we point out in Chapter 2, practicing urban wildlife biologists are still relatively rare, but demand is increasing. Sadly, few academic institutions are training new professionals to meet the current job market or the needs of an ever-increasing urban population.

Students with an interest in urban wildlife as a future profession will likely be confronted with some negativity by more traditional wildlife biologists. Hey, we've been there. Actually, often we're *still* there! As pointed out by Witter, Tylka, and Werner (1981:424) several decades ago, "Plying the wildlife trade in the world of high-rises and suburbia is [a career] with which most wildlifers are not overly comfortable. Conventional wisdom holds that wildlife management is most productive when conducted far removed from metropolitan environments and concentrations of people."

While others may regard your professional pursuits with a huge measure of naiveté and disdain, we encourage you to take heart. A shift in thinking has begun and is taking root. Back in 1997, The Wildlife Society (TWS), the professional organization for wildlife biologists in the United States, published an entire issue of the *Wildlife Society Bulletin* (previously published by TWS) devoted

Figure I.2 As the general population becomes more urbanized, wildlife biologists are being asked to have greater interaction with the people, particularly in the area of public education. (Courtesy Robin Graham)

to urban wildlife management issues—this is usually the first step in the professional recognition of a subdiscipline within the larger context of wildlife management. TWS now has both an Urban Wildlife and a Human Dimensions Working Group, and sessions on these topics are becoming more common at the organization's annual meeting. Additionally, there are now other professional environmental organizations and even entire peer-reviewed journals that focus on urban issues.

It is highly unlikely that state departments of natural resources (DNRs) and land grant universities will rush to embrace urban wildlife management in the immediate future. In large part, this is because they have fixed agendas, no budget for additional management and/or research obligations, no personnel who are trained in urban wildlife research and management, and little recognition of urban wildlife management as their responsibility. Where then does the future urban wildlife biologist find a job?

The lack of attention to urban wildlife issues by government agencies has spawned a grassroots movement of sorts in which responsibility for wildlife management is shifting away from state and federal agencies, with privatization becoming a growing trend. We believe future job opportunities will emerge in the private sector, with environmental consulting firms and private wildlife control businesses; with city and county government departments, including urban planning and recreation and parks; at the federal level in agencies such as the U.S. Department of Agriculture's Wildlife Services; and with nonprofit organizations, including urban nature centers, wildlife rehabilitation organizations, and environmental education centers. In addition, we have found that students trained in a wildlife and fisheries curriculum make excellent high school biology teachers (C.E. Adams and Greene 1990).

For now, college students will need to tailor their academic training around a holistic management paradigm. Traditional management practice focuses on either increasing (in the case of game species) or decreasing (in the case of "nuisance" species) animal populations. The focus of urban wildlife management, however, is in large part to provide urban residents with opportunities to enjoy the wild species that exist in their own backyards while also providing the knowledge and

Figure I.3 Aldo Leopold, the father of wildlife conservation in America, recognized that people are part of the wildlife management equation. (Courtesy Aldo Leopold Foundation)

skills required to avoid problems. The challenge is to manage both wildlife and human populations in urban ecosystems in a manner that allows for sustainable coexistence (Fox 2006). If urban residents recognize the potential for positive interactions with wild things in the city, it may help to ease the pressure on the last remaining true wild areas we have left.

In the late 1930s Aldo Leopold (Figure I.3) recognized that meaningful and significant wildlife management requires management of *people*. This fact is not often acknowledged within the more traditional wildlife management fields, but there is no way to escape it when dealing with urban wildlife. As we demonstrate in this book, urban areas are home to entire communities of wildlife. This affords future wildlife professionals the opportunity to better understand the human-wildlife relationship and make significant contributions to all areas of wildlife conservation.

SUMMARY: A PARADIGM SHIFT

One of the common problems associated with introducing urban wildlife management (UWM) as part of the curriculum in the wildlife sciences is the simplistic notions our colleagues, students, and others have regarding its conceptual framework. For example, a "raccoon in the garbage can" always seems to become the summative explanation of urban wildlife management. Other reductionist definitions include UWM as a subset of animal damage control, or that UWM is a particular suite of techniques peculiar only to urban areas. The latter problem is of our own making, given the inclusion of UWM in the *Wildlife Management Techniques Manual,* published by The Wildlife Society. Truth be known, wildlife management techniques consist primarily of catching, identifying, marking, following, and counting wild animals, flavored with a healthy dose of formulae and statistics to add scientific rigor to the first five activities. UWM is just one of the venues in which these activities take place; another example of the depth and breadth of human involvement with wild things. *Urban Wildlife Management* will explore an extensive range of concepts required to provide a more complete and accurate presentation of wildlife management to colleagues, students, and the general public. In order to do this it seemed appropriate to articulate the fundamental similarities and differences between wildlife management in urban vs. rural habitats (Table I.1). The differences identify the professional preparation, environment, limitations, public involvement, and professional isolation of urban when compared to traditional rural wildlife management programs. Table I.1 can also be used to identify and summarize the critical curriculum components for courses on UWM. The chapters that follow go into greater detail, with explanations and examples of how these differences need to be negotiated in order for wildlife biologists to operate effectively in urban habitats.

Table I.1   Wildlife Management Comparisons in Human-Altered Urban vs. Natural Rural Habitats

Similarities

1. Involves game, nongame, exotic, and/or threatened/endangered species.
2. Uses standard wildlife management procedures, e.g., the *Wildlife Management Techniques Manual*.
3. Action requires input, participation, and oversight by state or federal wildlife agencies.
4. Preparation in college-level wildlife management courses required.
5. Potential economic losses or gains are the primary catalysts for management action.
6. Wildlife management goals are both proactive and reactive.
7. Uses professional and popular outlets to disseminate the status of information to the *whole* community.

Differences

Urban	Rural
1. Lower diversity of native plant and animal species.[a]	1. Higher diversity of native plant and animal species.
2. Fewer sources of state and federal funding for management programs.	2. More sources of state and federal funding for management programs.
3. A new and developing focus for research, management, and education programs.	3. A large and established focus for research, management, and education programs.
4. Layers of jurisdiction increase with proximity to urban centers.	4. Layers of jurisdiction decrease with distance from urban centers.
5. Small scales of analysis with many legal and physical impediments in highly fragmented landscapes.	5. Large scales of analysis with few legal and physical impediments in less fragmented landscapes.
6. Requires extensive training and experience in the human dimensions of wildlife management.[b]	6. Requires less training and experience in the human dimensions of wildlife management.[b]
7. Limited academic and agency acceptance and participation.	7. Wide academic and agency acceptance and participation.
8. Residents have a more heterogeneous set of attitudes and expectations related to wildlife.	8. Residents have a more homogeneous set of attitudes and expectations related to wildlife.
9. Higher public demand for inclusion in the management process.	9. Lower public demand for inclusion in the management process.
10. Higher potential for threat to public health from zoonotic disease and parasites.	10. Lower potential for threat to public health from zoonotic disease and parasites.
11. Management to reduce artificially abundant wildlife populations.	11. Management to sustain artificially abundant wildlife populations.
12. Growing trend toward privatization and commercialization of wildlife management.	12. Majority of management efforts coordinated through state or federal agencies.
13. Exaggerated time frame for completion of management activities.	13. Significantly shorter time frame for completion of management activities
14. Managers may not have required training in wildlife management.	14. Managers have required training in wildlife management.

Table reviewed and edited by S. Ash, J. Davis, R. Denkhaus, S. Gehrt, S. Locke, and R. D. Slack.

[a] Deals primarily with a few species that are highly adaptable or fortuitously well suited to an urban environment.

[b] Includes conflict resolution; awareness of public attitudes, activities, knowledge, and expectations; public education; and identification and inclusion of all stakeholder groups.

SUGGESTED ACTIVITIES FOR THE INTRODUCTION

1. Evaluate how well your department's curriculum prepares its students to become a professional urban wildlife biologist (see Sidebar I.1 Job Description for an Urban Wildlife Biologist).
2. Take the Wildlife Hotline "Quiz" in Sidebar I.2. (Answers found in Appendix A).
3. Go to the U.S. Census Bureau website (www.census.gov). What percent of the population was classified as urban versus rural in the 2000 Census? How has this changed since the 1990 Census?
4. Go to the 2006 National Survey of Fishing, Hunting, and Wildlife-Associated Recreation Survey website (http://www.census.gov/prod/www/abs/fishing.html). How many residents are actively involved in wildlife watching? How does this compare to the number of residents involved in traditional hunting and fishing activities? How has this changed since the 2001 Survey? How much is spent on these activities?
5. Determine what kind of urban wildlife programs are available through your state department of natural resources and/or land grant university.

SIDEBAR I.1   Job Description for an Urban Wildlife Biologist

An urban wildlife biologist is an individual who works primarily in metropolitan (nonrural) environments, focusing on nondomestic vertebrate and invertebrate species and interactions between humans and wildlife (C.E. Adams 2003). During the course of a normal day, an urban biologist may be called on to handle anything from development of a plant or wildlife species conservation plan to assisting with conflict resolution issues between a developer and the local city planning board. Job titles and descriptions vary depending on the emphasis an employer places on specific tasks. Texas Parks and Wildlife Department, for example, states that the duties of its urban biologists include "providing opportunities for urban residents to reconnect with the natural systems, presenting educational programs for adults and students on a variety of habitat/wildlife issues, serving as technical advisors on multi-agency conservation planning initiatives, and assisting landowners with habitat restoration or enhancements" (http://www.tpwd.state.tx.us/expltx/ eft/careers/urban.htm, March 21, 2005). The North Carolina Wildlife Resources Commission, on the other hand, stresses "forging and working with partnerships of government and private entities to achieve conservation objectives" as well as wildlife damage management and public outreach in an announcement for a contract position (NCWRC, Urban Biologist Contract Position # 04-FD-JM1). The following job listing from the Missouri Department of Conservation provides some additional insight into the scope of knowledge and diversity of skills needed by those interested in this profession.

COMMUNITY CONSERVATIONISTS, MISSOURI DEPARTMENT OF CONSERVATION

Salary: $33,024–$58,764

Qualifications: Graduation from an accredited college or university with a Bachelor's Degree in Fisheries, Forestry or Wildlife Management, Environmental Planning, Biological Sciences, or applicable field of study and at least three (3) years of progressively responsible professional experience; or an equivalent combination of education and experience. **Experience in community development and natural resource planning is highly desirable.**

Duties and Responsibilities:

- Promotes the conservation of fish, forest, and wildlife in the urban environment through interaction with government officials, planning and zoning boards, land use committees, development committees, urban development industries, park boards, recreation planning committees, developers, public conservation groups, and others;
- Educates citizens about social, environmental and economic impacts of urban sprawl and the benefit of alternatives;
- Encourages the creation and support of local non-profit environmental groups; capitalizes on existing opportunities to deliver natural resource conservation messages by attending conferences, festivals, symposia, and workshops as presenters and attendees;
- Provides information, education, materials, research time, and coordination with citizen(s) that are willing to advocate for proper natural resource protection/management by local government; develops and assists with the development of educational materials, plans, grant applications, etc.;
- Provides technical assistance to local governments for natural resource management/protection and takes advantage of opportunities for involvement in community planning, assisting with plat review, planning and zoning code revisions, ordinance revisions, greenway establishment, green space conservation, aquatic resource management, storm-water master plans and long range land use master plans;
- Educates local governments of the negative impacts of sprawl, especially economic impacts through presentations to city councils, planning and zoning codes and ordinances;
- Utilizes cost share docket, other Missouri Department of Conservation (MDC) funds and outside financial resources to promote the establishment and/or maintenance of natural areas;
- Provides technical assistance on land such as parks, natural areas, riparian corridors, storm-water facilities, nature trails, institutional campuses, etc.;
- Builds strong working relationships with local governments;
- Advocates for and provides technical assistance on the use of conservation development techniques that maintain or restore healthy fish, forest, wildlife, or natural community resources;
- Shows professionals how conservation development can be an economically sound alternative;
- Assists with easements, deed restrictions, and other methods of insuring long-term environmental protection;

- Provides training to inter-agency personnel concerning urban natural resource management; completes reports;
- Performs other duties as required.

SIDEBAR I.2 Wildlife Hotline "Quiz"

To be an effective urban wildlife biologist you need to be versatile, have a broad range of knowledge, and think on your feet. Wildlife hotlines receive thousands of calls each year and each call is an opportunity to educate the public. In real life, the questions you ask of your caller are critical for determining what the caller is asking, and how you should answer. So look over the questions below, taken from actual calls to a wildlife hotline, and think about how you would respond. Suggestions for key points that need to be addressed can be found in Appendix A.

1. We have a neighborhood pond and it's overflowing with ducks—we've got a real population problem and the ducks are becoming very aggressive because there isn't enough to eat. What can we do?
2. There's some kind of animal in my attic and its making all kinds of noise during the night. What is it and why is it up there?
3. During the last two weeks we've been hearing birds in our chimney. They are calling throughout the day and it echoes up and down the flue and makes a terrible racket. What can I do to get rid of them? [This call was received in July.]
4. There's an armadillo tearing up my lawn at night. What kind of bait should I use to trap him?
5. Birds are always flying into my window and killing themselves. Why? What can I do to prevent it?
6. I found a baby porcupine in my [Houston] backyard. What should I do with it?
7. Do owls live around here [Albuquerque, New Mexico]? Could one have carried off my cat?
8. We have a cardinal that keeps attacking our bedroom window. Why is he doing that and how do I make him stop? He starts really early in the morning!
9. I just found a baby squirrel on the sidewalk. She seems okay but now that I've touched her I know the mom will kick her out of the nest if I try to put her back. How do I raise it?
10. A peacock is living in our neighborhood. He flies to the top of the houses and starts screaming at the crack of dawn. Can someone come out here and catch it?
11. I'm a golf course manager and we've got a problem with those Canada geese. They're all over the place and they poop on the greens and the cart paths. The golfers are really starting to complain. What should we do?
12. What kind of seed should I put in my bird feeder to attract a greater variety of species?
13. I found a snake and I need you to tell me if it's poisonous. Its brown and its BIG!
14. I'm from the city street department and we've got beavers flooding our roads. It damages the pavement and creates a hazard for drivers. Who do I need to talk to about having these things trapped and removed?
15. I've read about places where they give deer birth control pills when the population gets too big. Well, our deer population is too big—they're eating all of our landscaping plant and they're becoming a traffic hazard. Where can we get a supply of those pills? How do we get the deer to eat them?
16. I found a bat on the ground and it can't fly. How can I help it?
17. How can I get bats to live in my bat house? Is there someplace I can order bats?

APPENDIX A: ANSWERS TO THE WILDLIFE HOTLINE "QUIZ"

1. We have a neighborhood pond and it is overflowing with ducks. We've got a real population problem and the ducks are becoming very aggressive because there isn't enough to eat. What can we do?
   - **Teaching opportunity**: This question provides an opportunity to talk about how feeding wild animals, while usually well intentioned, often turns out to cause problems in the long run.
   - Suggest the community work with local and/or national organizations that have expertise in humane population control methods, including egg addling and egg oiling.
   - Discuss habitat modification techniques to make the area less attractive to waterfowl.
   - Explain that regardless of the type of population control the community chooses, they will need to post signs and educate their neighbors on the need to stop supplemental feeding.
2. There's some kind of animal in my attic and its making all kinds of noise during the night. What is it and why is it up there?
   - Depending on where the caller lives, there may be several wildlife species that are active at night and are likely to use attics when they are available; examples include raccoons, opossums, flying squirrels, tree squirrels, bats, and, yes, even rats.
   - A common first step is to figure out where the animal(s) is entering and exiting—this is important for determining exactly what type of animal is using the attic and the materials needed for exclusion.
   - Some methods for encouraging the animal to leave include placing a loud radio or strobe lights in the attic.
   - The homeowner may want to consider hiring a commercial wildlife control operator.

- Before taking any action, consider the time of year. If it's breeding season, extra steps may be needed to ensure that babies are not left behind before repairs are made to the entrance/exit site(s).

3. During the last two weeks we've been hearing birds in our chimney. They are calling throughout the day and it echoes up and down the flue and makes a terrible racket. What can I do to get rid of them? [This call was received in July.]
   - Most likely there are chimney swifts nesting in the flue.
   - Explain that all migratory birds, including swifts, are protected by the Migratory Bird Treaty Act. Harming the birds, their nest, eggs, or offspring is illegal.
   - **Teaching opportunity**: Point out that chimney swifts are nice to have around because they eat tons of insects during the summer—including mosquitoes.
   - If there is a lot of noise the nestlings will soon fledge (leave the nest)—at most, there will only be about two to three weeks of the noise. Once the young leave the flue the parents will too; at this point a chimney sweep can be hired to cap the chimney to prevent future nesting.
   - Meanwhile, a thick piece of Styrofoam®, cut to fit the fireplace opening, will dampen the noise.
   - One last thing—chimney swift nests create little risk of flue fires. Once the young fledge the nests tend to fall harmlessly down into the fireplace and can be discarded. Tests have shown that it's actually quite difficult to set a swift nest on fire.

4. There's an armadillo tearing up my lawn at night. What kind of bait should I use to trap him?
   - Ask if the caller has actually seen the animal—are they positive it is an armadillo? Are they able to convince you, based on their description?
   - Explain that armadillos dig to uncover grubs, which are a favorite food source. While the armadillo is undoubtedly damaging the lawn, so are the grubs.
   - Remove the food source by treating the lawn for grubs and the armadillo will move on.
   - **Teaching opportunity**: Trapping will not solve the problem. As long as there are grubs they will simply attract another armadillo (or another animal that eats grubs).
     - Moreover, armadillos are not easily baited, especially when there's a yard full of juicy grubs to be had without entering a strange metal contraption.
     - A trapped armadillo will attempt to dig its way out, often mutilating its feet in the process.
     - Finding an appropriate release site is difficult.
     - Most relocated animals do not survive for long after release; they are hit by cars trying to return to their home territory; they get into territorial fights with the resident wildlife; they become ill as a result of the stress of relocation; or they are unable to find food and water in the new location.
   - Treating for grubs takes longer initially than trap-and-remove, but it's a long-term solution that is also more humane.

5. Birds are always flying into my window and killing themselves. Why? What can I do to prevent it?
   - When a bird flies into a window hard enough to become injured or die, it's usually because the window is reflecting the sky (for all or some portion of daylight hours). The bird perceives the window as open space and attempts to fly through.
   - There are several ways to reduce the reflection hazard of your windows:
     - Mix up some soapy or salty water in a spray bottle and mist the windows, allowing the water to dry and create a haze on the glass. Depending on how much of the year the windows create a reflection hazard, this method may need to be repeated periodically as rain or other moisture removes the coating. This method does require a somewhat casual standard of housekeeping since it does not leave a crystal clear streak-free pane of glass—and that's the point.
     - There are commercially available products, such as fine screening or spider-web mimics, that can be placed on the outside of the window to reduce reflection.
     - Landscaping plants placed near the window may reduce reflection as well.

6. I found a baby porcupine in my [Houston] backyard. What should I do with it?
   - Members of the public don't always make accurate species identification. While it's not completely impossible for someone to find a porcupine in a Houston backyard, it is highly unlikely since this species' normal range does not include Texas.

- Chances are much greater that this individual has actually found an African pygmy hedgehog, an exotic pet that has escaped or has been intentionally released.
- In any case of uncertain identity, ask if the caller can safely take a digital photo and attach it to an email.
- Once the animal has been correctly identified, it may be necessary to contact a permitted wildlife rehabilitator or domestic animal shelter in the caller's area for assistance. Some animal shelters will accept exotic species under certain circumstances, either for use as an education animal or for placement in a new home.

7. Do owls live around here [Albuquerque, New Mexico]? Could one have carried off my cat?
   - Yes, there are many species of owls living in the Albuquerque area, including great horned owls (always have a trusted field guide nearby when taking hotline calls).
   - Yes, great horned owls are large enough to take a domestic housecat or even a small dog … and they've been known to do so.
   - However, there are many other wild predators that may prey on companion animals.
   - **Teaching opportunity**: Explain that domestic cats are not only at risk when allowed to roam freely outdoors, they are also predators that pose a risk to wild birds and mammals.
   - Don't forget to warn against pet owners taking action again a suspect owl, since doing so would be illegal (Migratory Bird Treaty Act), not to mention pointless.

8. We have a cardinal that keeps attacking our bedroom window. Why is he doing that and how do I make him stop? He starts really early in the morning!
   - When a bird repeatedly hits or "attacks" a window, it's responding to a reflection of itself and interpreting the image as an intruder on its territory. This behavior usually occurs only during the breeding season, when birds (especially male birds) are the most territorial.
   - The antireflection methods listed in Q5 above will work in this situation as well, but can usually be stopped or removed once the breeding season is over.

9. I just found a baby squirrel on the sidewalk. She seems okay but now that I've touched her I know the mom will kick her out of the nest if I try to put her back. Will you tell me how to raise it?
   - The short answer is "no." Most states require a wildlife rehabilitation permit to legally hold and care for wildlife.
   - The longer answer helps callers to understand why keeping it is not in the best interest of the animal.
     - Wild infants, just like humans, have very specific nutritional and behavioral needs.
     - Inappropriate diets can result in life-long disability or even death.
   - The best solution—assuming the squirrel actually needs assistance—is to contact a permitted wildlife rehabilitator in the caller's area.
   - **Teaching opportunity**:
     - Bird and mammal parents will NOT automatically reject their offspring once they've been handled by a human (or by our companion animals).
     - This tidbit of "common knowledge" may have originated from the fact that wild parents will, in some cases, remove offspring from the nest or den when there is something amiss that will keep that individual from surviving. Cruel? Perhaps, but by reducing competition between healthy and handicapped siblings the healthy offspring have an improved chance at survival (because even when you're healthy, survival isn't a sure thing for wild animals).
     - However, there are cases in which a healthy young animal is knocked out of a nest by wind or wanders off before it's old enough to survive.
     - A permitted wildlife rehabilitator can help the caller determine whether or not the animal actually needs assistance, and may be able to offer suggestions for returning the baby to its parent(s).

10. A peacock is living in our neighborhood. He flies to the top of the houses and starts screaming at the crack of dawn. Can someone come out here and catch it?
    - A bird that can fly up to the roof of a house is a bird that can fly away from anyone trying to catch it.
    - Sometimes peacocks can be captured after they have been habituated to enter a trap for food.

- Another option may be to wait until the bird is in its seasonal molt and therefore unable to fly.
- Suggest that the caller contact a game bird breeder for additional suggestions.

11. I'm a golf course manager and we've got a problem with those Canada geese. They're all over the place and they poop on the greens and the cart paths. The golfers are really starting to complain. What should we do?
    - Explain that a golf course is just about the most perfect Canada goose habitat that exists on this earth.
    - It may be possible to modify the course to create a less appealing habitat but these methods may be expensive and will probably change the look of the course significantly.
    - The PGA and LPGA have been involved in research related to wildlife and golf courses—suggest the caller contact these organizations for detailed suggestions.
    - Suggest the course manager contact an organization with expertise in goose population management (see Q1 above).
    - Lastly, while there are lethal removal methods, the caller needs to keep in mind that Canada geese are a protected species (there's that Migratory Bird Treaty Act again); it's important that the state wildlife management agency and the U.S. Fish and Wildlife Service be involved in any management plan.

12. What kind of seed should I put in my bird feeder to attract a greater variety of species?
    - Refer the caller to the Wild Birds Unlimited or similar website for information on specific seed suggestions for the birds in their area, but …
    - **Teaching opportunity**: Explain that while offering seed is one way to attract birds, a greater variety of species can be enticed to their yard with water, which all birds need, and by landscaping with fruit- and seed-bearing plants, which also provide shelter and nesting sites.
    - The caller might also like to know that native plants and water features are cheaper in the long run than feeders, require less maintenance, and allow birds to retain more of their natural behavior while also reducing the problems that occur when large numbers of animals congregate in a small area (including the spread of diseases such as finch conjunctivitis).

13. I found a snake and I need you to tell me if it's poisonous. Its brown and its BIG!
    - First order of business—warn the individual there's no guarantee of accuracy when an ID is made over the phone. A digital photo will increase the chance of an accurate identification but, again, no guarantees!
    - Ask questions about the snake's location, behavior, etc.
    - Remind the caller that even nonvenomous snake bites can be painful.
    - If the snake is not threatening a child or a pet, and it is not inside a building, chances are it will leave on its own if given the opportunity.
    - If the snake will not leave, consider contacting a wildlife control company for assistance.
    - **Teaching opportunity**: If there have been multiple sightings of this and/or other snakes, explain that snakes, like most wild animals, are attracted to areas that provide one or more of the following: food, water, shelter. Consider whether the attractant could be rodents.

14. I'm from the city street department and we've got beavers flooding our roads. It damages the pavement and creates a hazard for drivers. Who do I need to talk to about having these things trapped and removed?
    - Explain that trapping and removal (vie lethal means or for relocation) will not solve the problem. It will only open up habitat for other beavers to move in. Changes must be made to the habitat (see Q4 above).
    - There are a variety of methods and technologies that will allow the beavers to remain *and* stop the flooding, including "beaver deceivers."
    - Suggest the caller contact the state wildlife management agency or the Humane Society of the United States (www.hsus.org) for additional assistance.

15. I've read about places where they give deer birth control pills when the population gets too big. Well, our deer population is too big—they're eating all of our landscaping and they're becoming a traffic hazard. Where can we get a supply of those pills? How do we get the deer to eat them?
    - Begin by explaining that the use of immunocontraceptives for deer is an experimental method—one that has not proven to be very effective.

INTRODUCTION: A NEW WILDLIFE MANAGEMENT PARADIGM

- Additionally, this is an expensive population management method.
- Lastly, explain that any management of deer populations must be done with oversight by the state wildlife management agency … and then provide the agency's phone number and a contact name within the agency.

16. I found a bat on the ground and it can't fly. How can I help it?
    - Step 1 (and this is VERY important): Make note of the phone number on Caller ID and then ask if anyone has touched the animal. If there has been contact of any kind, inform the caller to immediately contact their local health department for instructions.
    - Explain that while rabies is relatively rare in bats, one should err on the side of caution. If treated immediately, rabies is preventable and treatment no longer involves a series of painful shots in the stomach—they are no worse than flu shots. Left untreated, however, rabies is fatal. Period.
    - One does not have to be bitten to contract rabies from an infected animal. If one has even a small cut on the hand, holding an infected animal can create a saliva-to-blood transmission of the virus. An infected bat may not exhibit overt signs of illness.
    - If no one has handled the bat, instruct the caller to contact a permitted wildlife rehabilitator in the area for instructions and assistance.
    - **Teaching opportunity**: Bats can't take flight from the ground like a bird—they must drop from some height (which varies depending on the species) in order to become airborne. If a bat is grounded—and this can happen for a variety of reasons—it must crawl up a tree or other vertical structure in order to fly away.

17. How can I get bats to live in my bat house? Is there someplace I can order bats?
    - Congratulate the individual on their progressive attitude toward bats—one of the most maligned species on the planet—and thank them for their efforts to keep their neighborhood mosquito population in check!
    - Explain there is no legal way to order bats for their house, and if they ever see an advertisement for bats they should consider it a scam.
    - The Bat Conservation International website (www.batcon.org) offers tips on how to site bat houses to improve the chance of attracting bats, and loads of other fantastic information on bats.
    - **Teaching opportunity**: North American bats are crucial to the success of the agricultural industry because they are able to consume huge quantities of pest insects in a single night.

LITERATURE CITED

Adams, C.E. 2003. The infrastructure for conducting urban wildlife management is missing. *Transactions of the North American Wildlife and Natural Resource Conference* 68:252–268.

Adams, C.E. and J. Greene. 1990. Perestroika in high school biology education. *American Biology Teacher* 52:408–412.

Adams, L.W. 1994. *Urban Wildlife Habitats: A Landscape Perspective*. Minneapolis, MN: University of Minnesota Press.

Adams, L.W. and D.L. Leedy, eds. 1987. *Integrating Man and Nature in the Metropolitan Environment*. Columbia, MD: National Institute for Urban Wildlife.

Adams, L.W. and D.L. Leedy, eds. 1991. *Wildlife Conservation in Metropolitan Environments*. Columbia, MD: National Institute for Urban Wildlife.

Conover, M. 2002. *Resolving Human-Wildlife Conflicts: The Science of Wildlife Damage Management*. Boca Raton, FL: Lewis Publishers.

Decker, D.J., T.L. Brown, and W.F. Siemer. 2001. *Human Dimensions of Wildlife Management in North America*. Bethesda, MD: The Wildlife Society.

Fox, C.H. 2006. Coyotes and humans: Can we coexist? *Proceedings 22nd Vertebrate Pest Conference* 22:287–293.

Hadidian, J. 2007. *Wild Neighbors: The Humane Approach to Living with Wildlife*. Washington, DC: Humane Society Press.

Johnson, A.A. 1970. *Biology of the Raccoon* (Procyon lotor *varius Nelson and Goldman*) *in Alabama*. Auburn University Agricultural Experiment Station Bulletin 402.Auburn, AL: Auburn University.

Landry, S.B. 1994. *Peterson First Guides: Urban Wildlife*. New York: Houghton Mifflin.

Manfredo, M.J., J.J. Vaske, P.J. Brown, D.J. Decker, and E.A. Duke. 2008. *Wildlife and Society: The Science of Human Dimensions*. Washington, DC: Island Press.

Platt, R.H., R.A. Rowntree, and P.C. Muick. 1994. *The Ecological City: Preserving and Restoring Urban Biodiversity*. Amherst: University of Massachusetts Press.

Rees, P. 2003. *Urban Environments and Wildlife Law*. Oxford, UK: Blackwell Publishers.

Shaw, W.W., L.K. Harris, and L. Vandruff, eds. 2004. *Proceedings of the 4th International Symposium on Urban Wildlife Conservation*. May 1–5, 1999. Tucson, AZ.

Shipp, D. 2000. *Urban Wildlife Spotter's Guide*. London: Usborne Publishing.

Tyldesley, D. 1994. *Planning for Wildlife in Towns and Cities*. Peterborough, UK: English Nature.

United States Census Bureau. 2006. *National Survey of Fishing, Hunting and Wildlife-Associated Recreation*. Washington, DC: United States Government Printing Office.

Whiston-Spirn, A. 1985. *The Granite Garden: Urban Nature and Human Design*. New York: Basic Books.

Witter, D.J., D.L. Tylka, and J.E. Werner. 1981. Values of urban wildlife in Missouri. *Transactions of the North American Wildlife and Natural Resources Conference* 46:424–431.

PART I

Urban Landscapes

CHAPTER 1

Wildlife Management
Past and Present

The forest stretched no living man knew how far.

—Willa Cather (1931)

KEY CONCEPTS

1. Identification of the key events in the history of wildlife management in North America.
2. A description of changes in the way wildlife is valued over the past 200+ years.
3. A definition of "urban wildlife" as the term is used in this text.
4. A list of three underlying rationales for categorizing wildlife.
5. A list of three types of wildlife behavior likely to be influenced by urbanization.
6. Adopting orphaned wildlife is not a wise practice and in many cases is illegal.

A BRIEF HISTORY OF WILDLIFE MANAGEMENT IN NORTH AMERICA

For hundreds of generations, conditions for wildlife populations in North America were nearly ideal, although, as with any healthy ecosystem, constantly in flux. Indigenous peoples used wildlife at sustainable levels for food, clothing, and shelter, and the animals played a large role in their culture and spiritual life (Decker, Brown, and Siemer 2001). This situation began to change with the arrival of European immigrants approximately 400 years ago. These new Americans carried with them a culture of human domination over nature … at least in theory; most Europeans did not have the legal right to exercise dominance over wildlife. Those privileged few who owned land also owned the wildlife living on that land.

Immigrants to North America found a seemingly infinite supply of natural resources. The abundance of wildlife during the 1600s and early 1700s must have been staggering to those who left behind lands that had been over-hunted for centuries. What's more, there were no legal restraints on the exploitation of these resources. In North America, wildlife was a *commons*, a resource owned by all, but, at that time, managed by no one. Although even the earliest settlers depended on domestic livestock and cultivated crops, wild game provided variety and an essential food source when crops failed (Root and De Rochemont 1994).

By the mid-1800s wildlife populations in the East were suffering under the combined effects of subsistence use, market hunting, and habitat loss. The migration of Americans westward during this time created a wave of similar pressures on wildlife. White-tailed deer (*Odocoileus virginianus*), elk (*Cervus canadensis*), black bear (*Ursus americanus*), most species of waterfowl, and wild turkey (*Meleagris gallopavo*) were extirpated in many areas of the East and Midwest (Decker, Brown, and Siemer 2001). The passenger pigeon (*Ectopistes migratorius*), a species once so abundant that when flocks flew overhead the sky grew dark, was effectively extinct in the wild by 1900.

RISE OF THE AMERICAN CONSERVATION MOVEMENT

As the extent of damage to wildlife became more apparent, concerned citizens spent much of the nineteenth century working to develop a meaningful and effective system for protecting this disappearing resource. In 1844, a group of about eighty militant conservationists formed the New York Sportsmen's Club, the sole purpose of which was "the protection and preservation of game."

The club had three primary targets: sale of game for market, spring shooting of game birds, and lax game laws (Figure 1.1). A majority of the club's members were attorneys; they developed an effective strategy of suing poachers, dealers, and hotel proprietors for the sale or possession of game killed out of season. They tracked down violators by following tips from informants and using private detectives. The group was so effective in its crusade against game-law violators their approach was adopted by newly formed sportsmen's clubs in Massachusetts, Rhode Island, and Ontario. Prior to 1870, however, most efforts were local rather than regional or national in scope (Trefethen 1975).

The American conservation movement began in earnest during the late 1800s with the emergence of two views of nature and the country's wildlife heritage. Sustainable-use advocates, often

Figure 1.1   Market hunting of waterfowl and other wildlife species was one impetus behind the enactment of the first wildlife laws. (Courtesy USFWS)

termed "progressives," included within their ranks Gifford Pinchot and Theodore Roosevelt, while John Muir and other "romantics" represented the preservationist standpoint [Lutts 1990; Figure 1.2 (a), (b), and (c)]. Originators of the conservation movement at this time often embraced both the sustainable-use and preservationist worldviews. It was due to the two major influences of this movement that the U.S. Congress created what would become the National Forest System (sustainable-use) and the first National Parks (preservation).

(a)

(b)

(c)

Figure 1.2 Conservation pioneers (a) John Muir, (b) Theodore Roosevelt, and (c) Gifford Pinchot.

During this period state agencies were created to manage wildlife for current and future citizens to use and enjoy. Laws to restrict harvest and protect habitat were passed by legislators across the country, but implementation of these laws proved more difficult. Reliable funding for enforcement often was limited or nonexistent for at least another 50 years.

In 1937 Congress took an unprecedented step to address the funding problems associated with wildlife management by passing the Federal Aid in Wildlife Restoration Act, also known as Pittman–Robertson for the legislators who sponsored it. Pittman–Robertson funnels an 11 percent federal user fee on hunting rifles, shotguns, and ammunition to the U.S. Fish and Wildlife Service. A 1970 law added a tax on handguns and archery equipment for wildlife management.

In 1950 Congress passed the Federal Aid in Sport Fish Restoration Act to address similar issues regarding fisheries management. Dingell–Johnson, as it is commonly known, collects a 10 percent manufacturers' user fee on fishing equipment and tackle. The 1984 Amendment to Dingell–Johnson expanded the tax to include new motorboat fuel taxes and duties on imported tackle and boats.

User fees established by the Restoration Acts, and collected by the U.S. Treasury from manufacturers, go into trust funds administered by the Department of the Interior. A maximum of 8 percent of the funds may be retained by the U.S. Fish and Wildlife Service for administration, and the rest is allocated to states based on a formula that considers the total area of the state and the number of licensed hunters in the state. These are cost-reimbursement programs; the states cover the full amount of an approved project and then apply for reimbursement through federal aid for up to 75 percent of the project expenses. States must provide at least 25 percent of the project costs from non-federal sources. Appropriate state agencies are the only entities eligible to receive federal aid grant funds.

The purpose of establishing the Restoration Acts was to provide funding for restoration, rehabilitation, and improvement of wildlife habitat and fisheries, wildlife and fisheries management research, and information distribution. Pittman–Robertson was amended in 1970 to include funding for hunter training programs and development, operation, and maintenance of public target ranges. Dingell–Johnson funds land for boating, fishing and fish production, research and inventory projects, and pay for public education about fish and their habitats. Both of these Acts have served the needs of wildlife management agencies and consumptive users well for decades.

CHANGING WILDLIFE VALUES

The "typical" American has changed dramatically over the past 50+ years. In contrast to the country's rural, agricultural heritage, 80 percent of Americans now live in areas classified as "urban" by the U.S. Census Bureau: a central city and its closely settled surrounding territory (including suburbs), regardless of legal status, that has a population of at least "50,000 and a density of at least 1,000 people per square mile" (U.S. Census Bureau 2000).

Life in cities and suburbs has changed Americans' attitudes and expectations concerning wildlife. Many, if not most, are several generations removed from a culture of living close to the land. They are more likely to value wildlife similarly to the way they value companion animals and people (Mankin, Warner, and Anderson 1999) than as a consumptive-use resource. Americans are now more likely to be involved in wildlife-related recreation such as observing, feeding, and photography, rather than traditional activities such as hunting and fishing [U.S. Census Bureau 2006; Figure 1.3 (a) and (b)].

Due in large part to a historic focus on consumptive issues, governmental agencies are not well positioned to address the concerns of an increasingly urbanized population. State wildlife agencies have a legislative mandate to manage all wildlife within their borders as a public resource, but in all but a few states funding for nontraditional wildlife management issues is extremely limited. As a result, a trend is developing toward increased privatization of wildlife management, particularly

(a)

(b)

Figure 1.3   Participation in traditional consumptive-use activities (a) has decreased over time while interest in nonconsumptive wildlife recreation (b) is on the rise. (Courtesy Bill Harper [a], USFWS [b])

in urban and suburban settings. Nongovernmental organizations, such as private wildlife control businesses, conservation groups, humane societies, and wildlife rehabilitators have stepped in to address public demand unmet by government agencies. This paradigm shift away from a system of managing wildlife as a commons for the good of the resource and toward private, profit-driven systems is not, by and large, the result of specific policy decisions. Rather, it has evolved as a grassroots response while agencies struggle to adapt to changing public expectations and funding limitations.

A NEW KIND OF WILDLIFE

While many Americans readily acknowledge that pigeons (*Columba livia*), starlings (*Sturnus vulgaris*), house sparrows (*Passer domesticus*), rats (*Rattus* spp.), mice (*Mus* spp.), and tree squirrels (*Sciurus* spp.) are "urban animals," few are aware that a wide diversity of other wild species live in human cities, towns, and suburbs. Burger (1999) provided a detailed and extensive compendium about animals in towns and cities, and studies in Tucson, Arizona, have described urban and suburban populations of coyotes (*Canis latrans*), burrowing owls (*Athene cunicularia*), Harris hawks (*Parabuteo unicinctus*), Cooper's hawks (*Accipiter cooperii*), javelina (*Tayassu tajacu*), deer (*Odocoileus* spp.), and bird communities, just to name a few (Shaw et al. 2003).

Categorizing Wildlife

Early in the development of management agencies, wildlife species were categorized into three groups based on their utility to people: game (including "furbearers" of interest to trappers), nongame, and nuisance. This approach reflected the interests of both agencies and stakeholders in wildlife as a consumptive-use resource. Game species were managed to produce the greatest possible harvest and nuisance species were targeted for lethal control. Nongame species received little attention.

In the early 1970s, growing public interest in the environment, concern over vanishing species, and legislation enacted as a result of these trends expanded the wildlife lexicon. The term *threatened or endangered* (T/E) allowed wildlife professionals to classify species based on their abundance or scarcity. In spite of this change, wildlife categories remained sharply drawn; for example, once a game species is listed as T/E, management strategies shift and hunting is prohibited … so for all practical purposes it is no longer a game species. If, however, the species recovers and is delisted it returns to its original classification as game.

With the publication of a book entitled *Urban Wildlife Habitats: A Landscape Perspective* (L.W. Adams 1994), the term *urban wildlife* was established. Rather than classifying wildlife based on their usefulness to humans or their population status, urban wildlife species are categorized as such based on whether populations can be found living in and around human settlements. Developing a definition of urban wildlife has been difficult, in large part because it attempts to categorize species in a way that is radically different from traditional classification methods. Some examples of definitions used by other authors include:

- All native, nondomestic, wild animals found in or around urban areas (Schaefer 2004).
- Any wild creature that lives in an urban environment or an urban-rural interface, including birds, reptiles, amphibians, mammals, fish, insects, and worms (U.S. Department of Agriculture Forest Service 2001).
- Any nondomestic animals that live in cities, suburbs, or other urban areas (Maurizi and Friedner 1997).
- Wild vertebrates local to the region that may occur in the urban area (Fletcher 1994).

For the purposes of this book, *urban wildlife includes all nondomestic vertebrate species, with populations in areas classified as urban*. Wild invertebrates will receive only cursory attention in this publication because management of these species is a specialty in its own right, considered to be the responsibility of entomologists (insects), arachnologists (spiders), and malacologists (mollusks) rather than wildlife biologists.

Categorizing wildlife as urban blurs the sharp distinctions established by traditional categories. For example, deer can be classified both as a game and as an urban species. To some degree, traditional categories are less applicable in urban ecosystems. While both game and nongame species can be found in urban habitats, game species are rarely hunted in commercial or residential areas. Conversely, nongame species, such as grackles (*Quiscalus* spp.), prairie dogs (*Cynomys* spp.), and bats *(Chiroptera* spp.*)*, often are the target of lethal control measures, individually or as communities. Species listed as threatened or endangered may experience localized abundance within urban habitats, thanks to greater abundance of food and water or lack of predation pressures.

The Unique Ecology and Behavior of Urban Wildlife

Wildlife populations adjusted to living in urban habitats may exhibit distinct behavioral differences when compared to animals living in rural environments. This adjustment to conduct the usual activities associated with species survival (e.g., breeding) within the specific conditions of the urban environment is called "synurbization" (L.W. Adams, VanDruff, and Luniak 2005). In other words, synurbization is the response of wildlife to the ecological changes associated with urban development. For example, urban development destroys the preexisting conditions required for survival in the natural environment by creating empty ecological niches that attract species preadapted to the specific conditions offered by the new urban niche. In general and when compared to rural counterparts, the behavior and ecology of species adapted to urban habitats are different. A summary of these differences is listed below. Note that these differences can be species-specific or generalizable to the whole assemblage of urban-adapted animals.

1. Movement and activity: Canada geese (*Branta canadensis*) have historically migrated up to 3000 miles to nest, but some individuals and flocks abandon migration altogether and become year-round urban residents (Hope 2000). Prolonged circadian activity is observed in urban birds singing earlier in the morning and later at night, and peregrine falcons feeding at night.
2. Reproduction: Changes in nesting locations were observed in Georgia, where 73 percent of the state's 1270 known breeding pairs of least terns (*Sterna antillarum*) nest on gravel roof tops compared with only 1 percent using traditional beach nesting grounds (Youth 1999). Prolonged breeding seasons are possible largely due to warmer winter conditions in urban habitats. A whole new assemblage of nesting sites and materials is available to wildlife in urban habitats (Figure 1.4). This can result in higher birth rates due to more reproductive cycles and greater offspring survival due to consistent supplies of anthropogenic food sources.
3. Tolerance of humans: Many urban species become habituated to human presence and lose at least some of their natural wariness. A survey of New Mexico urban and semirural residents found bobcat *(Felis rufus)* sightings were more frequent in areas of high-density housing than in "traditional" habitat, and that 70 percent of sightings were less than 25 meters from a house (Harrison 1998). Species living in urban areas often tolerate higher population densities than individuals living in rural habitats. Raccoon (*Procyon lotor*) density has been found to be higher in urbanized landscapes, possibly due to increased survival and reproduction rates and greater site fidelity in urban habitats (Prange, Gehrt, and Wiggers 2003).
4. Diet and nutrition: Feeding strategies also may change in urban habitats when animals make use of new sources of food (e.g., garbage, pet food, bird feeders, and hand-outs). Ring-billed gulls (*Larus delawarensis*) in Ohio, for example, have become more dependent on landfills than on fish from historic Great Lakes feeding areas (Belant, Ickes, and Seamans 1998; see also Chapter 8).
5. Survival and mortality: Unlike their rural counterparts, urban wildlife negotiates a whole new range of obstacles for survival. For example, birds and mammals in the urban landscape are often victims of collisions with vehicles, window panes, communication towers, wind generators, high-rise

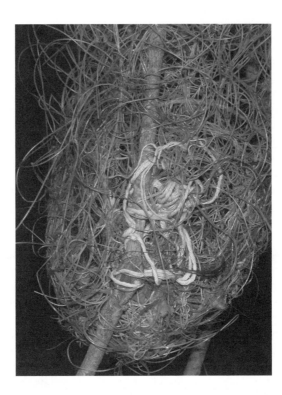

Figure 1.4   *(A color version of this figure follows page 158.)* Urban birds use a variety of human artifacts to build their nests. Pictured here is an oriole nest made entirely out of nylon fishing line and some yarn. (Courtesy Linda Causey; nest provided by Texas Cooperative Wildlife Collection)

buildings, and even with each other. Higher population densities increase intra-specific battles for food and shelter resources. Children and companion animals disturb or destroy adults and young in nests. Urban wildlife are exposed to many different forms of human abuse, entrapments, and entanglements not present in rural habitats. Diseases unique to urban habitats include backyard feeder diseases, such as salmonellosis, mycoplasma conjunctivitis, and trichomoniasis. Other diseases, easily proliferated in highly dense urban wildlife populations, include canine distemper virus, parasitic diseases, and mange. Urban wildlife are more likely to be exposed to the toxic effects of pesticides, heavy metals, and other pollutants resulting from human lifestyles in cities. Another obstacle to urban wildlife survival is to be "found" by a human as an infant and assumed to be abandoned. Wildlife rehabilitators across the country receive thousands of springtime calls from the general public concerning their discoveries of "orphaned" wildlife infants. Unnecessary human intervention or disturbance is driven by the same common misconceptions that motivate humans to adopt orphaned baby raccoons (see Sidebars 1.1 and 1.2; Burton and Doblar 2004; Ditchkoff, Saalfeld, and Gibson 2006).

SPECIAL CHALLENGES FOR WILDLIFE MANAGEMENT WITHIN URBAN SETTINGS

Wildlife management does not have an extensive history within metropolitan areas because prior to the mid-twentieth century America was a predominantly rural society. As a result, most wildlife management practices were developed with rural landscapes in mind. Urban and suburban environments present a host of special challenges for wildlife professionals (Decker, Brown, and Siemer 2001). A thorough discussion of the effects of urbanization on the wildlife management profession is presented in Chapter 2.

Urban Ecosystems

Urbanization is the process of transforming wild lands to better meet the needs and desires of humans. Usually this comprises clearing much of the native vegetation and replacing it with exotic species. Some, if not all, of the topsoil layer is removed and much of the remaining soil is either compacted, "waterproofed" by creating an impervious barrier to precipitation in the form of concrete or asphalt, or both. In the process, the land's habitat potential is significantly altered.

An *urban ecosystem* can be thought of as a system influencing, and being influenced by, human attitudes, behaviors, regulatory policies, and a sense of resource control throughout areas where humans live, work, and recreate at moderate to densely populated social scales (U.S. Department of Agriculture 1995). Certain species are naturally well suited to the altered environment, some are able to adapt to these changes, and others decrease in number or disappear through a combination of mortality and emigration. An extensive examination of urban ecosystems and population dynamics can be found in Part II: Urban Ecosystems.

Urban Habitats

Urban environments do not consist of one type of habitat. Examples of unique habitats within urban areas include parks, cemeteries, vacant lots, streams and lakes, residential yards, school grounds, corporate campuses, golf courses, airports, bridges, parking structures, and landfills. In addition, the extent of habitat diversity occurs within a much smaller area than would normally be found in rural landscapes. Part III: Urban Habitats and Hazards provides both large- and small-scale analyses of selected urban habitats.

Sociopolitical Factors

Most Americans like wildlife. They believe hearing birds sing and watching squirrel acrobatics add to their quality of life. However, this love affair with wildlife can quickly turn sour when animals "cross the line" from acceptable to unacceptable behavior (Schmidt 1997). The line is different for each individual. Human reactions to wildlife include a broad spectrum of emotions and reactions based on previous exposure to both formal and informal education programs and personal experience (Kellert 1980).

Natural resource agencies have a vested interest in shaping public understanding of wildlife management and conservation (Mankin, Warner, and Anderson 1999), and they have had success identifying and communicating with specific, traditional clienteles (Hesselton 1991; Kania and Conover 1991; Jolma 1994). However, the number of urban stakeholders who want explicit consideration in management has grown over the last several decades (Decker and Enck 1996).

While traditional stakeholders tend to have similar expectations for wildlife management, several regional studies suggest urban residents often have diverse and conflicting goals, such as the desire to reduce human-wildlife conflict *and* enhance wildlife viewing opportunities (Conover 1997). The wildlife profession has had difficulty communicating effectively with the general public (Decker et al. 1987; Gray 1993). Urban wildlife professionals must constantly consider both biological and sociopolitical factors when developing management strategies. Identifying stakeholders and attempting to understand their perspectives, expectations, and demands is a daunting task. Chapters 9 and 10 delve into the application of the human dimensions in urban wildlife management and the use of a stakeholder approach to management, respectively.

Whether we like it or not, wildlife management is driven by the bottom line. State and federal agencies prioritize wildlife research and management projects based on budgets. The primary driving force behind which wildlife species will be the focus of graduate student research at universities

Figure 1.5 Urban wildlife habitat may consist of a vast array of small plots of land classified as private property, easements, rights-of-way, and parks, and each with a different owner or regulatory jurisdiction. (Courtesy John M. Davis)

is the source and amount of funding. In this regard, literally millions of dollars have been spent
on deer, quail (*Colinus* and *Callipepla* spp.), and turkey research projects compared to the much
smaller amount to support urban wildlife management research projects. There is a bit of irony
here when one compares the shrinking population of hunters that benefit from game research with
the growing urban populations who desperately need a better understanding of the wildlife around
them. Within the wildlife profession, urban wildlife management is an empty niche that is quickly
being occupied by the private sector because there is money to be made by doing so. Furthermore,
many communities are generating additional revenue through various types of wildlife watching
programs, for example, Eagle Days in Missouri (see Chapter 9).

    Throughout most of the wildlife profession's history, management activities took place in
rural settings, primarily on public lands and large tracts of private agricultural and forest land.
Farmers, ranchers, and game managers have much in common; game management grew out of
an agricultural mindset of making land produce harvestable crops, including wildlife. Managers
have long worked to motivate private landowners in rural areas to enhance their properties for
wildlife (Decker, Brown, and Siemer 2001). In contrast, management of urban wildlife requires
working with many private owners holding small parcels of land. Consider for a moment the
logistics of gaining access to land in private subdivisions for management purposes—this would
involve the immensely complicated task of contacting possibly hundreds of property owners
(Figure 1.5). Public lands in metropolitan areas tend to consist of easements, rights-of-way, and
parks, all of which fall under the jurisdiction of a multitude of municipal and county governments
and private entities.

Wildlife professionals must learn to navigate a maze of legal considerations in urban and suburban areas. The sociopolitical landscape is cluttered with laws, regulations, ordinances, and policies, and varying levels of enforcement. Within city limits there exists a minimum of four layers of jurisdiction: federal, state, county, and municipal. Often there will be overlapping areas of responsibility within these jurisdictions. The issue of laws and jurisdiction, and their affect on urban wildlife management, are discussed in greater detail in Chapter 11.

Special Management Considerations

Ecological and sociological factors combine to create urban wildlife management challenges. On the wildlife side of the equation, a species' ecology and behavior can be used to predict generalities such as the presence and abundance of resident populations, while the specific circumstances often are tied to the geographic areas. On the human side of the equation, issues such as culture, economics, and politics can predict how humans will respond to a species' presence and abundance.

Chapter 12 examines three categories of wildlife management commonly encountered in the urban/suburban ecosystem, including management of threatened or endangered, introduced, and feral species. Chapter 14 takes an even closer look at the management implications of two prevalent urban species—resident Canada geese and urban deer.

No discussion of the overarching issues related to urban wildlife management would be complete without a consideration of zoonotic disease management (Chapter 13). In fact, the Department of Homeland Security has issued grants and contracts to research the possibility that certain types of zoonotic diseases and their vector species could be used for wide-scale disease transmission in urban communities. We present five categories of zoonotic diseases and discuss specific examples in terms of origins, carriers, life cycles, disease epidemiology, and control options.

O.K. let's get started. Recall an earlier quote by a professional biologist in the Introduction that reduced the whole concept to urban wildlife management to controlling raccoons in garbage cans? Inspired, we went to the literature and discovered an extensive body of information on urban raccoons. Needless to say, it is not the only focus of urban wildlife management, but the urban raccoon is definitely a much studied animal!

AN URBAN SPECIES OF SPECIAL INTEREST: THE RACCOON (*PROCYON LOTOR*)

The raccoon may be the wild mammal most suited to the urban environment. This might account for the common association of urban wildlife with a raccoon in a garbage can (Figure 1.6). For humans, raccoons have both aesthetic and utilitarian value. The raccoon has become a popular and charismatic animal in books, movies, and television programs. People will go to great lengths to invite raccoons into their backyards and homes, and raccoons are happy to accept the invitations. People are attracted to raccoons because they do not associate them with typical nuisance species (e.g., rats and mice). Raccoons are quite photogenic, with one possible exception. In Figure 1.6, one of the raccoons has a severe case of sarcoptic mange and has lost nearly all of its hair. Nevertheless, raccoons have playful mannerisms which people enjoy watching; even their clever thievery (i.e., the masked bandit) has endeared them to humans.

For some urban residents the raccoon is indeed one of the most beloved charismatic species. A television documentary many years ago showed a lady feeding a pair of raccoons Oreo™ cookies on her back patio. Soon the pair of raccoons brought their four offspring and they were also given cookies. As time went on, over a dozen raccoons showed up at her door wanting the cookies. The lady thought it would be interesting to see if the raccoons could find the cookies for themselves. She hid them in her den adjoining the back patio. The documentary ended showing about a dozen raccoons searching for cookies in her den. The raccoons probably left when no more cookies could be

Figure 1.6   Raccoons in a garbage can. Note one has mange causing the loss of most of its body hair. (Courtesy Justin Russell)

found, but the documentary left the viewer without any closure on this human-wildlife interaction. Supplemental feeding, intentionally or unintentionally, is a common type of interaction between humans and raccoons in urban environments.

Pet Raccoons

It was surprising to learn that First Lady Grace Coolidge had a pet raccoon, called "Rebecca." The raccoon was kept in a special house built by the president. Young raccoons are skilled at looking cute, so it's easy to see why people might think they would make enchanting (if potentially illegal) pets. In fact, we found many requests for information about pet raccoons on the Internet. However, it's a different story when that cute baby grows up to become a thirty-plus pound wild animal (see Sidebar 1.1). The "pet" raccoon, which by adulthood is habituated and unafraid of humans, does not learn proper raccoon behavior. Raccoons raised by untrained members of the public do not learn any of the life skills a raccoon mother provides, including foraging, avoiding dangerous situations (including humans), and proper raccoon behavior when encountering other raccoons or wildlife. Releasing such an animal "back to the wild" is tantamount to a death sentence.

Humans justify the adoption of young wild animals because of several misconception or myths they have about wild animals. The first, and one of the most common myths, is that "infant animals handled by humans will be immediately rejected by their parents." Second, that a solitary young animal has been orphaned—in truth, the parent is probably nearby. Third is the misconception that a Good Samaritan is an acceptable substitute for the natural wild parents. And lastly, the wildlife "rescuer" mistakenly perceives the wild infant as an acceptable long-term pet (Burton and Doblar 2004). The bottom line is that regardless of whether the animal is a raccoon, squirrel, white-tailed deer, or coyote (*Canis latrans*), from a moral, ethical, and humane standpoint, keeping a wild animal as a pet is never in the best interest of the animal.

Raccoon Economics

"The raccoon is unquestionably the most economically important furbearer in North America" (Zeveloff 2002). The raccoon pelts industry was valued at $100 million in 1982 (http://www.furs.

com/price.html), and even in the early twenty-first century, raccoon pelts are sought after by the garment industry. Raccoon trapping, and the associated pelt trade, is a growing business in the northern latitudes of the United States and Canada. Coats made with the raccoon's long-haired fur and signature gray and black shadings retail for between $2,000 and $6,000. In the 1950s, a popular television program featured Davy Crocket wearing a coonskin hat in every episode, which made coonskin hats all the rage for young viewers. Today, one can buy a coonskin hat made in Canada for $129.

The raccoon is also one of the most chronic human-wildlife conflict species in North America (Zeveloff 2002, see Table 2.1 in that publication). There is no published data on the magnitude of economic loss caused by raccoons in terms of property damage or disease transmission to humans, their pets, and other urban animals. However, it is unlikely that hunting could ever be used to control urban raccoon populations because of the methods used and the state regulations that govern the taking of game and fur-bearing animals.

Raccoons as Disease Vectors

There is a dark side to human-raccoon interactions as well. Raccoons are carriers of diseases (viral and bacterial) and parasites (external and internal) that can become a health hazard for humans and their companion animals (Table 1.1). Raccoon parasites run the taxonomic gamut, from protozoans to helminthes to arthropods. While not all of them have zoonotic implications, a wide variety of parasites have been found in the esophagus, stomach, intestines, pancreatic ducts, urinary bladder, subcutaneous tissue (e.g., in the skin of the feet), and lungs of raccoons (Johnson 1970). Of special concern to humans is the roundworm parasite, *Baylisascaris procyonis* (see Chapter 13). One study in Kansas found that up to 75 percent of the raccoons examined were infected with *B. procyonis* (Robel, Barnes, and Upton 1989). Raccoon ectoparasites include fleas, ticks, and sucking lice, all of which can be vectors of other diseases. In fact, it has been found that disease and parasites are the primary raccoon population control mechanisms.

Table 1.1   Some Infectious Diseases of Raccoons in the United States

Fungal diseases	Histoplasmosis
Viral diseases	Rabies
	Canine distemper
	Infectious enteritis
	Fox encephalitis
	Saint Louis encephalitis
	Eastern equine encephalitis
Bacterial diseases	Leptospirosis
	Tularemia
	Listeriosis
	Tuberculosis
Protozoan	Trypanosomiasis
	Coccidiosis
	Toxoplasmosis

Source: From A.A. Johnson. 1970. *Biology of the Raccoon* (Procyon lotor *varius Nelson and Goldman*) *in Alabama.* Auburn University Agricultural Experiment Station Bulletin 402. Auburn, AL: Auburn University.

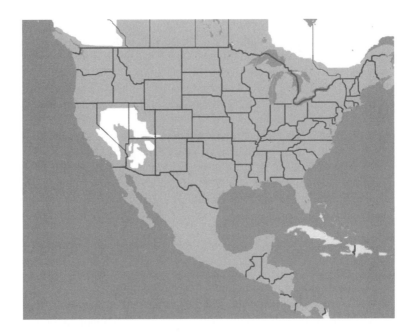

Figure 1.7   Raccoon (*Procyon lotor*) distribution in the Western Hemisphere. [Rendered from Zeveloff (2002) by Linda Causey]

The Urban Raccoon

Urban raccoon ecology, natural history, population dynamics, management, and examples of their interactions with humans have been given a fairly extensive treatment in the literature, much of which has been listed and summarized in Johnson (1970), MacClintock (1981), Zeveloff (2002), Defenders of Wildlife (2004), and Hadidian (2007). Thus, this section will touch on the highlights of these and other references to explain how the raccoon has become one of the most common animals found in urban environments. In fact, raccoons have the widest geographic distribution when compared to all other North American mammals (Figure 1.7). The wide distribution of the raccoon is due to the fact that it is not "picky" about anything, including: diet; den sites and types; climate and type of habitat; breeding seasons and mates; and proximity to other animals, including humans.

Diet

"Their diets include everything that is edible, a fact of life that accounts more than any other for their adaptability" (MacClintock 1981). "The raccoon's diet is so highly varied that it seems easier to describe the foods he doesn't eat, rather than those he does" (Hadidian 2007). Another author put it this way, "Given their broad omnivory, consummate opportunism, and wide-spread distribution, a complete listing of the raccoon's foods would be long, tedious, and perhaps impossible" (Zeveloff 2002). Even so, we felt up to the challenge to develop some type of synthesis of the raccoon's diverse diet, realizing what they eat is dependent on place, season, opportunity (access and availability), and learned behavior. In one study, information on the diet of raccoons was obtained from examination of the contents of 79 stomachs and 70 large intestines from 112 raccoons, and from the contents of 365 scats (fecal samples) (Johnson 1970). Other than direct observation of feeding behavior, these are the usual methods used to determine wild mammal diets. In Table 1.2 we have categorized the raccoon's diet into its basic food groups, including animals, plants, and anthropogenic sources.

Table 1.2   What the Raccoon Eats

Class	Example(s)
	Animals
Invertebrates	
Annelids	Earthworms
Crustaceans	Crayfish (their most favorite food)
Mollusks	Slugs, snails (land and water), oysters, clams
Insects	Grasshoppers, ground beetles, cut worms, beetle borers, crickets, cicadas, centipedes, millipedes, water beetles, crane fly larvae, caterpillars, wasp and hornet nests, bee hives, spiders and their egg cases
Vertebrates	
Fish	Minnows, trout, buffalo head, bull head, carp, sunfish, catfish, suckers, shiners, eels, shad, salmon, bass
Amphibians	Frogs (any catchable species), tadpoles, salamanders, toads
Reptiles	Turtles (yellow-bellied, mud, painted, soft-shell) or their eggs, sea turtle eggs, snakes (garter, red-bellied, copperhead, rattlesnake), skinks, fence lizard, alligator eggs
Birds	Eggs, nestlings, or adults, including woodpeckers, meadow larks, cardinals, song sparrows, blackbirds (yellow-headed, brown-headed, red-winged), pheasants, raptors, herons, egrets, cormorants, grackles, coots, rails, waterfowl, marsh wrens, wood ducks, herring gulls, sea birds, domestic poultry
Mammals	Young or adult muskrats, coypu, nutria, mice (several different genera and species), shrews, cottontail rabbits, pocket gophers, ground squirrels, chipmunks, squirrels (gray, fox, and flying), rats (cotton, wood, and Norway), mink, white-tailed deer, bats
	Plants
Fruits	Berries (any kind if available, edible, and sweet), grapes, cherries, plums, raspberries, watermelon, apples, figs, tomatoes, peaches, persimmons
Seeds	Field and sweet corn (in the milk stage), sorghum, peanuts, sweet potatoes, sunflower, grains (wheat and barley), acorns, nuts (beech, hickory, pecans, walnuts, chestnuts)
Stems	Sugarcane, cordgrass
	Anthropogenic
Garbage (homes, restaurants, grocery stores), pet food, bird feeders, road kills, intentional food supplements	

Source: Johnson (1970); Hoffman and Gottschang (1977); MacClintock (1981); Zeveloff (2002); Hadidian (2007).

Raccoons may also resort to cannibalism, with larger individuals consuming the smaller members of the population (Zeveloff 2002).

The trend in seasonal menus usually consists of an animal diet in the spring, fruits in the summer, and plants and seeds in autumn. Crayfish (Cambaridae) are a much sought after food regardless of season. The raccoon's diet, albeit seasonal, contains every component of the basic food groups found in the human food pyramid (Figure 1.8). This is an important comparison, demonstrating that the raccoon's food preferences make it well adapted to survive on the same foods that humans eat … and discard.

Denning

The wide geographic distribution of the raccoon indicates they live in habitats that are as varied as their diet. The only areas devoid of raccoons are the deserts of the western United States and the most extreme northern latitudes of Canada. Otherwise, they are found pretty much anywhere in the Western Hemisphere, and there are 25 subspecies based on geographic location (Zeveloff 2002). The basic raccoon habitat requirements are a permanent water supply, den sites (preferably trees),

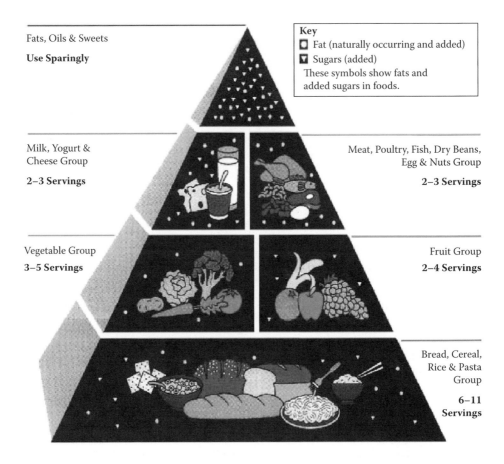

Figure 1.8 The basic food groups and how they should be distributed in the human diet. (http://www.healthcare. com/tag/basic-food-pyramid/)

and available food. On a smaller scale of analysis, ideal habitats for raccoons are wooded areas that have mature trees associated with a wetland (including swimming pools, fishponds, and bird baths in urban areas) of some type (Figure 1.9).

Trees are important for many reasons, including shelter from the elements, security from predators, a place to give birth, and as a source of food (nuts, berries, fruit). Raccoons are extremely adept tree climbers. They can negotiate vertical space in trees with the same agility as horizontal space on the ground. They climb up and down trees head first or tail first—doesn't matter. Raccoons have been observed jumping out of trees (from 30 feet up, in one case) to escape danger, without harming themselves in the process. They are also excellent swimmers. Many a 'coon hunter can attest to the animal's agility in water, having watched their 'coon hounds drowned by the same raccoon(s) they were trying to bring to bay. In areas that have few trees—grassland ecosystems—raccoons will den in ground burrows made and vacated by badgers, ground hogs, and armadillos; or they will den in brush piles (Zeveloff 2002).

Urban raccoons tend to seek out large trees for den sites when given a choice of alternatives. One study found that urban raccoons have many different tree dens that are shared by other raccoons although not at the same time (Hadidian, Manski, and Riley 1991). Being consummate opportunists, they also use several different areas in urban homes, including the attic, chimney, and crawl spaces within or under the house (Figure 1.11). They also use outbuildings such as garages, sheds, and abandoned doghouses as den sites. Raccoons have even been found in ventilation ducts of

Figure 1.9 Ideal raccoon habitat (e.g., water and woods) in an urban neighborhood. (Courtesy Clark E. Adams)

Figure 1.10 *(A color version of this figure follows page 158.)* Baby raccoons (kits) in a tree den located right outside the bedroom window of an urban resident. (Courtesy Wes Orear)

Figure 1.11   Where raccoons can live in urban neighborhoods. (Concept by Clark E. Adams, drawing by Linda Causey)

high-rise office buildings in the heart of downtown Cincinnati and probably in every other municipality in the United States. One female even commandeered an unused sofa on the third floor of a house to raise her litter of four (Cauley and Schiner 1973). In fact, any accessible structure or part of a structure can offer a safe shelter and den site. Other refuge sites include sewer and storm-water drainage tiles; abandoned houses; and church belfries (Hoffman and Gottschang 1977). The bottom line is that raccoons appear to have many addresses in urban neighborhoods. As such, a reasonable raccoon management option in urban areas would be to evict them from a residence, close all access points, and simply let them loose. They will probably seek out one of their alternative addresses in the neighborhood.

Population Densities

As we will discuss in Chapter 6, population dynamics is a study of factors that promote a species' presence or absence in various habitats. Raccoons have overcome the usual stresses on wildlife in urban ecosystems that affect "normal" movement and activity, diet and nutrition, reproduction, survival, and mortality (Ditchkoff, Saalfeld, and Gibson 2006). As such, urban raccoon population densities have ranged from 41 to 333/km²—values depending on the area being sampled and used for extrapolations (Smith and Engeman 2002). When compared to rural populations, urban/suburban raccoon populations exhibit unusually high numbers per unit area. Other studies isolated at least seven factors (listed below) that contributed to high raccoon populations in urban areas (Hoffmann and Gottschang 1977; Rolley and Lehman 1992; Gehrt and Clark 2003; Prange, Gehrt, and Wiggers 2003; Smith et al. 2003; Prange, Gehrt, and Wiggers 2004; Randa and Yunger 2006; Bozek et al. 2007).

1. Urban raccoons rely heavily on anthropogenic resources and are the most efficient of urban-adapted mesopredators at exploiting these resources.
2. Habitat use by raccoons in urban areas is concentrated in human-use areas.
3. Raccoons have nearly unrestricted movement throughout the urban matrix. As such, removal of resident raccoons from high quality habitat may result in rapid immigration by outsiders.
4. When compared to rural counterparts, urban raccoons have increased survival, high annual recruitment (i.e., larger litter sizes), smaller and more linear home ranges, and increased site fidelity. The linearity of raccoon home ranges in urban areas suggests they reduce travel time between food, water, and den areas.
5. Urban raccoons have few natural predators except for large owls, coyotes, bobcats, dogs, and humans. Telemetry studies of raccoon survival have consistently found that predation is a minor cause of mortality. However, the majority of raccoon deaths in urban neighborhoods are the result

of being run over by cars or from diseases such as distemper. Disease and parasites can be spread or transmitted quickly in highly dense urban raccoon populations.

6. Urban raccoons have few competitors for their habitat essentials of food, den sites, and water; a virtually unlimited supply of each limits intraspecific competition.

7. Concentration of anthropogenic food sources, high-density urban development, and greater vehicle traffic along roadways contribute to decreased movements and dispersal rates of raccoons in urban areas. Therefore the population's reproductive surplus is absorbed into the local population rather than dispersing into unoccupied or less densely populated raccoon habitats.

Given the above conditions for survival, it is probably safe to say that raccoon populations can be found in any urban or suburban community, large or small, in the contiguous United States. They are usually nocturnal but still have the highest rate of direct and indirect contact with urban residents. The downside of elevated raccoon densities in urban landscapes is altered community structure, effects on other species by monopolizing resources, spreading disease and parasites, and altering predator-prey relationships (Prange and Gehrt 2004).

SIDEBAR 1.1 Rascal

Prepared by Irene Reeb and Clark E. Adams

In the early 1970s, my neighbors were given the opportunity to adopt a baby raccoon. Its mother had been killed by a dog and the baby raccoon was left to survive on its own. Initially, the kit was nursed with an eye dropper until it was old enough to ingest solids, for example, dog food. The foster parents had to move out of the state and needed to find an alternative home for the young raccoon. My neighbors accepted the kit, which became the pet of their 8-year-old daughter, Dawn. She named the raccoon "Rascal" and kept it in a large cage under their patio deck. Dawn loved the raccoon. She carried it around and enjoyed sharing it with the neighborhood children. Rascal was fed dog food and table scraps. He was very tame and never bit the children. He did scamper through the house and tried to climb the house plants, so Dawn had to keep Rascal in her room when she wanted to play with him or wasn't holding him. When school began, Rascal stayed in his cage all day long, with play time limited to after school hours. It was decided he couldn't be happy living that way. He had reached adult size so he was taken down to Plum Creek about half mile from their home and released. Rascal returned several times to Dawn's home, especially in the evenings. He would scratch on or sit right outside the sliding glass door and wait for food. After a while, he returned only occasionally. After about one month he did not return at all.

Later that fall, another friend, Luther, who worked at the local college, called Dawn's mother. "Irene," he said, "come get your raccoon. He's scaring people in front of Dorcas Hall." Irene drove to the college campus and parked in front of the dormitory. She saw an obviously frightened raccoon walking stiff legged with its fur bristled being confronted with an audience of student spectators. She assumed it was Rascal because what other raccoon would come into such close association with humans? Irene put on a pair of leather gloves, opened the car trunk, and called out Rascal's name. This raccoon did not respond, and instead it ran under the car and then crawled up onto the motor. Luther opened the hood of the car. Irene grabbed the raccoon by the scruff of its neck. It turned and bit her on the hand through the glove. Even though it hurt, Irene hung on and tossed the raccoon in the trunk of her car and closed it.

When she got home and opened the trunk she knew the raccoon she had captured was not Rascal, based on its markings and visible fear of her. She closed the trunk and waited until her husband came home. He used a heavy bag to catch the raccoon and put him into the cage to be examined by a local veterinarian for symptoms of rabies. Irene received a tetanus shot and after a few days, the "mystery raccoon from the wild" was released to head for Plum Creek and freedom. It is unknown what really happened to Rascal.

SIDEBAR 1.2 Raccoon Survival

We've chosen to end this section by presenting some evidence of a human-raccoon encounter that resulted in dire, long-term consequences for the raccoon (C.E. Adams 1983, used with permission from the National Association of Biology Teachers).

One of my former students collected and prepared a study skin and cleaned the skull of a road-killed raccoon. Close examination of the skull revealed that the raccoon may have experienced a human encounter prior to the lethal event. Figure 1.12 shows the normal condition and relation of boney elements; postorbital process (PO), zygomatic arch (ZG), and coronoid process (CP). The skeletal aberrations appear on the left side particularly in the region of the coronoid process on the lower mandible, and postorbital process and zygomatic arch on the cranium (Figures 1.13 and 1.14). The coronoid process from the mandible was severed and refused onto the zygomatic arch which was also broken and rehealed as was the postorbital process. What happened?

The skull was that of an old male, as indicated by the well-developed sagittal crest on the top of the cranium, so it appears the bone damage occurred when this raccoon was much younger. It is entirely possible the animal was shot in the head, with the bullet passing first through the mandible, shearing off the coronoid in the process, severing the zygomatic arch, passing through the left eye, and glancing off the postorbital process of the cranium (Figure 1.14). The loose coronoid process and severed zygomatic arch, over a long period of time, refused as one boney mass (CP/ZG) during the healing process. The condition of the left eye, unfortunately, was not recorded on the specimen card.

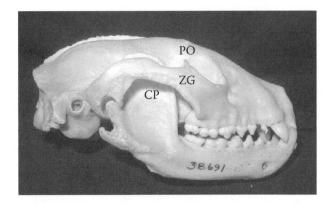

Figure 1.12   Undamaged right side of raccoon skull showing the normal condition of the postorbital process (PO), zygomatic arch (ZG), coronoid process (CP), and position of the lower mandible in relation to the upper maxilla. (Courtesy Walter Causey)

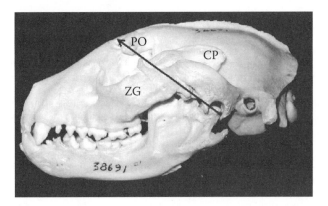

Figure 1.13   Damaged left side of the raccoon skull showing an abnormal postorbital process (PO), a fused coronoid process (CP) from the lower mandible with the zygomatic arch (ZG) on the upper maxilla, and extreme tooth wear on the upper and lower canine teeth. The arrow indicates the suspected path of the bullet. (Courtesy Walter Causey)

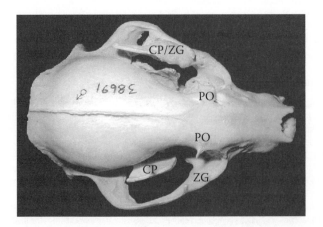

Figure 1.14   Dorsal view of the raccoon skull that illustrates the extent of the damage on the left side when compared to the normal right side. The fusion of the coronoid process with the zygomatic arch (CP/ZG) is particularly evident as is the damaged postorbital process (PO). (Courtesy Walter Causey)

Assuming the above explanation of the damaged skull is correct, one can only imagine the torturous pain and prolonged discomfort this raccoon experienced during the healing process. As is often the case, wounded animals will seek the shelter of their den. They can remain in their den for long periods of time until they feel they are absolutely safe. This raccoon was probably looking down at the hunters, giving them a clear head shot. Once he was hit, he may have quickly slipped into a tree den. It must have had to seek out incredibly soft foods given its inability to chew solid foods. It is unknown how long it took for the bones in this raccoon's skull to heal in the manner illustrated. This story exemplifies the resiliency of crippled animals' and their abilities to rehabilitate themselves—in this case, only to be harmed again by human actions.

SPECIES PROFILE: KILLDEER (*CHARADRIUS VOCIFERUS*)

They [plovers] sprint eight feet and stop. Like that.
They spring a yard (like that) and stop…
When they stop they, suddenly, are gravel.

—**Norman McCaig, Poet**

The killdeer is one of our more well-known shorebirds, possibly because you don't have to go to the shore to see one (Figure 1.15). Once in serious decline due to unregulated market hunting, killdeer can now be viewed in urban habitats. This North American plover, which has a cravat-like set of two dark bands against a white breast and a high-pitched, almost keening call, can be seen on gravel roads, parking lots, lawns, and golf courses. These flat surfaces must appear similar to the flat grasslands, sandbars, and mudflats used by their country cousins. Lighted parking lots attract plenty of insects, which make up about 75 percent of the bird's diet, along with some small invertebrates and seeds (Ehrlich, Dobkin, and Wheye 1988).

A killdeer's nest is not elaborate. Built on open ground or even flat rooftops, especially if either is gravel covered, nests may be unlined or lined with grass and other local materials. Both the male and female are active in chick-rearing. After courtship, the female will lay three to five buffy eggs with dark mottling in early spring (March to April, depending on the location). Chicks hatch about four weeks later and leave the nest within hours. Adult killdeer are known for their broken-wing distraction displays, in which they feign an injury to distract and lure predators away from their

Figure 1.15   Killdeer. (Courtesy Carl Schmidt)

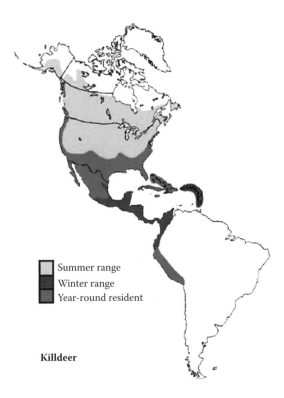

Summer range
Winter range
Year-round resident

Killdeer

Figure 1.16   Geographical distribution of the killdeer. http://www.fort.usgs.gov/products/Publications/555/ maps/range/range_tables/kill.htm.

offspring, then fly away at the last minute (Ehrlich, Dobkin, and Whey 1988). Rooftop nests present a different kind of threat to killdeer chicks. While they do afford some protection from common predators of ground-nesting birds, the downy, precocial killdeer chicks find it difficult to follow their parents away from the roof shortly after hatching to search for food. They will try, however, rather than be left behind. Sometimes the tiny chicks are unable to surmount the raised perimeter of the roof; in other cases, chicks have been observed leaping from multi-story buildings, and some— but certainly not all—do survive (Cornell Lab of Ornithology website).

Green roofs, depending on their size, may alleviate this problem somewhat, in part because they may allow the birds to find food without immediately leaving the roof. During 2007–2008, Swearingin et al. (2008) evaluated wildlife use of a newly constructed green roof approximately 320 m^2 in area. Wildlife surveys were conducted each week from January 2007 to March 2008. During the 13-month study, a total of 157 birds were observed flying over or using the green roof. Of the birds actually using the roof, 72 percent were killdeer, and these birds nested while others simply perched or loafed.

Despite their wide distribution (Figure 1.16), and though they are perceived to be common and abundant, even in altered or degraded urban habitats, several studies suggest killdeer populations may not be doing well in either rural or urban landscapes (Sanzenbacher and Haig 2001). The species is vulnerable to any number of twentieth-century problems, including domestic cats, pesticides, oil pollution, lawnmowers, and automobiles (Cornell Lab of Ornithology website).

Killdeer, particularly urban populations, are not well represented in the published research literature. The papers we were able to find focus on behavioral ecology, particularly as it relates to nesting. Other topics of research interest include social organization, vocalizations, and susceptibility to toxic agents. Sadly, it is urban "nuisance" species that tend to get attention (and research dollars), and since there are few, if any, human-wildlife conflicts associated with the presence of killdeer, the species does not appear to be a research priority (Cornell Lab of Ornithology website).

CHAPTER ACTIVITIES

1. Find out what portion of your state's annual DNR budget comes from either of the two Restoration Acts (Pittman–Robertson or Dingell–Johnson), then determine how much is spent on urban wildlife management issues.
2. Describe an important wildlife experience from your own childhood and how your relationship with wild things has changed since your childhood.
3. Keep track of urban wildlife issue stories covered by your local media—newspapers, local television and/or radio stations) over a two-week period. National news wire stories (AP, UPI, etc.) can be included as long as you found them through a local source. For each story you see, read, or hear, list the following information: publication name, article title, author, and publication date.
4. How would you check for the presence of raccoons in or around your home or in your neighborhood? Call your local wildlife control company to get information on raccoon activities in your neighborhood.

LITERATURE CITED

Adams, C.E. 1983. Road-killed animals as resources for ecological studies. *American Biology Teacher* 45:256–261.

Adams, L.W. 1994. *Urban Wildlife Habitats: A Landscape Perspective.* Minneapolis, MN: University of Minnesota Press.

Adams, L.W., L.W. VanDruff, and M. Luniak. 2005. Managing urban habitats and wildlife. In *Techniques for Wildlife Investigations and Management,* sixth edition, 713–734, ed. C.E. Braun. Bethesda, MD: The Wildlife Society.

Belant, J.L., S.K. Ickes, and T.W. Seamans. 1998. Importance of landfills to urban-nesting herring and ring-billed gulls. *Landscape and Urban Planning* 43:11–19.

Bozek, C.K., S. Prange, and S.D. Gehrt. 2007. The influence of anthropogenic resources on multi-scale habitat selection by raccoon. *Urban Ecosystems* 10:413–425.

Burger, J. 1999. *Animals in Towns and Cities.* Dubuque, IA: Kendall Hunt Publishing Company.

Burton, D.L. and K.A. Doblar. 2004. Morbidity and mortality of urban wildlife in the Midwestern United States. In *Proceedings of the 4th International Symposium on Urban Wildlife Conservation*, 171–181, ed. Shaw, W.W. et al. Tucson, AZ.

Cauley, D.L. and J.R. Schiner. 1973. The Cincinnati raccoons. *Natural History* 82:58–59.

Conover, M.R. 1997. Wildlife management by metropolitan residents in the United States: Practices, perceptions, costs, and values. *Wildlife Society Bulletin* 25:306–311.

Decker, D.J., T.L. Brown, B.L. Driver, and P.J. Brawn. 1987. Theoretical developments in assessing social values of wildlife: Toward a comprehensive understanding of wildlife recreation involvement. In *Valuing Wildlife: Economic and Social Perspectives,* 76–95, ed. D.J. Decker and G.R. Goff. Boulder, CO: Westview.

Decker, D.J., T.L. Brown, and W.F. Siemer. 2001. *Human Dimensions of Wildlife Management in North America.* Bethesda, MD: The Wildlife Society.

Decker, D.J. and J.W. Enck. 1996. Human dimensions of wildlife management: Knowledge for agency survival in the 21st century. *Human Dimensions of Wildlife* 1:60–71.

Defender of Wildlife. 2004. *People and Predators: From Conflict to Coexistence,* ed. N. Fascione, A. Delach, and M.E. Smith. Washington, DC: Island Press.

Ditchkoff, S.S., S.T. Saalfeld, and C.J. Gibson. 2006. Animal behavior in urban ecosystems: Modifications due to human-induced stress. *Urban Ecosystems* 9:5–12.

Ehrlich, P.R., D.S. Dobkin, and D. Wheye. 1988. *The Birder's Handbook: A Field Guide to the Natural History of North American Birds.* New York: Simon and Schuster.

Fletcher, D. 1994. Urban wildlife: An overview. *Urban Animal Management Conference Proceedings*, Canberra, Australia.

Gehrt, S.D. and W.R. Clark. 2003. Raccoons, coyotes, and reflections on the mesopredator release hypothesis. *Wildlife Society Bulletin* 31:836–842.

Gray, G. 1993. *Wildlife and People: The Human Dimensions of Wildlife Ecology*. Urbana, IL: University of Illinois.

Harrison, R.L. 1998. Bobcats in residential areas: Distribution and homeowner attitudes. *Southwestern Naturalist* 43:469–475.

Hesselton, W.T. 1991. How governmental wildlife agencies should respond to local governments that pass anti-hunting legislation. *Wildlife Society Bulletin* 19:222–223.

Hadidian, J. 2007. *Wild Neighbors*. Washington, DC: Humane Society Press.

Hadidian, J., D.A. Manski, and S. Riley. 1991. Daytime resting site selection in an urban raccoon population. In *Proceedings of the Second National Symposium on Urban Wildlife*, ed. L.W. Adams and D.L. Leedy. National Institute for Urban Wildlife,

Hoffmann, C.O. and J.L. Gottschang. 1977. Numbers, distribution, and movements of a raccoon population in a suburban residential community. *Journal of Mammalogy* 58:623–636.

Hope, J. 2000. The geese that came in from the wild. *Audubon* 102:122–126.

Johnson, A.A. 1970. *Biology of the Raccoon* (Procyon lotor *varius Nelson and Goldman*) *in Alabama*. Auburn University Agricultural Experiment Station Bulletin 402. Auburn, AL: Auburn University.

Jolma, D.J. 1994. Attitudes Toward the Outdoors: An Annotated Bibliography of U.S. Survey and Poll Research Concerning the Environment, Wildlife, and Recreation. Jefferson, NC: McFarland.

Kania, G.S. and M.R. Conover. 1991. Another opinion on how governmental agencies should respond to local ordinances that limit the right to hunt. *Wildlife Society Bulletin* 19:222–223.

Kazacos, K.R. and W.M. Boyce. 1989. *Baylisascaris larva migrans*. *Journal of the American Veterinary Medical Association* 195:894–903.

Kellert, S.R. 1980. American's attitudes and knowledge of animals. Transactions of the North American Wildlife and Natural Resources Conference 45:111–124.

Lutts, R.H. 1990. *The Nature Fakers: Wildlife, Science, and Sentiment*. Golden, CO: Fulcrum.

MacClintock, D. 1981. *A Natural History of Raccoons*. New York: Charles Scribner's Sons.

Mankin, P.C., R.E. Warner, and W.L. Anderson. 1999. Wildlife and the Illinois public: A benchmark study of attitudes and perceptions. *Wildlife Society Bulletin* 27:465–472.

Maurizi, R. and T. Friedner. 1997. The effect of city living on urban wildlife. *ASCI*. http://www.uvm.edu/~rmaurizi/oldStuff/urban.html, accessed July 7, 2004.

Prange, S. and S.D. Gehrt. 2004. Changes in mesopredator-community structure in response to urbanization. *Canadian Journal of Zoology* 82:1804–1817.

Prange, S., S.D. Gehrt, and E.P. Wiggers. 2003. Demographic factors contributing to high raccoon densities in urban landscapes. *Journal of Wildlife Management* 67:324–333.

Prange, S., S.D. Gehrt, and E.P. Wiggers. 2004. Influences in anthropogenic resources on raccoon (*Procyon lotor*) movements and spatial distribution. *Journal of Mammalogy* 85:483–490.

Randa, L.A. and J.A. Yunger. 2006. Carnivore occurrence along an urban-rural gradient: A landscape-level analysis. *Journal of Mammalogy* 87:1154–1164.

Robel, R.J., N.A. Barnes, and S.J. Upton. 1989. Gastrointestinal helminthes and protozoa from two raccoon populations in Kansas. *Journal of Parasitology* 75:1000–1003.

Rolley, R.E. and L.E. Lehman. 1992. Relationships among raccoon road-kill surveys, harvests, and traffic. *Wildlife Society Bulletin* 20:313–318.

Root, W. and R. De Rochemont. 1994. *Eating in America: A History*. New York: Ecco Press.

Sanzenbacher, P.M. and S.M. Haig. 2001. Killdeer population trends in North America. *Journal of Field Ornithology* 72(1):160–169.

Schaefer, J. 2004. *Florida's Definition of Urban Wildlife*. http://www.wec.ufl.edu/faculty/ SchaeferJ/3401/November24.html, accessed July 7, 2004.

Schmidt, R. 1997. Drawing the line for wildlife. *Wildlife Control Technology* 4:6–7.

Shaw, W.W., L. Harris, M. Livingston, J.-P. Charpentier, and C. Wissler. 2003. Wildlife habitats in urban environments. Paper presented at *Effects of Urbanization Symposium: Wildlife Habitats in Urban Environments*. University of Arizona, School of Renewable Natural Resources, Tucson, AZ.

Smith, H.T., R.M. Barry, R.M. Engeman, S.A. Shwiff, and W.J.B. Miller. 2003. Species composition and legal economic value of wildlife road-kills in an urban park in Florida. *Florida Field Naturalist* 31:53–58.

Smith, H.T. and R.M. Engeman. 2002. An extraordinary raccoon, *Procyon lotor*, density at an urban park. *Canadian Field Naturalist* 116:636–639.

Swearingin, R. M., C. Pullins, T. Guerrant, and B. Washburn. 2008. Green roofs in the airport environment: Pleasant dreams or nightmares? Poster presentation at the 10th Annual Bird Strike Committee Meeting, Orlando, FL.

Trefethen, J.B. 1975. *An American Crusade for Wildlife*. Alexandria, VA: Boone and Crockett Club.

U.S. Census Bureau. 2000. *Urban and Rural Classification Census 2000*. Washington, DC: United States Government Printing Office.

U.S. Census Bureau. 2006. *National Survey of Fishing, Hunting and Wildlife-Associated Recreation*. Department of the Interior, Fish and Wildlife Service. Washington, DC: United States Government Printing Office.

U.S. Department of Agriculture. 1995. *Urban Community Ecosystems: A National Action Plan*. Washington, DC: United States Government Printing Office.

U.S. Department of Agriculture. 2001. *The Urban Forestry Manual*. Athens, GA: Southern Center for Urban Forestry Research and Information, USDA Forest Service.

Youth, H. 1999. These rare birds flock to a shopping mall in search of home furnishings. *National Wildlife* 37(4):18–19.

Zeveloff, S. 2002. *Raccoons: A Natural History*. Washington, DC: Smithsonian Institution Press.

CHAPTER 2

The Changing Landscape of Wildlife Management

There are two spiritual dangers in not owning a farm. One is the danger of supposing that breakfast comes from a grocery store, and the other that heat comes from the furnace."

—Aldo Leopold (*A Sand County Almanac*, 1949)

KEY CONCEPTS

1. Four societal events set the stage for urban wildlife management.
2. A description of how human-wildlife interactions changed with the societal shift from agrarian to urban life styles.
3. Four lines of evidence justify the need for wildlife management in urban areas.
4. There are several justifications for educational programs about urban wildlife.
5. How urban residents reconnect with the natural world around them.
6. Three criteria designate "nuisance" wildlife.
7. The scope of urban wildlife problems compared in terms of damage, offending species, and cost to the urban resident.
8. Examples are provided to show how urban wildlife management can become an integral part of the training and experience of wildlife professionals now and in the future.

DEMOGRAPHIC FACTORS THAT SET THE STAGE FOR URBAN WILDLIFE MANAGEMENT

The convergence of three societal events at the end of World War II set the stage for urban wildlife management. First, automobiles and homes became more affordable and available, and returning soldiers had the means to purchase both. Second, Cold War mania provided a stimulus for the passage of the Highway Revenue Act of 1956, which created the Highway Trust Fund. This legislation enabled the development of the network of super highways through, around, and out of the cities. The original political justification for extensive highway construction was to provide city populations a rapid escape mechanism in the event of nuclear attacks. Lastly, America began to shift from a largely agrarian to a primarily urban society around 1945, as people moved away from family farms to the city for work (Figure 2.1). Highways, affordable automobiles, really cheap gas (about ten cents/gallon), and an increasingly urban population created the opportunity for mass human migration from self-contained communities in cities and small towns to suburban developments several miles from the city core. In fact, urbanization was the most massive and sudden shift of humanity in its history.

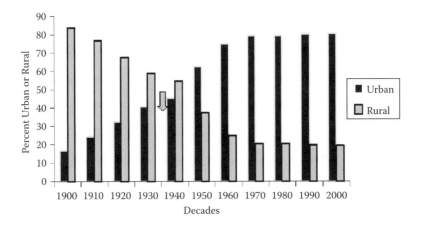

Figure 2.1 Change from rural to urban residence in the United States 1900 to 2000. (Courtesy Clark E. Adams)

These three events led to a phenomenon called *urban sprawl,* "a process where the perimeters of the city are extended outward into the countryside, one development after the next, with little plan as to where the expansion is going and no notion as to where it will stop" (Wright 2004). Urban sprawl set the stage for a human-wildlife interface that had not existed before, and reinforced a separation between people and wildlife (Figure 2.2). However, the trend toward urbanization is not limited to the United States, or even North America. According to a 2003 study, 74 percent of people living in developed countries could be classified as urban. The same study found that in less developed

Figure 2.2 Urban sprawl in El Paso, Texas. (Courtesy John M. Davis)

regions, 42 percent of the population qualified as urban, and this number was expected to reach 57 percent by 2030 (United Nations website).

THE SEPARATION OF PEOPLE AND NATURE

One might conclude if people moved to the city fringe they would take advantage of the opportunity to reconnect with nature and wild things. However, the opposite reaction seemed to occur; people brought the structure of the city ecosystem to the country, simplifying and destabilizing their surroundings. This was accomplished by changing the natural landscape, and removing and replacing natural vegetation and endemic wildlife with exotic and domesticated species.

A detailed description of the structural, biotic, and socioeconomic features of urbanization was developed by McDonnell and Pickett (1990). Structural features consist of dwellings, factories, office buildings, warehouses, roads, pipelines, power lines, railroads, channelized waterways, reservoirs, sewage disposal facilities, dumps, gardens, parks, cemeteries, and airports. Crops, ornamentals, domesticated pets, pests, and disease organisms are the dominant biotic features of urbanization. Socioeconomic features include changes in human values, wealth, lifestyles, resource use, and waste.

Urban dwellers are usually unaware that by changing the structural and functional components of natural ecosystems they are creating alternative habitats for a wide array of invertebrate and vertebrate animals that are quick to take advantage of the altered landscape. This commonly results in increased human-wildlife interactions (e.g., coyotes in city parks), whether intended or not. A variety of these wildlife encounters are discussed in detail throughout this book.

Aldo Leopold's prophetic quote at the beginning of this chapter is a succinct evaluation of how an urbanized society runs the risk of using the most simplistic explanations for complex phenomena. Decades and generations of human isolation from the natural world through urbanization have produced a society that lacks a connection to the natural world and a relationship with that world. For example, few people in contemporary society appreciate what their last breakfast cost wildlife in terms of food production, transportation, packaging, and distribution. Leopold points out a second disconnect between society and nature in term of their understanding of the costs to the environment and wildlife just to provide the energy required for home heating (e.g., surface coal mining).

In order to more fully understand the separation between people and wildlife that occurs with urban sprawl, one needs to compare the interactions between people and wildlife in a rural community with what most readers experience now on a daily basis in their urban communities. Perspective Essay 2.1 was written by Clark Adams, who recalls human-wildlife interactions in a small agrarian community in Iowa in the early 1950s. Of particular importance in this essay is the description of people's attitudes, activities, knowledge, and expectations concerning wildlife in their day-to-day lives at that time. Furthermore, it is instructive to reflect on the past from time to time in order to recall where we have been in relation to where we might be heading in our associations with wild things. It is unlikely that urban societies will ever return to the condition described in this essay. Furthermore, it seems imperative that urban communities understand their impact on species diversity, interrelationships with wild things and natural habitats, natural cycles (water and biogeochemical), and how the production, provision, and utilization of energy resources affect wildlife and their habitats. Urban society should also realize that unsustainable use of natural resources endangers human society as well as wildlife. These comparisons and impacts will be addressed in subsequent chapters.

THE NEED FOR WILDLIFE MANAGEMENT IN URBAN AREAS

There is strong evidence justifying the need for wildlife management in urban areas. This need is based on decades of population shifts from rural to urban areas (Figure 2.1), a lack of wildlife management paradigms focused on urban people and wildlife, changing animal damage control issues, a growing body of literature (professional and popular) about wildlife in urban areas, career opportunities in the private sector, and generations of human isolation from their natural environments. Furthermore, from 1985 to 2005, the frequency of the U.S. population sixteen years of age and older that reported taking time to observe (35 to 66 percent) and/or feed (46 to 79 percent) wildlife has nearly doubled (U.S. Department of the Interior 1985, 1991, 1996, 2001, and 2005).

The expansion of urban areas into formerly natural environments has caused an increase in human-wildlife encounters, resulting in a variety of human emotions, explanations, and reactions, mostly conceived in an intellectual and experiential vacuum. Some state agencies are now advertising available positions in various aspects of urban wildlife management (e.g., wildlife damage biologist, urban wildlife biologist, and urban outreach supervisor), but most wildlife biologists employed by state and federal agencies and the universities have been caught off guard. The traditional wildlife management curricula produce wildlife biologists who focus their attention on game, nongame, or threatened and endangered species in nonurban habitats.

One study found the degree to which contemporary wildlife curricula, nationally, are used to train urban wildlife biologists was, at best, a token effort (L.W. Adams, Leedy, and McComb 1987). Criteria that defined the training and tasks expected of an urban wildlife biologist emphasized the human dimensions of wildlife management (Tylka, Shaefer, and Adams 1987). Results from a more recent national study on the degree to which land grant universities and state agencies are addressing urban wildlife management are presented in the next section of this chapter.

The number of peer reviewed manuscripts on urban wildlife management research in the two major journals of The Wildlife Society (TWS), the *Journal of Wildlife Management* (JWM) and, until its demise, the *Wildlife Society Bulletin* (WSB), has grown steadily since 1997. One entire issue—58 articles—of the WSB (1997: 25, no. 2) was dedicated to research on white-tailed deer (*Odocoileus virginianus*) overabundance in urban communities. A later issue of WSB (2005: 33, no. 2) contained ten articles on golf courses and bird conservation. Research focused on urban wildlife management issues became a more common occurrence in WSB from 2005 until the last issue in December 2006. Both the fifth and sixth editions of the *Wildlife Management Techniques Manual* have sections devoted to urban wildlife research. Newspaper articles about wildlife in urban areas such as coyotes (*Canis latrans*) in city parks, cougars (*Puma concolor*) in people's backyards, and urban deer (*Odocoileus* spp.) herds are becoming more frequent. The complex nature of conducting field research in urban settings was explained by VanDruff, Bolen, and San Julian (1996). Again, the emphasis was on the considerations of habitat structure in urban compared to rural environments, and the problems resulting from frequent contacts with humans during the research process.

Finally, the typical urban resident is unable to identify common wildlife species, does not know why particular species of wildlife occur in their backyards or how to deal with a problem species, and lacks an understanding of interrelationships between people and wildlife. Typical misconceptions about urban wildlife include:

- All snakes are "poisonous"
- Solitary fawns have been abandoned and need human care (Figure 2.3)
- Missing pet cats and puppies have strayed off or have been stolen
- Feeding wildlife is a necessary activity (see Case Study 2.1)
- Wild predators do not exist in urban habitats

THE NEED FOR PUBLIC EDUCATION PROGRAMS ABOUT URBAN WILDLIFE

Several indicators justify the need for educational programs that inform urbanites about the wildlife around them. One example is found in an examination of what formal public school curricula

Figure 2.3 Deer fawn lying in the grass waiting for its mother to return. (Courtesy Wildlife Services)

teach students about how the natural world works. An analysis of urban high school students' knowledge of wildlife conducted by C.E. Adams et al. (1987) provided guidance in educational program development. Observations of how people react to wildlife in their backyard provided further insight. Furthermore, an inventory of the types of programs offered by state, federal, and private agencies and how they were delivered was considered.

One might expect the biology class to be the most logical place in the public school curricula to learn something about urban environments, plants, and animals. C.E. Adams and Greene (1990) suggested an alternative biology curriculum built around a conceptual framework that provided both teachers and students with a socially relevant approach to biology education. Many of the changes they suggested would convert the traditional content of biology courses into material that promotes student awareness, knowledge, and actions concerning societal (urbanite implied) impacts on biological resources. Their suggested revisions would have made the biology curriculum, in part, a course in survival training, life-long learning, and the connection between people and wild things. In the same issue of the *American Biology Teacher*, C.E. Adams (1990) provided examples of classroom activities involving urban wildlife, for example, tracking radio-collared fox squirrels (*Sciurus niger*) around their urban habitat. This is exactly what happened eighteen years later at Texas A&M University. McCleery et al. (2005) designed educational opportunities for undergraduate students to participate in on-campus wildlife research projects. One study involved tracking several radio-collared squirrels that lived on campus. This was an experiential learning program that gave students a hands-on approach in capture and handling techniques, radiotelemetry, habitat measurements, and population estimation.

Human reactions to wildlife in their backyard represent a broad spectrum of emotions and perceptions based on previous information (formal and folklore), experience, the kind of animal involved, and what it is doing. It is difficult to categorize this spectrum of emotions and reactions based on a static taxonomy (Kellert 1980) that ignores the dynamics of human-wildlife encounters in urban environments. Clearly, a person's emotional response and reactions will be different when they encounter a snake than when they encounter a cliff swallow (*Petrochelidon pyrrhonota*); a seemingly helpless animal or one that appears to be aggressive; a juvenile or an adult.

Examples such as these provide some guidance for the types of educational programs needed. One of the most basic lessons urbanites need regarding the wildlife around them is how to determine

whether the animal is behaving normally. However, humans may apply an anthropogenic interpre-
tation of unusual behavior; for example, "the animal wants to be my friend" or "it's in trouble and
knows I want to help." Cases in point include feeding wildlife, rescuing "orphaned" baby deer, and
encouraging the nearly exponential proliferation of some charismatic species (e.g., resident Canada
geese (*Branta canadensis*) and urban white-tailed deer (*Odocoileus virginianus*); see Chapter 14).
Schmidt (1997) suggested that human reactions to wild animals can be based on their perceptions
of whether the animal has "crossed the line," that is, whether the human-wildlife encounter might
result in the real or perceived potential for damage.

Since the mid-1980s, there has been a proliferation of educational programs about urban wildlife
and habitats by state and federal agencies and private organizations. Nearly all of the programs were
designed for informal methods of information transfer that can occur in any setting to individuals
for whom participation is a personal choice (C.E. Adams and Eudy 1990). Some programs were
designed to address the educational needs of teachers and their students, including Project WILD,
Aquatic WILD, Flying WILD, Project WET, and Project Learning Tree. These programs continue
to be offered as activity-based learning systems to supplement the state-mandated public school curricula. Community-based programs on urban wildlife and their habitats include Master Naturalist,
Master Gardner, and Watchable Wildlife. The latter program led to an annual event called Eagle
Days in several communities throughout Missouri, Crane Mania in Grand Island, Nebraska, and
the Gulf Coast Birding Trail in Texas. Nearly all of the programs available at the time this book
was written relied on volunteer facilitators and participants to conduct training workshops. The
long-term success of these programs is based on a steady recruitment and retention of qualified
volunteers who have high achievement and altruistic values, identify with program goals, and have
a high interest in and concern for the environment (Greene and Adams 1992).

The findings of C.E. Adams et al. (1987) discouraged any assumptions of an enlightened public
concerning wildlife, particularly among urban high school students. The students could not correctly identify many common urban wildlife species (i.e., opossum, *Didelphis marsupialis* vs. rat,
Rattus norvegicus), the relative numbers of selected mammals within Harris County, Texas (i.e.,
raccoons, *Procyon lotor*, were considered rare while cougars were thought to be abundant), the eating habits of eight of sixteen mammals, and the effect of human presence on the relative abundance
of twelve of sixteen mammals. The students' lack of knowledge might be attributed to a lack of
contact with wildlife given their lifetime residency in an urbanized environment and related lifestyle. One needs to know how urban residents learn about wildlife in order to understand some of
the study results.

The majority of urban residents learn both fact and folklore about wildlife through television
programs filmed in exotic locations around the globe, and under circumstances quite different from
those described in Perspective Essay 2.1. Why wouldn't young people consider cougars abundant in
urban settings when large predators are so common in the popular media? Why wouldn't they consider these animals to be harmless when they see, on television, a former Secretary of the Interior
sitting right beside a seemingly unrestrained adult cougar? Other examples of how urbanites learn
about wildlife in the utilitarian worlds of advertising, the fashion industry, and shopping malls have
been provided by Price (1999). C.E. Adams et al. (1987) suggested that urban wildlife education
programs focus on the basic principles of diversity, interrelationships, cycles, and energy (DICE; see
discussion in Chapter 3) exemplified with human-wildlife interactions, and include wildlife-related
activities using species common to the urban environment.

OUTCOMES OF HUMAN-WILDLIFE ENCOUNTERS

Stories of human-wildlife encounters in urban areas are increasingly newsworthy in both print
and electronic media. Particular attention has been given to human encounters with predatory

mammals, such as coyotes, cougars, and bears (*Ursus* spp.). For example, the occurrence of a coyote in New York City's Central Park, cougars in the backyards of El Paso, Texas, and coyote, cougar, and bear attacks on humans in several locations in the United States, have all received media attention. The particular importance of these observations is the realization that urban residents lack the tools required to make informed decisions about how to avoid, interpret, and react to wildlife encounters.

The best people to interview concerning how urbanites respond to the particular wildlife species they encounter, their knowledge and attitudes about wildlife based on the questions they ask, and their frustration with their inability to find information quickly may be wildlife rehabilitators and animal damage control (ADC) specialists. In many states, these are the individuals the public most often turns to for answers to their questions about the wildlife around them.

URBANITES NEED TO RECONNECT
WITH THE NATURAL WORLD

The predominant method used by urbanites to reconnect with the natural world around them is by constructing environments, installing feeders, and wildscaping their yards (Damude et al. 1999). *Wildscapes* are habitats that provide the essential ingredients for a variety of wildlife: food, water, shelter, and space. This is accomplished by planting and maintaining native vegetation, installing birdbaths and ponds, and creating structure. Feeders can supplement native vegetation but should never replace it. The goal is to provide places for birds, small mammals, and other wildlife to feed and drink, escape from predators, and raise their young.

Nearly all of the constructed environments lead to a paradox in urban wildlife management; programs like Backyard Wildlife and Wildscapes are designed to invite wildlife and, once the habitats are provided, the animals will come. But when too many animals respond to the invitation they are seen as a nuisance and become the target for management efforts designed to reduce their numbers (see Chapter 14). In spite of this problem, wildscaping programs tend to be viewed by departments of natural resources (DNRs) and nonprofit organizations as unequivocally positive. Some alternative considerations concerning the value of constructed habitats for urban wildlife are discussed in Chapter 9.

Another dilemma is caused by supplemental feeding of wildlife and the provision of alternative housing. Urbanites spend billions of dollars annually to purchase food and accoutrements to facilitate the wildlife feeding process (U.S. Department of the Interior 2005). The majority of feeders are designed to attract seed-eating birds, but other species are attracted to the seed, including squirrels (*Sciurus* spp.), mice (*Mus* spp.), rats (*Rattus* spp.), opossums, raccoons, even bears. Additionally, some species are attracted to the animals feeding on the seed, such as snakes (Serpentes), domestic cats (*Felis catus*), skunks (Mephitidae), owls (Strigiformes), and hawks (Falconiformes) (see discussion of food chains in Chapter 3). It is reasonable to expect that providing animals with a continuous, reliable, high-calorie food source will lead to higher than normal birth and survival rates. High recruitment rates add to population increases that exceed the carrying capacity of a natural environment and create other concerns (see Chapter 6).

An additional way for urbanites to reconnect with the natural world around them is to utilize the benefits provided in urban open areas (e.g., parks, vacant lots, cemeteries, greenways, abandoned railroad tracks, lakes, streams, and rivers). The structure of urban habitats already provides plenty of food, water, and shelter for many wildlife species. In fact, peregrine falcons (*Falco peregrinus*) have been brought back from virtual extinction because of the resources provided in urban environments (see Chapter 8). Many urban open spaces provide excellent birding opportunities during the height of migration. As discussed in Chapter 7, such places offer urban birders the opportunity to see a great variety of birds without ever leaving the city limits.

URBAN WILDLIFE SPECIES ARE INCREASING,
SOMETIMES TO NUISANCE LEVELS

Many wildlife species have adjusted so well to the urban lifestyle they are now perceived by some segments of the population as a nuisance. Warranted or not, nuisance wildlife control (NWCO) technicians often are called in to address the problems associated with these species. NWCOs may be employed by USDA-Wildlife Services (WS), local governments, or by private pest control companies.

Some variation in the common nuisance species should be expected given the difference in types of resources (food, water, shelter) provided by urban communities, patterns of urban sprawl into natural habitats, age of communities, and patterns of land fragmentation during development. The types of species that provoke large numbers of calls to NWCOs will vary regionally; for example, alligators (*Alligator* spp.) and nutria (*Myocaster coypus*) in the Southeast, bear and cougar in the West, polar bears (*Ursus maritimus*) in the extreme Northwest, and cattle egrets (*Bubulcus ibis*) in the South Central United States.

SOME INSIGHTS INTO THE MAGNITUDE OF URBAN WILDLIFE PROBLEMS

We felt an investigation of the common nuisance species within urban areas in each state would also provide some interesting comparisons. In addition, we thought knowing what these animals do that qualifies them as a nuisance species would provide some extremely interesting examples of how these species have adapted to the urban environments. So, as part of our research for this book, we requested and obtained data from a national database maintained by USDA-Wildlife Services (WS). These records (Management Information System, MIS Database) are reports completed by WS personnel. The MIS Database contains records of the number of public inquiries (requests for assistance) that came from urban and rural residents, identification of nuisance species, and the economic impact (losses) reported to WS by the public resulting from damage caused by all species in four damage categories, including agriculture, human health and safety, property, and natural resources. The MIS Database provided longitudinal (Fiscal Years 1994 to 2003) and regional (Table 2.1) scales of analysis of the magnitude of urban wildlife issues across the continental United States.

Analysis of the data yielded 350,661 projects conducted by WS in response to requests for assistance by the public with urban wildlife damage problems. In addition, MIS data provided nearly 1.2 million records of damage occurrence that were used to evaluate urban damage losses. Each record represented one occurrence of a human-wildlife conflict in the form of actual damage or a perceived threat observed by Wildlife Services or reported to them by the public. These data are a good index for determining which wildlife species are causing, or are perceived as causing, damage to resources considered important to the public, but may not represent population sizes of those damaging species. However, empirical observations by WS employees suggest that damage levels in specific habitats or environments appear to correlate as directly proportional to relative population sizes within these habitats for the species or species groups causing damage.

It should be noted that the resource subcategory "human health and safety" is always underreported for loss values in the WS MIS Database. Difficulty in deriving actual damage losses for this resource resulted in a value of zero being reported frequently. In addition, even when some values can be determined, all primary and collateral damage costs can rarely be assessed. This is especially true where human lives have been lost. Economic damage data from the MIS Database was collected from October 1, 1993, through September 30, 2003.

The MIS Database was examined for the urban species of most concern by various regions (as defined in the 2005 National Survey of Fishing, Hunting, and Wildlife Associated Recreation) (Figure 2.4) in the United States. There were some disclaimers for using the regional breakout in

Table 2.1 National and Regional Records of Ten Species Most Recorded as Causing Damage to Urban Environments by Wildlife Services during FY 1994–2003.

Region	Top 10 Species — Number of Records									
Nationwide	Raccoons 73,084	Coyotes 56,081	Skunks 54,198	Beaver 45,958	Deer 36,943	Geese 34,808	Squirrels 28,975	Opossums 21,871	Foxes 20,635	Blackbirds 19,228
Pacific	Skunks 26,476	Raccoons 22,871	Coyotes 18,964	Opossums 8,972	Coots 5,614	Foxes 4,801	Beaver 4,479	Geese 3,804	Bobcats 3,670	Cougar 3,508
Mountain	Skunks 3,202	Woodpeckers 2,919	Raccoons 2,537	Pigeons 2,244	Gophers 1,994	Coyotes 1,429	Squirrels 1,128	Beaver 1,061	Mourning doves 1,036	Foxes 935
West North Central	Skunks 2,144	Beaver 1,861	Geese 1,404	Hawks/owls 835	Raccoons 770	Blackbirds 588	Coyotes 526	Pigeons 514	Mink 497	Squirrels 462
West South Central	Coyotes 29,227	Raccoons 15,654	Skunks 8,635	Beaver 8,545	Squirrels 8,241	Opossums 7,012	Black rats 6,299	Bobcats 4,759	Armadillos 4,357	Pigeons 3,755
East South Central	Blackbirds 2,707	Geese 1,808	Pigeons 1,680	Beaver 1,518	Hawks/owls 1,271	Raccoons 905	Bats 616	Vultures 520	Skunks 520	Ducks 502
East North Central	Deer 29,149	Raccoons 13,276	Hawks/owls 10,322	Black bears 10,089	Beaver 7,808	Geese 6,847	Skunks 6,632	Squirrels 5,537	Woodchucks 5,349	Crows 4,315
New England	Beaver 8,491	Raccoons 8,289	Skunks 4,673	Foxes 3,491	Geese 3,212	Bats 3,193	Squirrels 2,889	Black bears 2,766	Woodchucks 1,827	Deer 1,811
Middle Atlantic	Geese 9,393	Crows 2,675	Blackbirds 1,572	Gulls 1,037	Sparrows 582	Raccoons 404	Blue jays 395	Vultures 393	Hawks/owls 378	Pigeons 330
South Atlantic	Beaver 12,190	Raccoons 8,378	Woodchucks 7,911	Geese 7,601	Squirrels 7,577	Blackbirds 6,889	Foxes 5,853	Deer 4,594	Snakes 4,013	Vultures 3,185

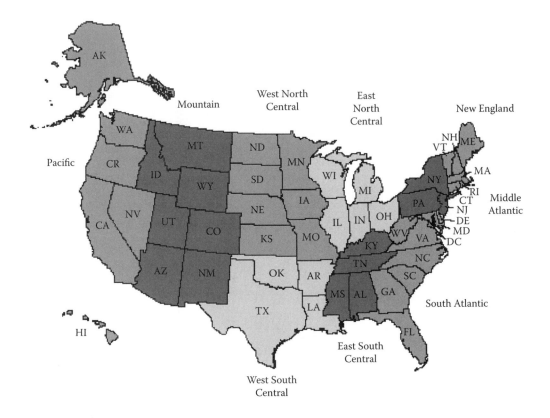

Figure 2.4 Regions used in the analysis of urban species of most concern. (Courtesy U.S. Department of the Interior)

Figure 2.4 as one scale for data analysis. For example, there were regional differences in geographic size, the number of WS offices, and the level of WS personnel reporting activity. However, the size of the MIS data source may offset these problems.

The national and regional reviews of the urban species of most concern were not provided in the form of scientific names; rather, they are presented as common names or species groups, for example, deer, blackbirds (Icteridae), geese. Such "lumping" is sometimes desirable because several species collectively cause identical damage in the same habitats to the same kinds of resources, and evaluation of the magnitude of damage to these resources is more accurately reflected when the impacts of these combined groups are considered. For instance, data reveal that, collectively, puddle ducks (dabbling species) cause significant damage to some urban resources, but when one species of dabbling waterfowl is analyzed alone, it does not appear to cause damage of any real concern. Again, if only white-tailed deer are analyzed for damage impacts, it may appear that there's little damage by deer in western states. However, when this species is lumped with mule deer (*Odocoileus hemionus*), exotic deer (e.g., red deer [*Cervus elaphus*], fallow deer [*Dama dama*]), and all other deer species, we have a more accurate picture. In this evaluation scheme, only the top ten species or species groups were presented in both national and regional analyses even though all species reported were used in the analysis (Table 2.1).

Urban Species of Most Concern: National Analysis 1994 to 2003

The national overview of the top ten urban species of most concern provided a baseline to compare on the regional level. Listed in order of greatest to least magnitude of damage caused by each

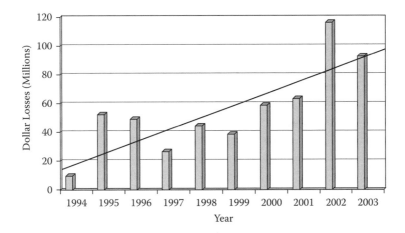

Figure 2.5   Trends in reported dollar losses caused by urban wildlife, FY 1994 to 2003. (Courtesy Clark E. Adams)

species on a national scale were raccoons, coyotes, skunks, beaver, deer, geese, squirrels, opossums, foxes, and blackbirds (Table 2.1).

Urban Species of Most Concern: Regional Analysis 1994 to 2003

As might be expected, the species of special concern were different by region and did not always reflect the nationwide list. For example, birds rather than mammals were the species of special concern in the Middle Atlantic region, and no reptiles were reported in the top ten list except for the South Atlantic region. In the Pacific region, black bears were reported (3,225 records) nearly as often as cougars, but they were not included in the top ten list.

There were differences in number of public reports (records) for each region. For example, the level of public reports in the Pacific, West South Central, East North Central, and South Atlantic regions were nearly ten times greater than in the other regions. This difference might be attributed to the size of the urban population, and those factors listed above that might bias the MIS data set. Nevertheless, there is no other national synthesis that demonstrates which species are causing damage to resources in urban environments.

Economic Impact of Damage to Resources by Urban Wildlife: National Overview

There was a reported economic damage to resources by urban wildlife of $550.8 million from FY 1994 to 2003. Nearly 75 percent of the damage was to personal property ($397.3 million) followed by agricultural losses (e.g., gardens; $88.7 million), human health and safety ($50.2 million), and natural resources (e.g., loss of native flora or fauna; $14.1 million). The level of economic loss increased steadily during the ten-year period (Figure 2.5).

Economic Impact of Damage to Resources by Urban Wildlife: Regional Overview

The increasing level of reported economic loss may be correlated with a higher level of public reports of urban wildlife damage in years succeeding 1994, for example, public awareness of whom to report to, rate of urban sprawl into natural habitats with increased human/wildlife conflicts and consequent damage, and more rigorous pursuit of these data by WS personnel. Figure 2.5 does not include a FY 1995 outlier that was a $1.2 million single occurrence plane crash.

Table 2.2   Sum of Dollar Losses Due to
Damage Caused by Various
Species of Wildlife in Nine U.S.
Regions (1994 to 2003)

Region	Sum of Dollar Losses
South Atlantic	119,096,379
New England	102,597,126
Pacific	92,143,657
West South Central	57,274,703
East North Central	55,931,311
Mountain	38,542,420
Middle Atlantic	33,888,330
East South Central	31,092,058
West North Central	20,208,391

There was an increasing trend in dollar losses in every region from FY 1994 through 2003. At the end of the ten-year period, the South Atlantic, New England, and Pacific regions reported the highest dollar losses due to damage caused by various species of urban wildlife (Table 2.2).

There can be no doubt that there is a growing trend of urban human-wildlife conflict management issues that will need professional attention. The degree to which the public can expect professional attention to these issues is addressed later in this chapter.

A nuisance species could be defined as any that causes negative impacts on human health or economics but, in truth, "nuisance" is in the eye of the beholder. For example, an urban coyote that preys on urban rodents will probably attract little attention from its human neighbors, especially if it stays out of sight. But small dogs and cats provide an easy meal for coyotes, and at this point the predator is likely to be viewed as a nuisance, or worse, by pet owners. Large flocks of birds may produce droppings that cover trees and buildings or pollute water impoundments. Raccoons, opossums, squirrels, skunks, armadillos (Dasypodidae), foxes (*Vulpes* spp.), and many other species find den spaces for living and raising their young in houses and other buildings. In short, if there is a resource that can be exploited, urban wildlife will find and utilize it, becoming a nuisance to some humans in the process. More often than not their exploitive behavior is aided and abetted by humans who, consciously or inadvertently, invite wildlife into their backyards or homes.

URBAN HABITATS AS A DOMINANT FOCUS OF WILDLIFE PROFESSIONALS

There is great satisfaction in knowing, or at least assuming, that one's efforts are on the cutting edge, presenting a new paradigm for action. The issues addressed above represent a current wildlife management need that is sure to exist into the foreseeable future. If there was ever a time to conduct urban deer management, it is now. Nearly every urban community within the range of white-tailed deer has a deer problem. There is probably a greater need for wildlife biologists with state and federal agencies to apply their management skills to urban rather than rural deer herds. Interestingly, some developers of residential areas have marketed the concept of close proximity of deer as a benefit to prospective customers. It may be an attractive advertising ploy, but the wildlife management and conflict resolution problems that ensue are formidable. Conversely, management of urban wildlife may focus on species restoration of threatened or endangered species (e.g., the peregrine falcon, Chapter 8).

Urbanization and the encroachment of humanity into undeveloped wildlife habitats will continue. Every state wildlife agency has been given the legislative mandate to manage its wildlife resources, regardless of whether the wildlife resides in the city or country. There is some overlap of jurisdiction for certain species (e.g., migratory species, threatened/endangered species, and marine mammals;

see Chapter 11). Still, the degree to which these agencies are prepared to embrace the challenges of urban wildlife management is a matter for speculation. Few of the wildlife professionals within these agencies have been trained specifically to manage wildlife in an urban setting. Few land-grant universities that offer degrees in wildlife sciences include even one course in urban wildlife management. Traditional wildlife management strategies applied in rural areas are not always appropriate in urban areas where people become a larger part of the equation (VanDruff, Bolen, and San Julian 1996).

The Infrastructure for Urban Wildlife Management Is Missing

In 1999 the Urban Wildlife Working Group of The Wildlife Society conducted a national survey of state departments of natural resources and land-grant universities that offered at least a bachelor's degree in Wildlife Science (Management) (Adams, 2003). The survey was conducted by faculty and students associated with the Human Dimensions in Wildlife Management laboratory in the Department of Wildlife and Fisheries Sciences at Texas A&M University (College Station, Texas). The e-mail survey was designed to determine how well these universities, and the forty-eight state wildlife agencies in the continental United States, were prepared to address urban wildlife management issues.

Survey questions were designed to determine the status of urban wildlife management in each respondent's state related to:

Relevant urban wildlife management issues;
Number of urban wildlife biologists;
Qualifications and tasks that differentiate urban from other wildlife biologists;
Number of urban wildlife biologists, degree of respondent (university or state agency) responsibility for urban wildlife management;
How urban wildlife management issues are addressed;
Number of universities and colleges that offer at least a B.S. degree in wildlife sciences and/or courses in urban wildlife management;
Future need for urban wildlife biologists.

Response rates were 80 percent from universities ($n = 37$) and 90 percent from state wildlife agencies ($n = 46$), respectively. Survey results can be considered as an accurate portrayal of national trends.

Only one university respondent reported there were no urban wildlife management issues relevant to his or her state. Of the remaining questions, there were small frequency differences in university and state agency responses. For example, 85 percent admitted that urban wildlife was a growing management concern, and 90 percent agreed that urban residents needed educational programs about the wildlife around them.

Two-thirds (62 to 67 percent) felt there was a growing human curiosity about the wildlife in their urban habitats, yet 60 to 83 percent of the respondents agreed that:

1. There is a growing concern about dangerous human-wildlife encounters in urban environments.
2. There is a need to show urban humans how to reconnect with the natural world around them.
3. Urban habitats provide plenty of food, water, and shelter for many wildlife species.
4. The number of many species of urban wildlife is increasing to nuisance levels.

The lower levels of agreement with the above issue statements were always by the university respondents. Finally, only 10 to 11 percent of the respondents agreed that urban wildlife management may become the dominant future focus in their states.

Based on the definition of an urban wildlife biologist provided in the Introduction, university respondents reported the employment of 7 urban wildlife biologists out of a total of 545 traditional wildlife biologists. Agency respondents reported 46 urban wildlife biologists out of a total of 5,409 traditional wildlife biologists.

Nearly half (44 percent) of the respondents said there were few to no qualifications that differentiated urban from other wildlife biologists. However, given a list of qualifications provided by Tylka, Shaefer, and Adams (1987), respondent groups had different levels of agreement. For example, 65 percent of the university compared to 29 percent of the agency respondents felt an urban wildlife biologist should be able to evaluate effects of urbanization on habitat. Only 43 percent of university respondents and 32 percent of agency respondents said the ability to solve wildlife damage problems should be a qualification. Few of the university (27 percent) and agency (15 percent) respondents identified the ability to evaluate public attitudes as a necessary qualification.

Thirty-six percent of the university and 29 percent of the agency respondents said there were "few to no" tasks that differentiated urban from other wildlife biologists. Based on a list of tasks provided by the Tylka, Shaefer, and Adams (1987) study, respondent groups had different levels of agreement. For example, 35 percent of the university respondents compared to 21 percent of the agency respondents felt that a task of an urban wildlife biologist should be animal damage control. Forty-one percent of the university respondents and 27 percent of the agency respondents identified the establishment of urban wildlife habitats as a task unique to urban wildlife biologists. Sixty-two percent of the university respondents and 31 percent of the agency respondents identified conducting research on urban wildlife management as a task appropriate for an urban wildlife biologist.

Respondents were asked to identify the level of responsibility (all, some, none) their university or agency had for urban wildlife management. None of the university respondents and 54 percent of the agency respondents said all urban wildlife management was their responsibility. Sixty percent of the university respondents and 46 percent of the agency respondents said that they had some responsibility for urban wildlife management. Forty and zero percent of the university and agency respondents, respectively, said urban wildlife management was not their responsibility.

The degree to which there is an academic infrastructure in place to train urban wildlife biologists was determined by asking survey respondents how many colleges or universities in their state offered at least a bachelor of science degree in Wildlife Science, or courses in wildlife management or urban wildlife management. University respondents could identify only sixty-seven universities or colleges that offered at least a BS in wildlife sciences compared to seventy-eight identified by agency respondents. Both respondent groups said that there were 111 colleges and universities that offered courses in wildlife management, but only six or seven that offered courses in urban wildlife management. These were national totals!

The estimated future need in 2004 for urban wildlife biologists employed in land-grant universities was 20 compared to 170 by agencies. However, less than 25 percent of the respondents expected their needs to be met in the next five years and, unfortunately, we're unaware of any follow-up study to determine whether or not this projection was accurate.

This study showed that the infrastructure for urban wildlife management was not well developed within academia or wildlife agencies, and it agrees with the findings of L.W. Adams et al. (1987). The degree to which land-grant universities and state agencies were prepared to embrace the challenge of urban wildlife management was less than encouraging. Few of the wildlife professionals within these agencies have been trained to manage wildlife in an urban setting, that is, qualify as urban wildlife biologists. Few universities or colleges offer even one course in urban wildlife management. Overall, study results pointed out that universities and agencies:

1. Are relying on conventional management philosophies and skills to address urban wildlife management issues.
2. Are applying token efforts to enormous if not insurmountable urban wildlife management problems.
3. Do not recognize urban wildlife management problems as their responsibility.
4. Are somewhat oblivious to present and emerging urban wildlife management problems.

It would be instructive to repeat the above analysis now, over 10 years later.

PERSPECTIVE ESSAY 2.1   Human-Wildlife Interactions in the 1950s

I spent the first fourteen years of my life, 1942 to 1958, in Algona, Iowa, which, at that time, had a population of about 4000 people. A half-hour walk would put one into the country to conduct any number of activities, including hunting; fishing; catching insects; building forts, rafts, and dams; looking for rocks or Indian artifacts; collecting mushrooms, nuts, or berries; making tools, musical toys, or weapons from the surrounding vegetation; visiting your special place for solitude and reflection; hiking; snow sliding; and a host of other activities that cost nothing but your time.

We fished year-round in the Des Moines River nearly every weekend and after school. Fishing included so much more than catching fish! River fishing usually resulted in wet, hot, dirty, insect-bitten, barefooted, and sun-burned anglers (Figure 2.6), which was a far cry from fishing programs currently popular in urban areas (Figure 2.7). The spring rains caused night crawlers (*Lumbricus terrestris*) to emerge after dark, and it was common to collect a gallon can full of worms to be used or sold as fish bait during the summer. Night crawler husbandry was a familiar practice. In addition, youngsters knew how to raise tadpoles to adult frogs, care for and feed baby owls until they fledged, catch pocket gophers for ten cents bounty, and live capture thirteen-lined ground squirrels (*Citellus tridecemlineatus*) with string.

Rabbits (*Sylvilagus floridanus* and *Lepus californicus*), fox squirrels (*Sciurus niger*), ring-necked pheasants (*Phasianus colchicus*), and waterfowl were common fare on our dinner table. During open season we hunted nearly every weekend and after school. In fact, the importance of hunting in our community was demonstrated each year during the opening day of pheasant season. Work stopped at noon. My mother made a special meal after which all the men and boys headed for the corn-fields to hunt pheasants. Hunting was so universally accepted as a recreational and/or subsistence pursuit that we often took our 22-caliber rifles or 410 shotguns to school and stored them in our lockers so we could head to our hunting areas as soon as the last bell rang.

Figure 2.6   Fishing in the 1950s. (Courtesy USFWS)

Figure 2.7 Fishing from a well-stocked swimming pool is common in urban cities. (Courtesy Clark E. Adams)

We learned the behavior patterns of the animals in our community, including their escape routes, how they tried to avoid detection, defense mechanisms, where they slept, and what they ate. Trapping wild animals for the fur trade was a common fall and winter enterprise by both youngsters and adults. This meant checking trap lines at 4:00 a.m., dispatching the captured animals, skinning them, and stretching the pelt on the appropriate board for the species, storing them for future sale and still making it to school or work on time. Trapping animals also led to lessons in animal anatomy. For example, knowing the location of the scent gland in skunks (*Mephitis mephitis, Mephitis putoris*) and mink (*Mustela vison*) was critical! Other things we learned included special physical adaptations, for example, badger (*Taxidea taxus*); the differences between carnivores, herbivores, and omnivores; and predator-prey relationships. Pet ownership was as much a matter of function as companionship. Dogs were expected to perform specific duties beyond eating and sleeping.

Unlike contemporary urbanized societies, the human-wildlife interface that existed during this time in my life was based on a utilitarian human attitude of domination and sustained yields. The extrinsic value of wildlife was a far more important consideration than their intrinsic value. Our decisions concerning wildlife were not constrained by physical, psychological, societal, economic, political, or structural barriers. There were nearly daily interactions with wildlife, but people seemed more aware of the reasons behind these interactions and what actions, if any, were appropriate and necessary to promote or prevent the interaction.

Clark E. Adams

CASE STUDY 2.1: NEIGHBORHOOD MOOSE KILLED BY KINDNESS

An Alaska state biologist says the people who feed wild animals in their backyards are responsible for the death of a moose (*Alces alces*) in early March 2005. On March 3, police shot and killed a moose after a 6-year-old Anchorage boy was stomped on the head. Rick Sinnott says he's seen too many people feeding animals and not thinking about what could happen. At the same time, a wildlife conservation group says it is an example of how things need to change.

It happened in a Muldoon neighborhood where trash cans without lids can tempt a hungry moose. But officials say the moose that hung out in the neighborhood wasn't just dumpster diving, it had been hand-fed.

"We know that scores, if not hundreds, of people are feeding moose every day in Anchorage and they're creating dangerous situations," Sinnott says. "But for us, we've got other jobs other than just being the moose monitors. And we do our other jobs. If we can, we try to bust these people. But it's almost impossible."

Shannon Eldridge lives just a few homes away from where it happened. "Pretty scary because I have two little kids that play outside. It's just a sad situation that the little boy got hurt. They're pretty dangerous now. There's a lot of them on the road. They're causing accidents, they're tromping on kids and people. They need to be taken care of a little bit, I think."

The group Defenders of Wildlife says they may have an answer to that. At this week's Board of Game meeting, they'll testify against a proposal to create a limited moose hunt in Anchorage. Instead they want to create an Anchorage moose committee to find other ways for city residents to coexist with moose.

Karen Deatherage says a similar idea for creating an Anchorage bear committee has been successful. "And it was really a successful program because we sat down at a table, we rolled up our sleeves, and we came up with long-term solutions to some of the conflicts in Anchorage with wildlife."

State officials agree that the Anchorage bear committee has helped residents better understand the issues raised when so many people live so close to bear habitat. There's plenty of agreement that the moose issue is just as serious.

Jeffrey Hope
Excerpted from an article at http://www.ktuu.com
March 3, 2005

SPECIES PROFILE: BOBCAT (*LYNX RUFUS*)

She's acting normal for a bobcat, which is aggressive. They don't usually respond like your average housecat.

—Bob Horvath, Wildlife Rehabilitator

Bobcats (Figure 2.8) are known to be secretive animals … so secretive, it seems, they can den in high-density housing areas and most of the human residents will be none the wiser (Harrison 1998). They are also solitary, with male and female coming together only to mate (Whitaker 1998). Unlike another North American felid, the lynx (*Lynx lynx*), bobcats don't spend most of their time in trees, preferring to rest and hide in rocky outcrops and thickets … or under the foundation of a suburban three bedroom, two bath, single family ranch-style home.

These tawny cats have a stubby tail with a black tip (the common name refers to this "bobbed" tail), slightly tufted ears, and mottled spotting. Adults weight between 15 and 30 pounds (6.8 to 13.6 kg) and are roughly the size of a large cocker spaniel. Prey include small to medium-sized vertebrates, including reptiles, birds, rabbits (*Sylvilages* spp.) and hares (*Lepus* spp.), mice, squirrels, opossums, and domestic cats. They are capable of killing larger prey, including deer, but do so rarely, especially in an urban habitat where easier prey is readily available.

Breeding occurs in early spring, beginning at two years of age. Females raise their young alone. The kittens begin exploring at about one month and are weaned at two months. By fall of their birth year they are hunting on their own but remain with their mother for about one year (Whitaker 1998). Young bobcats are quite cute and are sold as pets, even though it may be illegal to keep them as

Figure 2.8   Bobcat. (Courtesy Dave Menke/USFWS)

companion animals in some places. As so often happens with exotic pets, once the animal reaches maturity or the newness wears off, the owners may have a change of heart and release the animal, rationalizing that because it is a wild animal it will instinctively know how to survive. Since the Internet has increased access to any number of exotic pets, one can't help but wonder if increased bobcat sightings in urban and suburban areas may have less to do with wild populations adapting to the presence of humans than with the adaptation of former pets to life outside the living room (personal observation).

Riley et al. (2003) found that age and sex played a role in the sensitivity of bobcats to urbanization. They suggest areas frequented or modified by humans may be perceived by female bobcats as less desirable for rearing young. Female bobcats' use of only the interior of California's Golden Gate National Recreation Area suggests a higher sensitivity to urbanization and/or human presence (Riley 2006). The 2003 study found young female bobcats were more urban-associated than adult females, while the opposite was true for males. Female bobcats are territorial so young females may be willing or be forced to accept marginal habitats. Adult male bobcats have larger home ranges that increase over time (Riley et al. 2003).

So while there are reports of bobcats living in human developments, this species is not likely to exploit the urban landscape to the same degree as raccoons and coyotes. The prevalence of small habitat patches is a challenge to animals with extensive space requirements, low reproductive rates, or susceptibility to human persecution; larger mammalian carnivores tend to be negatively impacted by development and other types of human activity associated with urbanization (Matthiae and Stearns 1981, Dickman 1987, Mackin-Rogalska et al. 1988).

CHAPTER ACTIVITIES

1. It seems that in every community experiencing urban sprawl there are specific factors that determine why and how the process took place. Examine the rate, direction, reasons for, and repercussions of urban sprawl in your community. Aerial photo records of the history and direction of urban sprawl within each student's community may be available from county government sources.

2. Identify the "problem" wildlife in your community, including the factors that distinguish them from other wildlife. Compare your results with Table 2.1. Determine the economic losses attributed to "problem" wildlife within your community.

3. Investigate how community residents are attempting to reconnect with the natural world around them; for example, feeding, photographing, observing wildlife; participating in educational programs about wildlife if they are available; consumptive activities; and "citizen scientist" projects.

4. Describe a day in the life of an urban wildlife biologist by contacting one working for Wildlife Services, your state's DNR, or other levels of government. How do you find one?

LITERATURE CITED

Adams, C.E. 1990. Resource ecology activities for introductory high school biology. *American Biology Teacher* 52:414–418.

Adams, C.E. 2003. The infrastructure for conduction urban wildlife management is missing. *Transactions of the North American Wildlife and Natural Resources Conference* 68:252–265.

Adams, C.E. and J.L. Eudy. 1990. Trends and opportunities in natural resource education. *Transactions of the North American Wildlife and Natural Resource Conference* 55:94–100.

Adams, C.E. and J. Greene. 1990. Perestroika in high school biology education. *American Biology Teacher* 52:408–412.

Adams, C.E., J.K. Thomas, P.-C. Lin, and B. Weiser. 1987. Urban high school students' knowledge of wildlife. In *Integrating Man and Nature in the Metropolitan Environment,* 83–86, ed. L.W. Adams, and D.L. Leedy. Columbia, MD: National Institute for Urban Wildlife.

Adams, L.W., D.L. Leedy, and W.C. McComb. 1987. Urban wildlife research and education in North American colleges and universities. *Wildlife Society Bulletin* 15:591–595.

Damude, N., K. Bender, D. Foss, and J. Gowen. 1999. *Texas Wildscapes: Gardening for Wildlife*. Austin: University of Texas Press.

Dickman, C.R. 1987. Habitat fragmentation and vertebrate species richness in an urban environment. *Journal of Applied Ecology* 24:337–351.

Greene, J.S. and C.E. Adams. 1992. An evaluation of volunteerism in selected conservation education programs. *Transactions of the North American Wildlife and Natural Resource Conference* 57:175–184.

Kellert, S.R. 1980. American's attitudes and knowledge of animals. *Transactions of the North American Wildlife and Natural Resources Conference* 45:111–124.

Harrison, R.L. 1998. Bobcats in residential areas: Distribution and homeowner attitudes. *Southwestern Naturalist* 43:469–475.

Leopold, Aldo. 1949. *A Sand County Almanac*. Oxford, UK: Oxford University Press.

Mackin-Rogalska, R., J. Pinowski, J. Solon, and Z. Wojcik. 1988. Changes in vegetation, avifauna, and small mammals in a suburban habitat. *Polish Ecological Studies* 14:293–330.

Matthiae, P.E. and Stearns, F. 1981. Mammals in forest islands in southeastern Wisconsin. In *Forest Island Dynamics in Man-Dominated Landscapes,* ed. R.L. Burgess and D.M. Sharpe. New York: Springer-Verlag.

McCleery, R.A., R.R. Lopez, L.H. Harveson, N.J. Silvy, and R.D. Slack. 2005. Integrating on-campus wildlife research projects into the wildlife curriculum. *Wildlife Society Bulletin* 33:802–809.

McDonnell, M.J. and S.T.A. Pickett. 1990. Ecosystem structure and function along urban-rural gradients: An unexploited opportunity for ecology. *Ecology* 71:1232–1237.

Price, J. 1999. *Flight Maps: Adventures in Nature in Modern America*. New York: Basic Books.

Riley, S.P.D. 2006. Spatial organization, food habits, and disease ecology of bobcats and gray foxes in urban and rural zones of a national park. *Journal of Wildlife Management* 70(5):1424–1435.

Riley, S.P.D., R.M. Sauvajot, T.K. Fuller, E.C. York, D.A. Kamradt, C. Bromley, and R.K. Wayne. 2003. Effects of urbanization and habitat fragmentation on bobcats and coyotes in southern California. *Conservation Biology* 17(2):566–576.

Schmidt, R. 1997. Drawing the line for wildlife. *Wildlife Control Technology*, December, 6–7.

Tylka, D.L., J.M. Shaefer, and L.W. Adams. 1987. Guidelines for implementing urban wildlife programs under state conservation agency administration. In *Integrating Man and Nature in the Metropolitan Environment*, L. W. Adams and D. L. Leedy, eds. Columbia, MD: National Institute for Urban Wildlife.

U.S. Department of the Interior, Fish and Wildlife Service and U. S. Department of Commerce, Bureau of the Census (1985, 1991, 1996, 2001, and 2005). *National Survey of Fishing, Hunting, and Wildlife-Associated Recreation.* Washington, DC: United States Government Printing Office.

VanDruff, L.W., E.G. Bolen, and G.J. San Julian. 1996. Management of urban wildlife. In *Research and Management Techniques for Wildlife and Habitats,* 5th edition, ed. T.A. Bookhout. Bethesda, MD: The Wildlife Society.

Whitaker, J.O., Jr. 1998. *National Audubon Society Field Guide to North American Mammals.* New York: Alfred. A. Knopf.

Wright, R.T. 2004. *Environmental Science: Toward a Sustainable Future,* 9th edition. Upper Saddle River, NJ: Prentice-Hall.

PART **II**

Urban Ecosystems

CHAPTER 3

Ecological Principles in the Urban Context

The famous balance of nature is the most extraordinary of all cybernetic systems. Left to itself, it is always self-regulated.

—**Joseph Wood Krutch,** *Saturday Review*, **June 8, 1963**

KEY CONCEPTS

1. The four basic principles of ecology (DICE) are listed, with examples.
2. Natural and urban ecosystems are compared in terms of:
   a. Structure: flora and fauna.
   b. Function: nutrient cycling and energy flow.
   c. Food webs and food chains.
   d. Species diversity and numbers of each species.
   e. Ecosystem services that are preserved and lost due to urbanization.
3. Ecological succession is defined and illustrated in the context of urban communities.

The four chapters included in this section—the ABCs of ecosystems—are intended to be an introduction for some readers, and a review for others.

ECOLOGICAL PRINCIPLES

Ecology is the study of relationships between the biotic and abiotic components of an ecosystem. It's a complex and complicated subject, but for our purposes we can distill it down to four fundamental ecological principles: Diversity, Interrelationships, Cycles, and Energy (DICE).

Diversity

Diversity is defined as the variety within ecosystems (Wilson 1992, cited in Folke, Holling, and Perrings 1996). Natural ecosystems tend to be complex, highly diverse, and highly stable. Urban ecosystems, on the other hand, are the products of human manipulation, which tends to simplify and destabilize natural ecosystems. Figures 3.1 and 3.2 demonstrate habitat simplification in suburban and metropolitan environments, respectively, and in Perspective Essay 3.2 we provide an example of how the typical urban lawn is living testament to habitat simplification and destabilization. The

Figure 3.1   Typical suburban landscaping causing simplification of habitat. (Courtesy Clark E. Adams)

remainder of this chapter points out several ways human manipulation of natural habitats destabilizes ecosystem structure and function.

To better understand the value of diversity, think of an ecosystem as a Boeing 747 jet. The structural components of our 747 ecosystem include sheet metal, rivets, wires, and a host of other materials. Now, imagine you've entered this ecosystem (boarded the plane) and have claimed your territory (found a seat). You look out the window and notice someone is removing rivets (*Rivetus solidus*) from the wing of the plane. You call this to the attention of the flight attendant who assures

Figure 3.2   Typical metropolitan area with high-rise buildings, concrete surfaces, and automobile traffic. (Courtesy J. Davis)

Table 3.1 Interrelationships between Ecosystem Components

Interrelationships	Example
Abiotic × abiotic	Temperature and rainfall = climate
Abiotic × biotic	Water temperature predicts fish species
Biotic × abiotic	Human pollution of the environment
Biotic × biotic	Predator/prey relationships, life cycles

you there's nothing to worry about. There are thousands of rivets in the plane and the person is just harvesting a few of them. You continue your trip without any mishaps.

On your return trip you notice that the same person is removing even more rivets from the plane. Again you voice your concern to the flight attendant, who again assures that there are more than enough rivets left. But how many members of the rivet species can be removed before the entire structure collapses?

Perhaps the 747 is a diverse ecosystem with several kinds of fasteners (species) whose function is to hold the plane together. If so, the loss of many or even most rivets might not be disastrous. But what if rivets are the only species performing this function on the 747? What if rivets are a *keystone species*, a part without which the 747 ecosystem cannot exist? Apply this same analogy to the ecosystem of your choosing, or even the whole planet, and you'll have a good idea of why diversity is so important.

Some of the most compelling arguments for preserving biodiversity are those concerning the effects of eroding biodiversity on human health. Both scientists and public health advocates have asserted that the preservation of biodiversity is crucial for present and future human health (Ostfeld and Keesing 2000). On the other hand, human sources of diversity in the urban landscape include introduction of exotic species, modification of landforms and drainage networks, control or modification of natural disturbance agents, and the construction of massive and extensive infrastructures (Pickett et al. 1997, as cited by Zipperer et al. 2000).

Interrelationships

"Everythingisconnectedtoeverythingelse" is a run together sentence used as an analogy for ecological interrelationships. Ecosystems are composed of biotic (living) and abiotic (nonliving) components. Table 3.1 provides several illustrations of how these components interact. Humans, in general, do not recognize how much our presence affects ecological relationships. Urban development, for example, increases the amount and type of pollutants released into the environment, changing the biotic and abiotic structure of aquatic and terrestrial habitats, and altering plant and animal relationships. Without question, ecosystems are complex. In order to really understand this complexity one needs to integrate population, community, and ecosystem processes across multiple scales (Ostfeld et al. 1998). This integrative approach is illustrated by the interrelationships that exist between oaks, white-footed mice (*Peromyscus leucopus*), deer (*Odocoileus* spp.), ticks (Ixodida), and gypsy moths (*Lymantria dispar*), and their link to Lyme disease, discussed in Chapter 14.

Cycles

Cycles are the mechanisms that make a finite amount of substance infinitely available in natural ecosystems. Nitrogen, carbon, hydrogen, oxygen, phosphorus, and sulfur (N-CHOPS) are the fundamental elements required by all living organisms, but only so much of each element exists on the planet Earth. Later, we'll explain trophic levels and nutrient cycling in terms of the major actors, explain the roles they play in the cycle, and human impacts on the nutrient cycles. Each of these elements is recycled through producers, consumers, and decomposers, making them available forever in just the right amounts to sustain plant and animal life.

In contrast, urban systems were designed around the principles of "too much of a good thing" and "one-way trips." Examples, respectively, are found in the overuse of fertilizers on lawns and gardens, and the huge landfills for urban garbage where many recyclable products are thrown away, including metals, glass, plastics, and organic materials.

The water cycle has a somewhat different nature; it provides a continuous supply of *purified* water using a system of three filters: air, soil, and plants. Urban development affects several aspects of the water cycle, including infiltration and runoff, and the purification process (e.g., removal of pollutants from air and soil). The water cycle is discussed in greater depth in Chapter 4.

Energy

Energy becomes available to every aquatic and terrestrial ecosystem by trapping sunlight energy through the process of photosynthesis. Sunlight energy is free, everlasting, and nonpolluting. In general, urban developments do not take advantage of sunlight energy. Instead, the primary energy sources for urban ecosystems are fossil fuels (e.g., coal and gas) that have adverse long-term effects on terrestrial and aquatic ecosystems in both urban and rural settings.

ECOSYSTEM STRUCTURE

Ecosystems, no matter how unique, have certain basic characteristics in common. All ecosystems are *biotic* communities (living plants, animals, and microbes) interacting with one another and their environment. When we use the term "environment" we're really talking about *abiotic* or nonliving factors that act upon the biotic community. To put it even more simply, an ecosystem is the marriage between biotic communities and the abiotic conditions they live in (Wright 2004).

The abiotic factors—including temperature, water, chemicals, and soils—are what cause the differences we observe between various ecosystems. For example, in regions with a high average annual temperature and rainfall we can expect to find the plant, animal, and microbe species typical of a tropical rainforest. The opposite end of the ecosystem spectrum would be a region with low average temperatures and rainfall, commonly referred to as tundra (Odum and Barrett 2004). Additionally, the primary difference between freshwater and marine ecosystems is the absence or presence, respectively, of certain chemicals—in this case, sodium chloride (NaCl, salt).

Natural ecosystems are inherently stable, self-contained units that derive energy from the sun, recycle matter (e.g., nutrients), and support complete food webs. Urban ecosystems, on the other hand, are inherently unstable, derive little of their energy directly from the sun, remove rather than recycle matter, and have incomplete food webs (Adams 1994). "Realistically, urban-industrial environments are parasites on the biosphere in terms of life-support resources" (Odum and Barrett 2004).

In order to fully understand the challenges of urban wildlife management, we first need to explain the individual components of natural ecosystem structure and function, and how they compare to those of urban ecosystems. Ecosystem structure refers to basic components and how they fit together. As was explained above, these components can be organized into two categories that affect the structure of every ecosystem—the abiotic components and the biotic community.

Abiotic Structure

The abiotic structure of an environment, sometimes referred to as its inorganic components, is made up of various physical and chemical factors (Table 3.2). It can be helpful to divide abiotic factors into two types: *conditions* and *resources*. Conditions represent the state of the habitat for

Table 3.2   Examples of Abiotic Factors

Conditions	Resources
pH (acidity vs. alkalinity)	Chemical nutrients (N, P, CO_2)
Salinity	Light (intensity and wavelength)
Temperature	Oxygen
Turbidity (cloudiness of water)	Space to live, feed, reproduce
Wind	Water or moisture

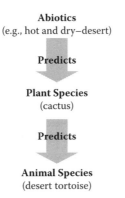

Abiotics
(e.g., hot and dry–desert)

Predicts

Plant Species
(cactus)

Predicts

Animal Species
(desert tortoise)

Figure 3.3   How ecosystems are formed.

a given abiotic factor. Examples include climate (temperature and moisture), pH (acid or alkaline), turbidity (cloudiness), and wind (speed and direction). Resources determine the habitat conditions. As a result, the resources that are available or absent will determine which plant and animal species exist in a particular area (Begon, Harper, and Townsend 1996).

For example, abiotic components predict the types of plants that will grow in an area and the plants, in turn, predict the types of animals that will exist there (Figure 3.3). Temperature and moisture are the abiotic factors that most often define ecosystems. As we discussed at the beginning of this chapter, it is the combination of these two components that determine the dominant vegetation type (e.g., trees, shrubs, grasses, or succulents), while temperature determines differences within related vegetation types (e.g., rainforest, temperate forest, tundra).

Interestingly, it's rare to see a reference to regional climatic differences between urban environments (e.g., desert-urban, temperate-urban, tropical-urban) discussed in the literature, even though these variations certainly exist. Perhaps this is because the term "urban ecosystem" is relatively new. One need only visit the cities of Phoenix and Seattle to recognize that temperature and moisture variations are at work even in the city, resulting in at least some variation in the species living in each.

Often cities and even suburbs become microclimates, with weather that is significantly different from that of the surrounding area. Cities become heat islands—areas in which local temperatures are measurably higher than those in the nearby countryside. Variations in relative humidity, precipitation, and wind speed also are found even within municipalities (Table 3.3). The increased temperatures are the result of human modification of the landscape, particularly an increase in thermal mass through the use of brick, concrete, and asphalt. Higher temperatures result in a longer growing season for plants. Additionally, city lakes and ponds may remain ice-free longer than in nearby rural bodies of water, which may influence the use of habitat by wildlife. Some studies have shown that cities have more cloud cover and more precipitation than surrounding areas (although most of this moisture is quickly lost due to increased runoff—see Chapter 5). Wind speed can be affected by the architectural features within human-created landscapes (Adams 1994).

Table 3.3   Effects of Cities on Urban Climate

Climate Element	Compared to Countryside
Temperature	1°F to 6°F (3.3°C) higher
Relative humidity	6% less
Precipitation	5% to 15% more
Wind speed	20% to 30% less
Contaminants	5 to 25 times more
Clouds	5% to 10% more
Sunshine	5% to 15% less

Source: Landsberg (1981) and Herold (1991), cited in Adams (1994).

Biotic Structure

Biotic structure refers to the way living organisms interact with one another. Biotic communities are composed of plants (everything from algae to trees), animals (mites to monkeys and anything in-between), and microbes (bacteria, fungi, and protozoans). Ecosystem organisms interact with each other through feeding and nonfeeding relationships called *trophic levels* and *food webs.*

Members of the biotic community can be placed into three trophic (feeding) categories: (1) producers, (2) plant and animal consumers, and (3) decomposers. These three groups work together to produce food, move it through the food chain, and return the abiotic components to the environment.

Life on earth depends on energy from the sun. Energy is a one-way trip! *Producers,* also known as *autotrophs* (self-feeders), are able to transform the sun's energy into organic matter. Green plants are the primary producers (Table 3.4). All other organisms feed on organic matter produced by plants as their source of energy. Through a process called *photosynthesis,* plants capture the sun's kinetic energy using specific wavelengths of light, and chemically convert it into potential energy (e.g., sugar). A variety of molecules are used by plants during photosynthesis, but the most important one is the chlorophyll molecule, which has the capability to capture various wavelengths (e.g., blue, red, yellow, orange) of light energy. Interestingly, the reason chlorophyll is green in color is because plants do not absorb the green wavelength of light for photosynthesis. Rather, green light is reflected. All terrestrial ecosystems depend on green plants, and each ecosystem has specific producers that carry on the crucial work of photosynthesis. The potential energy stored in leaves, stalks, and stems becomes available to other organisms when they consume plants, or when they eat the animals that consume plants. All energy is ultimately lost from the ecosystem in the form of heat.

The *consumers* or *heterotrophs* (other-feeders) are a diverse group of organisms—including everything from bacteria and fungi to shrews (Soricidae) and sharks (Superorder Selachimorpha) (Table 3.4). Because this is such an extensive group biologists have found it helpful to divide the consumers into subgroups based on their food source. The animals that feed directly and only on plants are called

Table 3.4   Matching Organisms to Trophic Levels

Organism	Producer	Primary Consumer	Secondary Consumer
Plants	X		
Rabbits		X	
Foxes			X
Opossums		X	X
Snakes			X
Decomposers		X	X
Detritus feeders		X	X
Parasites	X	X	X

primary consumers or *herbivores* (herb eaters). Mice (*Mus* spp.), rabbits (*Sylvilagus* spp.), and deer (*Odocoileus* spp.) are examples of this group. The animals that feed on herbivores are referred to as *secondary consumers* or *carnivores* (meat eaters). Coyotes (*Canis latrans*), cougars (*Puma concolor*), and domesticated cats (*Felis catus*) are examples of this group. *Decomposers* (e.g., bacteria and fungi) recycle the essential nutrients of life by decomposing dead plant and animal materials, and their wastes. These nutrients are returned to plants to use again in another round of photosynthesis. An alternative method of recycling nutrients is through a group of organisms called *detritus feeders*. The prime example of a detritus feeder is the earthworm (*Lumbricus terrestris*, see Perspective Essay 4.1 on Darwin's Earthworms). The primary diet of these organisms is dead plant or animal material. Both decomposers and detritus feeders occupy all trophic levels because their services are needed in each one. Decomposers and detritus feeders aren't the most glamorous members of the biotic community, but without their ability to recycle nutrients the whole system would come to a grinding halt.

Herbivores and carnivores usually feed only at one trophic level while other animals eat a little of everything (Table 3.4). The smorgasbord samplers are *omnivores* (all-eaters). Raccoons (*Procyon lotor*), opossums (*Didelphis virginiana*), and humans are generally recognized as omnivores. However, some carnivores such as coyotes, cats, and bears (*Ursus* spp.) will abandon their carnivorous habits to feed on more abundant food sources such as garbage.

Parasites are a special group with feeding adaptations that allow them to feed on a "host" species without killing it (at least not immediately) or share what the host has eaten (Table 3.4). Parasites can feed on either plants or animals. Parasites can feed on (leeches and ticks) or in (tapeworms and roundworms) their host species. Some plant parasites (e.g., mistletoe [*Phoradendron serotinum*]) also carry on the process of photosynthesis, which places them in a producer category.

Food Chains and Webs

Trophic levels and how they affect the interrelationships between the organisms in an ecosystem can be examined in greater detail by exploring the concepts of *food chains* and *food webs*. A food chain is a simplistic, straight-line model of the way organisms feed on one another. A typical grassland ecosystem food chain is illustrated below. Grass, as a producer, is the starting point of the chain, followed by various consumers:

$$\text{Food Chain} = \text{Grass} \rightarrow \text{Grasshopper} \rightarrow \text{Frog} \rightarrow \text{Snake} \rightarrow \text{Owl}$$

Of course, ecosystems are not this simple. Owls do not just eat snakes—they also prey on lower-order consumers such as rabbits and mice, as do snakes! Virtually all food chains are interconnected, forming complex feeding relationships, also known as food webs. The food web is a convenient method of demonstrating the complexity of natural ecosystems.

Figure 3.4 illustrates a typical food web within a natural ecosystem. Mice eat grass seeds and the mice are then consumed by owls. Of course, mice feed on other types of seeds as well as the occasional insect, just as owls may consume rabbits in addition to mice. Insects are even an important food for some smaller owl species (e.g., screech owls).

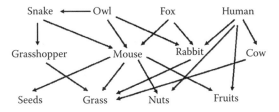

Figure 3.4 Natural ecosystem food web based on three trophic levels. (Courtesy Clark E. Adams)

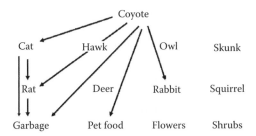

Figure 3.5   Urban ecosystem food web based on four trophic levels. (Courtesy Clark E. Adams)

Urban food webs paint quite a different picture, as can be demonstrated by constructing a web of common urban producers and consumers, both native and introduced. For example, common representatives of the producer level in a suburban area might include flowers, trees, and shrubs used in landscaping; as well as sunflower seed mix, sugar water, edible garbage, and pet food. Urban herbivore representatives would be rabbits, mice, deer, Canada geese (*Branta canadensis*), tree squirrels (*Sciurus* spp.), hummingbirds (*Archilochus colubris*), and several species of granivore (seed-eating) birds. Humans, raccoons, and opossums are all urban omnivores, while urban carnivores include domesticated dogs (*Canis familiaris*) and cats, coyotes, foxes, raptors (e.g., meat-eating birds such as hawks and owls) and, in some parts of the United States and Canada, black bears (*Ursus americanus*), polar bears (*Ursus maritimus*), and cougars. Figure 3.5 illustrates a typical urban food chain—garbage fed on by rats that are fed on by cats that are fed on by coyotes (it may come as quite a shock to many urban residents to find out their pets have been preyed upon by coyotes!). Note the food web of the coyote can also include the garbage and the rat, as well as rabbits and pet food. Cases have even been reported of rats and cats feeding side by side on garbage (Sullivan 2004).

Symbiotic Relationships

Symbiotic (which translates as "living together") relationships are common in natural ecosystems. Species interact in ways that are mutually beneficial, neutral, or detrimental to one or both. There are four types of symbiotic relationships, based on whether the results are positive or negative: *mutualism, commensalism, parasitism,* and *competition.*

Mutualism is a form of symbiosis in which both species benefit from the interaction. A fine example of mutualism is the relationship between the pronuba moth (*Tegeticula* spp.) and the soap tree or yucca plant (*Yucca elata*). In exchange for the pollination services that the moth performs as she deposits her eggs inside the flower, the yucca provides shelter and food for the moth larvae. Both species not only benefit from this arrangement, their survival is now dependent on the existence of one another.

Commensalism is the symbiotic relationship whereby one species benefits from the interaction while the other is unaffected. This can be demonstrated using the example of the remora (*Remora* spp.) and larger fish. The remora is a long, slender fish that has a dorsal fin modified as a sucker-like attachment organ, with which it grasps the side of another fish or a turtle. Remoras use their host as transportation while at the same time taking advantage of food fragments left behind as the host eats. The host animal is not harmed by the remora's activities.

Parasitism is the symbiotic relationship in which one species benefits and the other is adversely affected. Technically, predation and disease are forms of parasitism, but for most people the term parasite calls to mind creatures like ticks and fleas (*ectoparasites*—external) or tapeworms (*endoparasites*—internal). As long as the host animal is generally healthy, this type of parasite doesn't usually kill the host. However, if the host animal's health is compromised in some way (e.g., an injury or disease), or if the parasite is carrying a pathogen (e.g., ticks carrying Lyme disease or mosquitoes carrying West Nile virus) death can result (see Chapter 14). Some readers might not be aware that their companion animals can harbor several different types of parasitic diseases that use humans to complete part of their lifecycle. This is one reason veterinarians strongly recommend that puppies be tested and treated for parasitic nematodes such as roundworms (*Ascaris* spp.) and hookworms (*Ancylostoma* spp. and *Necator* spp.).

When both species in a symbiotic relationship are adversely affected the interrelationship is called *competition*. Competition for food, habitat, and mates is part of the natural order of life, but it always results in an inequitable distribution of available resources among the competing species. However, "all out war" between species is reduced by specialization and adaptation to a particular *niche*. An animal's ecological niche refers to what the animal feeds on, where it feeds, when it feeds, where it finds shelter, where it nests, and how it responds to abiotic factors. Niches allow potential competitors to coexist within the same ecosystem. Both plants and animals are involved in competitive relationships.

Biotic Communities

The first step to take when investigating a biotic community may be to catalog all the species present. By doing so, you're likely to notice that the individuals of any particular species in the ecosystem are part of a *population*—the interbreeding group that lives within a given area. Natural ecosystems are wonderfully diverse, and it's their diversity that creates stability. For example, most herbivores consume many different kinds of plants, so if one particular plant species becomes temporarily or even permanently scarce, other food sources are still available. The same applies for omnivores and carnivores—because they are able to feed on many different species of plants and/ or animals, and because there is a great variety of species available upon which to feed, fluctuations in populations are generally minor and, to some degree, predictable.

Once again, the situation is quite different in urban ecosystems. The species diversity, sometimes called *richness*, is much lower in urban areas than in natural ecosystems. Urban and suburban residents who have observed large flocks of birds at feeders or roosting on buildings, or have marveled at the number of squirrels that can be seen in one small city park might dispute this statement, but it is true nonetheless. While it is accurate to say that wildlife *densities* tend to be significantly higher in urban areas, the individual animals represent a relatively small number of different species (*diversity*). To put it more simply, there are lots of plants and animals living in the city, but there aren't lots of different *kinds* of plants and animals living there.

Urban Flora

As an area undergoes urbanization, a pattern begins to emerge that categorizes urban plant communities into three broad groups: *natural remnants*, *derelict lands*, and *planted communities* (Adams 1994). This chapter introduces these plant communities, but they are discussed in more detail in Chapter 7.

Natural remnants of vegetation persist even in the midst of heavy development. These remnants retain the characteristics of the native plant community that existed before urbanization. Remnants face many threats once they become surrounded by urbanization. As the patches become fragmented—smaller in size and further apart—pollination and dispersal of seeds becomes increasingly difficult for the plants contained within. As forest remnants are carved up, species once found inside the relatively protected interior are forced to deal with the harsher conditions found at the edge of the habitat (e.g., increased wind, greater temperature fluctuations). Urbanization affects the local hydrology, which inevitably affects the plant communities in the remnant patches. Species that are able to tolerate the changes may persist; those that can't disappear. Introduced exotic species create yet another threat. Table 3.5 provides examples of both native plants and the exotics that may take their place as development occurs.

Table 3.5   Exotic Plants and the Native Species They Replace

Vegetation Type	Exotic Species	Native Species
Forest systems	Chinese privet (*Ligustrum sinense*)	Green ash (*Fraxinus pennsylvanica*)
	Chinaberry (*Melis azedarach*)	Flowering dogwood (*Cornus florida*)
	Mimosa (*Albizia julibrissin*)	American elm (*Ulmus americana*)
	Siberian elm (*Ulmus pumila*)	Sweet gum (*Liquidambar styraciflua*)
	Chinese tallow (*Sapium sebiferum*)	Live oak (*Quercus virginiana*)
	Japanese honeysuckle (*Lonicera japonica*)	Maple (*Acer* spp.)
Prairie systems	Tall fescue (*Festuca arundinacea*)	Big bluestem (*Andropogon gerardii*)
	Johnsongrass (*Sorghum halepense*)	Indian grass (*Sorghastrum nutans*)
	Purple scabies (*Scabiosa atropurpurea*)	Switchgrass (*Panicum virgatum*)
	Sweetclover (*Melilotus officinalis* & *M. alba*)	Rose gentian (*Sabatia campestis*)
	Japanese brome (*Bromus japonicus*)	Prairie coneflower (*Ratibida pinnata*)
Aquatic systems	Giant salvinia (*Salvinia molesta*)	Pondweed (*Potamogeton* spp.)
	Water hyacinth (*Eichhornia crassipes*)	Water stargrass (*Heteranthera dubia*)
	Hydrilla (*Hydrilla verticellata*)	Arrowhead (*Sagittaria* spp.)
	Parrot's feather (*Myriophyllum aquaticum*)	Spatterdock (*Nuphar luteum*)
	Eurasian watermilfoil (*Myriophyllum spicatum*)	Pickerel weed (*Pontederia cordata*)

Source: Brockman (1986); Cushman and Jones (2004); Borman and Korth (1998); and Davis (2003).

Derelict lands, sometimes referred to as vacant lots, are characterized by having been disturbed by construction at some point. Depending on how long it has been since the disturbance, the plant species that are likely to be found in these patches include invasive exotic species, including some listed in Table 3.5, and native pioneer species (e.g., ragweed). Like remnants, these small areas of habitat are fragmented and disconnected from one another. The biodiversity within these patches tends to be low, although those species represented may exist in large numbers.

Planted communities make up the majority of urban areas. They are largely artificial, making extensive use of high-maintenance species in a highly controlled manner. Many of the species found in planted communities come from Europe, Asia, and South America—boxwood (*Buxus* spp.), nandina (*Nandina domestica*), crepe myrtle (*Lagerstroemia* spp.), photinia (*Photinia* spp.), ornamental pear (*Pyrus calleryana*), Bermuda grass (*Cynodon dactylon*), and Asian jasmine (*Trachelospermum asiaticum*).

Urban Fauna

Changes in soil, hydrology, and vegetation created by urbanization exert pressure on wildlife populations. Shifts in species composition are, at least to some extent, predictable. For example, urban ecosystems select for small- to medium-sized, highly adaptable predators. Large predators require large territories within which to hunt and, as mentioned earlier, urban habitat patches tend to be small and fragmented. The coyote is a perfect example of a predator that does well in urban ecosystems. Coyotes actually increase in numbers living in close proximity to humans; they don't require large amounts of space, and they find ample and diverse prey (rats, mice, rabbits, cats, and small dogs) within the city limits.

As a general rule, specialists decline in urban areas, while generalists thrive. Specialists are those species that have very strict survival requirements regarding diet, habitat, and/or nesting sites. Urbanization, by its very nature, significantly alters the existing habitat, creating more edge habitat while decreasing available interior habitat. Edge-loving species are attracted to these areas; generalists, with their wider range of acceptable living conditions, adapt; and the interior specialists disappear. As the specialists leave, the decrease in interspecies competition for resources allows the generalists to not only survive, but to increase in numbers.

Humans play an important role in which species are able to make the transition from life in the country to city living. Garbage is an important food source for many urban generalists, particularly omnivores like opossums and raccoons. Providing supplemental food for wildlife has become a popular hobby and lucrative business in the United States. (Hope 2000). The number of seed-eating (granivore) birds rises in an urban environment due to the availability of bird feeders that provide an endless supply of easily attainable food.

Urbanization, and associated human actions, affects the behavior of wild species living in this type of system, too. Artificial food supplies (e.g., birdseed, garbage, pet food) reduce the need for individual animals to move around in search of something to eat. The result is smaller home ranges for urban animals compared to their rural cousins. The heat islands discussed earlier in this chapter cause a number of behavioral changes, from longer breeding seasons (Adams 1994) to the disappearance of traditional migration activity (Hope 2000). Other impacts of urbanization and human activities on the population dynamics of urban wildlife are discussed in Chapter 6.

The wildlife species found in cities and towns across the United States are remarkably similar (Burger 1999), in spite of significant regional variations in climate and terrain. Table 3.6 lists a

Table 3.6   Common Urban Wildlife Species

Invertebrates	Birds
Black widow spider (*Latrodectus* spp.)	Northern cardinal (*Cardinalis cardinalis*)
Garden spider (*Argiope* spp.)	Blue jay (*Cyanocitta cristata*)
Silverfish (*Lepisma saccharina*)	Mourning dove (*Zenaida macorura*)
American cockroach (*Periplaneta americana*)	Ruby-throated hummingbird (*Archilochus colubris*)
House cricket (*Acheta domestica*)	American robin (*Turdus migratorius*)
Praying mantis (*Mantis religiosa*)[i]	European starling (*Sturnus vulgarus*)[i]
Mosquito (*Aedes* spp. and *Culex* spp.)	Killdeer (*Charadrius vociferus*)
House fly (*Musca domestica*)	Chimney swift (*Chaetura pelagica*)
Cabbage butterfly (*Pieris rapae*)	Purple martin (*Progne subis*)
Tent caterpillar moth (*Malacosoma* spp.)	Pigeon (aka rock dove, *Columba livia*)[i]
Pharaoh ant (*Monomorium pharaonis*)	House sparrow (*Passer domesticus*)[i]
Paper wasp (*Polistes* spp.)	Mallard (*Anas platyrhynchos*)
	Canada goose (*Branta canadensis*)
Fishes	American coot (*Fulica americana*)
	Great blue heron (*Ardea herodias*)
Common carp (*Cyprinus carpio*)[i]	American kestrel (*Falco sparverius*)
Goldfish (*Carassius auratus*)[i]	Peregrine falcon (*Falco peregrinus*)
Catfish (*Ameiurus* spp.)[i]	Screech owl (*Otus* spp.)
Green sunfish (*Lepomis cyanellus*)[i]	Vulture (*Cathartes aura* and *Coragyps atratus*)
Bluegill (*Lepomis macrochirus*)	
Largemouth bass (*Micropterus salmoides*)[i]	**Mammals**
	Opossum (*Didelphis virginiana*)
Amphibians	Eastern mole (*Scalopus aquaticus*)
	Little brown bat (*Myotis lucifugus*)
Spotted salamander (*Ambystoma maculatum*)	Red bat (*Lasiurus borealis*)
Bullfrog (*Rana catesbiana*)[i]	Cottontail rabbit (*Sylvilagus* spp.)
Green frog (*Rana clamitans*)	Norway rat (*Rattus norvegicus*)[i]
Leopard frog (*Rana pipiens*)	House mouse (*Mus musculus*)[i]
Woodhouse's toad (*Bufo woodhouseii*)	Tree squirrel (*Sciurus* spp.)
	Nutria (*Myocastor coypus*)[i]
Reptiles	American beaver (*Castor canadensis*)
	White-tailed deer (*Odocoileus virginianus*)
Snapping turtle (*Chelydra serpentine*)	Red fox (*Vulpes vulpes*)[i]
Red-eared slider (*Trachemys scripta*)[i]	Gray fox (*Urocyon cinereoargenteus*)
Mediterranean gecko (*Hemidactylus turcicus*)[i]	Raccoon (*Procyon lotor*)
Green anole (*Anolis carolinensis*)	Striped skunk (*Mephitis mephitis*)
Fence lizard (*Sceloporus* spp.)	Coyote (*Canis latrans*)
Five-lined skink (*Eumeces inexpectatus*)	Mountain lion (*Felis concolor*)
Common garter snake (*Thamnophis sirtalis*)	
Gopher snake (*Pituophis catenifer*)	
Rattlesnake (*Crotalus* spp.*)*	

Sources: Landry (1994); Davis (2003); [i] Introduced in some or all regions of the United States.

selection of animal species—native and introduced—commonly found in and around human-created landscapes. Additionally, people may introduce exotic wildlife into the system, accidentally or intentionally, which can have profound effects on the introduced and native species (see Perspective Essay 3.1).

ECOSYSTEM FUNCTION

In order for ecosystems to function properly they need to include a critical number of community members (e.g., producers, consumers, and decomposers) as well as the basic abiotic factors, so that the relationships that drive the system can develop. With rare exceptions, every natural ecosystem depends on solar energy, and recycles matter. In contrast, urban ecosystems depend primarily on fossil fuels while directing matter into landfills.

Matter is defined as anything that occupies space and has mass—that is, it can be measured where gravity is present. Matter can be changed from one form to another and it can be recycled, but it cannot be created or destroyed. *Energy*, which has no mass and does not occupy space, has the ability to move matter. Like matter, energy can be changed from one form to another, it can be measured, and it cannot be created or destroyed. Unlike matter, energy cannot be recycled.

The basic building blocks of all matter are atoms. Only 108 different kinds of atoms occur in nature, known as the naturally occurring *elements*. Of these, only seventeen—called the essential elements—are directly involved in sustaining all plant and animal life on earth (Table 3.7). Like building blocks, the elements can be combined and disassembled to build different kinds of matter, but the atoms themselves do not change during this process. This constancy of atomic structure is referred to as the law of conservation of matter, which states, "during an ordinary chemical change, there is no detectable increase or decrease in the quantity of matter." Matter is constantly being recycled, and natural ecosystems do a good job of this through decomposers and detritus feeders, but urban ecosystems do not.

Biogeochemical Cycles

One of the most remarkable functions of ecosystems is to provide a process where a finite amount of matter, for example, carbon, hydrogen, and nitrogen, becomes an infinite resource for all living organisms. This is done by an extremely simple process called recycling. The simplicity of the process becomes evident when one studies how biogeochemical cycles work. First, the actors in the process are always the same, that is, producers, consumers, decomposing bacteria and fungi or other microbes that carry out specific functions (Figure 3.6). What is recycled is one of the essential nutrients (nitrogen, carbon, hydrogen, oxygen, phosphorus, and sulfur). Living organisms only borrow nutrients for a period of time, eventually to be released back into the ecosystem to be picked up by another organism. This is why it is possible that the carbon atoms in your body may have been in

Table 3.7   Essential Elements for Sustaining Life

Carbon (C)	Nitrogen (N)	Manganese (Mn)
Hydrogen (H)	Sulfur (S)	Copper (Cu)
Oxygen (O)	Calcium (Ca)	Chlorine (Cl)
Phosphorus (P)	Iron (Fe)	Molybdenum (Mo)
Potassium (K)	Magnesium (Mg)	Zinc (Zn)
Iodine (I)	Boron (Bo)	

Note: The pnemonic for remembering these elements is: "See (C) Hopkins (H) café (Ca Fe) managed (Mg) by (Bo) mine (Mn) cousin (Cu) Clyde Mo Zinc (Cl Mo Zn)."

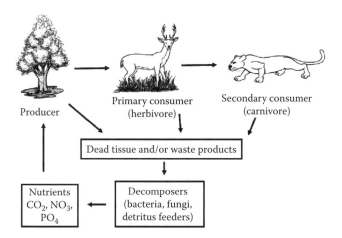

Figure 3.6 Trophic levels and recycling of nutrients. (Courtesy Clark E. Adams)

the body of a different animal or a plant a few months ago. It is important to understand how matter is processed in an urban compared to a natural environment.

The human impact on any biogeochemical cycle is to overload it with too much of the raw material, for example, carbon, nitrogen, or phosphorus. How is this done? In the nitrogen cycle, plants can only accept elemental nitrogen in the form of nitrate (NO_3). This is done by combining three atoms of oxygen with one atom of nitrogen. A particular assemblage of nitrogen-fixing bacteria is responsible for the union of oxygen with nitrogen or it can be accomplished by lightening or volcanic action. Plants then use this molecular form of nitrogen to make proteins. Human agricultural and lawn care practices require large inputs of nitrate fertilizer—more than required for plant growth. This overindulgence in the use of nitrate fertilizer results in quantities of unused fertilizer being left on the land, which eventually washes into natural waterways. Over-fertilizing also puts the nitrogen-fixing bacteria out of work.

The human impact on the carbon cycle is to discharge large quantities of CO_2 into the atmosphere by burning fossil fuels to produce electricity and to run the transportation industry. Plants cannot assimilate the excess CO_2. The excess CO_2 in the atmosphere is suspected of causing climatic change, often referred to (somewhat misleadingly) as "global warming." Recall that temperature is one of the primary abiotic predictors of the type of plant community that will develop in an area (Figure 3.3). Increased atmospheric CO_2 may also be the cause of higher than normal temperatures in urban compared to rural environments.

Urban wildlife does its part to recycle matter—food wastes are a heavily exploited resource, while birds and rodents often use discarded cloth and paper as nesting materials (recall Figure 1.4). Humans, on the other hand, are part of a throw-away culture in which matter is used and then deposited in landfills. The introduction of additional quantities of the essential elements into ecosystems throws the system out of balance and overloads the ability of plants and animals to incorporate these materials. For example, a city with one million human residents consumes an estimated 25,000 tons of water and 2,000 tons of food per day, and produces 50,000 tons of effluent water and 2,000 tons of waste daily (Deelstra 1989). Landfills are the antithesis of functioning ecosystems. They represent a one-way-trip, a use-it-and-lose-it mentality. Given enough time the matter contained in these mountains of waste will be broken down into other forms, but the process takes much longer than under more natural conditions.

Many city governments have begun to realize the value of landfill garbage and have developed waste-to-energy (WTE) facilities to extract the recyclables and use combustible materials to produce electricity. In fact, several states import large quantities of garbage from elsewhere just to maintain operations at their WTE facilities. Chapter 8 offers a detailed account of how landfill waste or garbage left within urban ecosystems becomes a valuable resource and attractant for wildlife.

Another way in which humans impact natural systems is by introducing elements that are not included in the list of essential elements, such as mercury (Hg), lead (Pb), aluminum (Al), as well as chemical compounds like pesticides (e.g., DDT, PCB) that cannot be assimilated or tolerated by living organisms. Chapters 5 and 6 examine the impacts of these selected environmental pollutants on the population dynamics of wildlife populations.

Energy

We owe our lives to the sun… How is it, then, that we feel no gratitude?

—**Lewis Thomas,** *Earth Ethics* **(Summer 1990)**

Of the solar energy that is intercepted by the earth's atmosphere, only 0.8 percent is used for photosynthesis (Odum and Barrett 2004). Solar energy absorbed by green plants is transformed into chemical energy by photosynthesis. Photosynthesis allows sunlight energy to become available to almost every aquatic and terrestrial ecosystem. While sunlight is the foundation of natural ecosystems, urban environments make little use of this infinite energy supply. Instead, the primary sources of energy for man-made ecosystems are fossil fuels (e.g., coal and gas), finite energy sources that have adverse effects on terrestrial and aquatic ecosystems around the globe.

As was mentioned earlier in this chapter, the fundamental difference between matter and energy is that energy cannot be recycled. No matter what path it takes through the ecosystem, ultimately energy is lost from the system in the form of heat. This fact is in keeping with two natural laws: (1) the first law of thermodynamics states that energy can be changed from one form to another; and (2) the second law of thermodynamics states that there will be a loss of energy during each energy conversion. Without a steady supply of energy (kinetic or potential) in the form of sunlight, every biotic component in natural and urban ecosystems would spontaneously move toward increasing *entropy* (disorder, death, and decay).

Imagine you have two wooden desks. One desk is kept inside, protected from factors that cause entropy (and imagine, while you're at it, the amount of energy required to do this!). The other desk is left out in your backyard with no protection from weather, wildlife, children, etc. Can you predict what would happen to the two desks over time? Of course, both desks will eventually succumb to entropy, but the one in the yard will do so much more quickly. In fact, no structural component of natural and urban ecosystems will last forever without some form of energy intervention to sustain its form.

Urban ecosystems are unsustainable in the long term because, in addition to their need for a steady supply of solar energy, they can only function with a continuous new supply of matter. Urban systems quickly decay without heavy subsidies of energy, water, and nutrients in the form of fertilizers provided by urban residents. Urban decay is prevalent in any large metropolitan area where an eroding tax base has left certain areas nearly devoid of the essential goods and services required to sustain the structural and functional integrity of the neighborhood.

Thermodynamics and Conservation of Matter

One way to illustrate how matter and energy are processed in natural ecosystems is to examine what happens with coal when it is ignited. A lump of coal represents a stored form of potential energy resulting from photosynthesis millions of years ago. By applying a catalyst (fire) the potential energy inside the coal is transformed into kinetic energy, as well as the original chemical elements of carbon, hydrogen, oxygen, and sulfur:

$$\text{Coal (CSHO)} + \text{Fire} \rightarrow CO_2\uparrow + SO\uparrow + H_2O\uparrow + \text{Ash}\uparrow + \text{Light} + \text{Sound} + \text{Heat}$$

The kinetic energy from burning the lump of coal consists of heat, light, and sound. The thousands of carbon, hydrogen, sulfur, and oxygen atoms contained in this coal sample is transformed by fire into carbon dioxide (CO_2), sulfur oxide (SO), water (H_2O), and the incompletely burned residue is ash. Note that energy has been lost but matter was conserved, albeit in the form of other chemical compounds. The excess CO_2 is released into the atmosphere and may be picked up by plants in another round of photosynthesis, or the molecules may accumulate and cause the climate change phenomenon mentioned earlier. The SO radical can combine with H_2O in the atmosphere and cause a serious abiotic change called "acid rain." Acid rain is known to cause the destruction of aquatic and terrestrial ecosystems in urban and rural areas. The ash, unless captured, becomes part of the particulate matter in the atmosphere, which is believed to cause some of the urban climatic effects illustrated in Table 3.3.

ECOLOGICAL SUCCESSION

It is said that there are two constants in life—death and taxes. Well, there is another one that should be added to the list, and that is "change." Ecosystems have a built-in response to disturbance called ecological succession. *Ecological succession* is defined as the "orderly and progressive replacement of one community type by another until a climax stage is reached." Climax communities are defined by the dominant vegetation, predicted by ambient temperatures and precipitation (see Figure 3.3), in a specific region. For example, the climax community in the desert southwest (high temperatures, low precipitation) would be cactus (Cactaceae), creosote bush (*Larrea tridentata*), and other forms of drought-tolerant plants On the other hand, in the southeast (high heat, high precipitation), the climax plant community is categorized as subtropical, consisting of lush broad-leaved plants capable of rapid transpiration. Soil formation, another example of ecological succession, is discussed in Chapter 4.

Human activities can dramatically change a natural ecosystem within a few hours and encompass extremely large areas in the process. A common example of this is clearing the land to develop another shopping mall, interstate highway, airport, or subdivision. Ecological succession can be illustrated beginning with a deciduous forest climax community in the northeastern United States (Figure 3.7). The development process begins when the forest is bulldozed down and the area cleared of all vegetation, thus interrupting the existence of the natural ecosystem. The third step in the process is to build a new shopping mall with all of its buildings, streets, and parking lots. The continued existence of the shopping mall is only possible if there are enormous investments of matter and energy, but one day this investment stops. Ecological succession begins at the fourth step, with the deterioration (decomposition) of the shopping mall infrastructure. For example, cracks in the concrete expose the soil, which allows grasses to return (step 5). The grasses contribute to more disruption of the shopping mall structure, allowing shrubs to get a toe-hold (step 6). Grasses and shrubs, together, contribute to the conversion of the shopping mall into suitable habitat for the growth of small trees (step 7). In time, the shopping mall is entirely gone, replaced by the deciduous hardwood forest climax community. Now there is an enormous time scale required to make ecological success work in the manner illustrated in Figure 3.7, but it does happen. Later chapters discuss the impacts of urban development on wildlife.

There are several real-life illustrations of ecological succession in urban communities. For example, imagine what would happen to an urban community if all of the people suddenly and permanently left the area, similar to what happened in Chernobyl, Russia, after the nuclear reactor meltdown in April 1986. Before and after pictures show how the human structures in Chernobyl

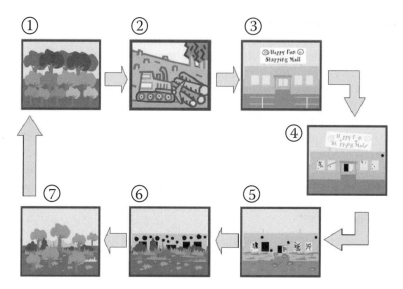

Figure 3.7 Ecological succession from forest to shopping mall and back to forest. See text for explanation. (Concept by Clark E. Adams, diagram by Linda Causey)

are deteriorating due to ecological succession that, in time, will turn the area back into the natural habitat that existed before human settlement. Another example is how abandoned residences ultimately convert back into the climax community of the geographic region, for example, grassland or forest. One of the most obvious examples of ecological succession within the urban ecosystem is successional habitat patches, such as vacant lots (see Chapter 7). Each example demonstrates the resiliency of nature to restore the natural habitat, plant and animal diversity, and ecosystem stability and services that existed prior to human interventions.

ECOSYSTEM SERVICES

> The belief that we can manage the Earth and improve on Nature is probably the ultimate expression of human conceit, but it has deep roots in the past, and is almost universal.
>
> **—Rene J. Dubos, *The Wooing of the Earth* (1980)**

Ecosystem services are ecological processes that produce, directly or indirectly, goods and services from which humans benefit. Nature provides "life support services" at virtually every scale, and many are free and irreplaceable by technology. The list of ecosystem services is extensive. Norberg (1999) listed 29 specific goods or services provided by nature. These goods and services were categorized according to certain ecological criteria, including:

1. Goods or services associated with certain species or a group of similar species, and for which the goods or the target of the service is internal to the ecosystem; for example, foods, timber, pharmaceuticals, pollination, and predators.
2. Services that regulate some exogenous chemical or physical input, that is, the system itself cannot alter the magnitude of the input; for example, processes that drive material and energy flows in ecosystems.
3. Organization of biological processes at virtually all scales, that is, from the way gene sequences are organized to networks of energy and material flows at the level of ecosystems (Norberg 1999).

In addition to housing the genetic library for all life on earth, organisms help to sustain a flow of ecological services that are prerequisites for human economic activities (Folke, Holling, and Perrings 1996). There is some survival value for the urban resident in knowing what types of goods and services are provided by nature (Table 3.8), because there are losses or changes in these goods and services during the urbanization process. Losses or changes in ecosystem services (Table 3.9) can have significant negative impacts on the quality of life for humankind on local and global scales. However, awareness of the existence of a particular good or service is not sufficient. "The dynamics of the ecosystem(s) generating and sustaining a good or service, along with links to other systems, to energy, biogeochemical, and hydrological flows, and to human activities have to be addressed" (Limburg and Folke 1999). In addition, McDonnell et al. (1997) stated that "To understand the ecology of urbanization, the individual components of ecosystem structure, physical and chemical

Table 3.8   The Tangible and Intangible Contributions of Nature

Goods

Breathable air	Building materials (e.g., timber, stone)
Drinkable water	Fuel/energy (e.g., oil, natural gas, wood, nuclear)
Nutrition	Pharmaceuticals
Wild foods (terrestrial and aquatic)	Other chemicals and minerals
Cultivated foods (terrestrial and aquatic)	Art/ornamental objects
Plant and animal species for domestication	Genetic information
Clothing (both plant- and animal-derived)	

Services

Tangible	Intangible
Biotic	Enjoyment/recreational opportunities
Photosynthesis	Educational opportunities
Pollination	Aesthetic appreciation
Recycling of nutrients	Cultural inspiration
Decomposition and waste assimilation	Historic appreciation
Population growth and control	Religious inspiration
Natural selection	
Adaptation	
Resistance to invasion of foreign species	
Succession	
Denitrification	
Erosion prevention	
Soil generation and preservation	
Abiotic	
Maintenance of ecosystem structure	
Nutrient regeneration/recycling	
Water purification	
Detoxification/filtering of pollutants	
Maintenance of atmospheric composition	
Climate moderation	
UV protection by ozone	
Mitigation of floods and droughts	
Resilience after catastrophic events	

Sources: Norberg (1999); Folke, Holling, and Perrings (1996).

Table 3.9   Environmentally and Biologically Relevant Effects of Urbanization

Physical and Chemical Environment	Population and Community Characteristics	Ecosystem Structure and Function
Air pollution	Altered assimilation rates	Altered decomposition rates
Hydrological changes	Altered biological cycles	Altered nutrient retention
Local climate change	Altered disturbance regimes	Greater nutrient flux
Soil changes and movement	Altered reproductive rates	Increased habitat patches
Water pollution	Altered succession rates and direction	Habitat fragmentation and simplification
	Altered survival rates	Increased debris dams
	Changed growth rates	Increased sediment loading in streams
	Genetic drift and selection	Loss of biological diversity
	Introduced species	Loss of biological organization
	Landscape fragmentation	Loss of forest understory
	Population size and structure	Loss of plant productivity
	Reduced species diversity	Loss of redundant pathways for nutrient and energy transfer
	Social and behavioral changes	

Source: McDonnell and Pickett (1990).

environments, populations, communities, and ecosystems need to be studied in order to appreciate the ecologically important impacts of urban development and change on natural areas."

However, in urban landscapes, humans (*Homo sapiens*) are the dominant species, perhaps even a keystone species, whose existence determines the continued existence and/or population dynamics of all other species of plants and animals in that community. The human impact on nature's services cannot be fully understood without the integration of ecology, sociology, political agendas and the decision-making process, and natural resource economics. This integration can be explained in terms of how people prioritize their resource and service needs, and then act on them through patterns of consumption, investment, resource use, waste production, and disposal decisions—which in turn affect ecosystem energy and mass flow into, through, and from urban ecosystems. In this regard, consideration of the effects of local economics, public policy, law, diversity of stakeholder perspectives, human dimensions, land-use planning decisions, and problem-solving skills become important. This is the conceptual framework we used to develop the remaining chapters in this book.

This book is not just about human and wildlife interactions in the city. The tentacles of the urban core extend well beyond the city limits by way of highways and associated structures, power production and transmission, waste disposal services, transportation, parks and recreational services, food production and distribution, and the mass communication industry (Figure 3.8). As such, these impacts of urbanization on the exurban environments, habitats, and wildlife need to be part of the story told in this book. Examples of this extended impact are provided in subsequent chapters on urban green and gray spaces, population dynamics, and urban streams and soils, among others.

PERSPECTIVE ESSAY 3.1   People and Wildlife:
The Lesser Anteater (*Tamandua tetradactyla*)

The animal shown in Figure 3.9 is an anteater and lives in the Amazon basin and the forests of northern South America. It goes by several common names. Its primary food sources are arboreal ants and termites. Lesser anteaters rip open bark with a third toe adapted for this function and use their sticky, long tongues to lap up their food.

Given this animal's normal geographic range and food habits, it was quite a surprise to find one hiding under a bush of a residential home in Seward, Nebraska, in September 1968; Seward

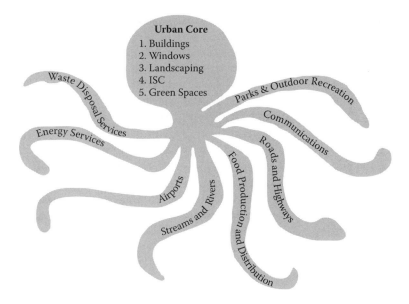

Figure 3.8   The resource demands of the urban core extend far beyond the city limits, with consequent impacts on wildlife and natural habitats. (Concept by Clark E. Adams, diagram by Linda Causey)

Figure 3.9   The lesser anteater (*Tamandua tetradactyla*). (Source unknown)

is several thousand miles north of the anteater's normal habitat. The events that transpired after the discovery of the anteater proved to be interesting, informative, and even a little bizarre.

The story began on a Sunday afternoon when my neighbor, a music teacher, called me, a biology teacher, and told me there was a female anteater in his backyard. I was skeptical to say the least. I went over to his house and sure enough there was an anteater in his backyard!

It was cold outside so I quickly picked up the anteater to take it to my laboratory at Concordia College. In so doing, I lost the left sleeve of my coat to the animal's strong prehensile third toe.

I placed the anteater in a chicken wire cage only to observe her tear it apart with her front toes. I then placed her in a more secure cage made of steel.

Now I had to figure out what to feed her. I called the Henry Dorley Zoo in Omaha, Nebraska. I told them I had an anteater—I even gave them the scientific name—and explained that I needed to know what to feed it. I also asked if they would be interested in taking the animal off of my hands. They were skeptical of my identification, but they gave the recipe for anteater gruel (which consisted of a mixture of egg yolk, dog food, milk, and a few other ingredients tossed into a blender to form a soup-like concoction)—which the anteater subsequently ignored. The Henry Dorley Zoo referred me to the Children's Zoo in Lincoln, Nebraska.

I called the Children's Zoo and told them I had an anteater. They were skeptical of my identification also—even though I was a biology teacher I did not command a great deal of respect with zoo personnel in the taxonomy department. Nevertheless, they agreed to pick up the animal and I warned them to bring a strong cage. It took about a week for personnel from the Children's Zoo to arrive. When they did, they were quite surprised to find out that I really did have an ANTEATER!

The fact that an anteater was found in Seward, Nebraska, was big news locally and state-wide. The local newspaper did a story on her, statewide news services picked up the story, and television stations in Lincoln and Omaha reported on the event. I believe that nearly every child attending elementary school in Seward had a field trip to my lab to see the anteater. I was holding the anteater for the children so they could see how strong she was. I forgot about that third toe and again lost a sleeve on my suit coat.

The nagging question of how the anteater ended up in Seward, Nebraska, all the way from South America remained unanswered. My neighbor's house was just across the street from a new church, and the anteater was found on the same day the building was being dedicated. Many people from all over Nebraska and other states were at the event, so maybe the anteater hitched a ride!

Eventually, the local police department received a call from a lady in Chicago, Illinois, who claimed the anteater was hers. She had left it in her car at the Lincoln airport to fly to Omaha for a one-day meeting. The police asked for her address in Chicago. They then contacted the Chicago police to verify the address—they found a vacant lot! They never heard from the lady again. Anteaters are not on the list of exotic pets desired by humans so the incident with the lady just made no sense.

The Children's Zoo adopted the anteater and called her "Connie," but Connie died soon after her adoption. An autopsy determined the cause of death to be malnutrition and heavy parasite loads in her intestines.

Readers may assume this story is a bizarre and uncommon tale, but it isn't all that unusual. My co-author, who worked at a wildlife rehabilitation center in Houston, Texas, has many stories to tell of the exotic animals that often ended up at the facility. There seems to be no limit to the appeal that exotic pets have for so-called animal lovers, but the novelty can wear off quickly as unexpected expenses rise and the wild "pets" refuse to settle into domestic bliss. Some of the species that were found by the public and delivered to the wildlife shelter during Kieran's two-year tenure with the nonprofit organization included African pygmy hedgehogs (*Atelerix albiventris*), green iguanas (*Iguana iguana*), red-tailed boa constrictors (*Boa constrictor imperator*), Burmese pythons (*Python molurus bivittatus*), a Moluccan cockatoo (*Cacatua moluccensis*), a coatimundi (*Nasua nasua*), and a pet black bear (*Ursus americanus*).

One of three things usually happens when exotic animals are introduced to a new habitat, on purpose or unintentionally. In one case the animal is completely unsuited for the environmental conditions it finds itself in, and it either dies of starvation, predation, or disease. In the second case, the animal has lost its fear of humans during previous captivity, and may even associate humans with food and shelter, so it finds its way into someone's yard. At this point, the local wildlife rehabilitation center or nature center or zoo or biology teacher is likely to get a call from an astonished citizen who starts the conversation by saying something like, "You're not going to believe this, but there's an anteater in my backyard!"

The last scenario is somewhat less common, but has a greater long-term impact. Some translocated animals find themselves ideally suited to their new habitat, so much so that they may even begin to out-compete some native species. Examples of wild species that have been intentionally or accidentally introduced to North America include everything from the house sparrow (*Passer domesticus*) and the European starling (*Sturnus vulgaris*) to Mediterranean geckos (*Hemidactylus turcicus* Linnaeus) and fire ants (*Solenopsis wagneri*).

Clark E. Adams
Kieran J. Lindsey

SPECIES PROFILE: CHIMNEY SWIFT (*CHAETURA PELAGICA*)

… a shift of wing, and they're earth-skimmers, daggers
Skillful in guiding the throw of themselves away from themselves.

—Anne Stevenson, Poet

Chimney swifts are so ungainly when clinging to the inside of a chimney you might never guess they are some of the most acrobatic flyers in the bird world. They are members of the Apodidae family; Apodidae means "without feet." This isn't technically true, but the legs and feet are small and weak. Still, their cigar-shaped bodies and boomerang wings allow them to dance through the air, diving and soaring, and they do nearly everything but incubate their eggs on the wing. Prior to the arrival of Europeans—and their chimneys—in North America, these swifts probably nested in caves and hollow trees. Chimneys initially increased the number of nest sites, and now help to

Figure 3.10 Chimney swift on nest. (Courtesy of Marilyn St. Louis)

mitigate the loss of hollow trees, which are often removed for safety or aesthetic reasons. However, recent changes in chimney design and covered flues may be a factor in declining population numbers (National Geographic 1999; Cornell Lab of Ornithology website).

Chimney swifts feed exclusively on flying insects and, like our North American bats, are invaluable for controlling mosquitoes. Estimates place the number of insects consumed by two adults and their chicks at 12,000 per day. They nest in colonies of at least a few pairs. Nests consist of a half-saucer of twigs, glued together and to the chimney wall with saliva. Eggs may be laid before the nest is completed. The female broods for nineteen to twenty-one days. Breeding pairs may be assisted by a male or female helper that feeds young and may even incubate the eggs. Chicks are altricial, fledging about a month after they hatch by climbing up the vertical walls with sharp claws and tail bristles before taking their first flight (Ehrlich, Dobkin, and Wheye 1998).

Although few people hope to make life easier for mosquitoes, conflicts can arise between humans and chimney swifts. The noise—and referring to it as anything else is gilding the lily—swift chicks make during the last two weeks in the nest isn't on anyone's Top 40 list. When the parents return to the nest to feed the youngsters, the chicks begin to clamor for mom's or dad's attention (and insects). As these cries echo up and down the flue it can create quite a din in the den below. Swifts, and other migratory birds, are federally protected and it's illegal to disturb the nest or the chicks. One trick to dampen the noise is to cut a thick slab of Styrofoam to fit into the fireplace opening (personal observations). Chimney swifts aren't raising babies during the seasons in which fireplaces get a workout, and when the swiftlets fledge the homeowner can legally install a chimney cap to prevent future nesting (Georgia Department of Natural Resources website).

Another cause for homeowner alarm is the fear that chimney swift nests will create a flue fire hazard after the birds leave on their annual migration to South America. Swift nests usually fall off the wall shortly after the young fledge (more motivation to make that climb and take wing!). A soiled, creosote-filled flue is a much greater hazard, and this kind of build up prevents chimney swift nests from attaching to the walls (Missouri Department of Conservation Website).

While there can be concern among homeowners hosting chimney swifts, there's very little published research on human-swift conflicts. Studies of this species tend to focus on population dynamics and, to a lesser extent, behavioral ecology.

CHAPTER ACTIVITIES

1. Determine the size of your ecological footprint by testing yourself at the following website: http://www.earthday.net/footprint/index.asp. If this site is no longer available conduct a Google search using "ecological footprint" as your search term.
2. Visit the website of Friends and Advocates of Urban Natural Areas (FAUNA, http://www.urbanfauna.org/). Examine how the city of Portland, Oregon, is attempting to reestablish a sustainable urban ecosystem in terms of the actions citizens are taking to accomplish this goal, and in terms of what your community is doing to design human developments that include natural things.
3. Conduct an inventory of urban wildlife (vertebrates) in your neighborhood using the safari technique, road-kill inventory, personal encounters, neighborhood informants, written and/or published inventories. Which species are the most abundant? Which species are endemic or exotic?
4. Find examples of ecological succession at work in your community.
5. Construct a list of ecosystems services that at one time were free, but now community residents have to pay for them.

LITERATURE CITED

Adams, L.W. 1994. *Urban Wildlife Habitats: A Landscape Perspective*. Minneapolis: University of Minnesota Press.

Begon, M., J.L. Harper, and C.R. Townsend. 1996. *Ecology: Individuals, Populations, and Communities*. Oxford, UK: Blackwell Science.

Borman, S. and R. Korth. 1998. *Through the Looking Glass: A Field Guide to Aquatic Plants*. Stevens Point, WI: University of Wisconsin-Stevens Point Foundation.

Brockman, C.F. 1986. *Trees of North America: A Field Guide to the Major Native and Introduced Species North of Mexico*. Racine, WI: Golden Press.

Burger, J. 1999. *Animals in Towns and Cities*. Dubuque, IA: Kendall/Hunt Publishing Company.

Cushman, R.C. and S.R. Jones. 2004. *Peterson Field Guides: The North American Prairie*. New York: Houghton Mifflin.

Davis, J.M. 2003. Urban systems. In *Texas Master Naturalist Statewide Curriculum,* 1st edition. Haggerty, M. M., ed. College Station, TX: Texas Parks & Wildlife Department.

Deelstra, T. 1989. Can cities survive? Solid waste management in urban environments. *AT Source* 18:21–27.

Ehrlich, P.R., D.S. Dobkin, and D. Wheye. 1988. *The Birder's Handbook: A Field Guide to the Natural History of North American Birds*. New York: Simon and Schuster.

Folke, C., C.S. Holling, and C. Perrings. 1996. Biological diversity, ecosystems, and the human scale. *Ecological Applications* 6:1018–1024.

Hope, J. 2000. The geese that came in from the wild. *Audubon* 102:122–126.

Landry, S.B. 1994. *Peterson First Guides: Urban Wildlife*. New York: Houghton Mifflin.

Limburg, K.E. and C. Folke. 1999. The ecology of ecosystem services: Introduction to the special issue. *Ecological Economics* 29:179–182.

McDonnell, M.J. and S.T.A. Pickett. 1990. Ecosystem structure and function along the urban-rural gradients: An unexploited opportunity for ecology. *Ecology* 71:1232–1237.

McDonnell, J.J., S.T.A. Pickett, P. Groffman, P. Bohlen, R.V. Pouyat, W.C. Zipperer, R.W. Parmelee, M.M. Carreiro, and K. Medley. 1997. Ecosystem processes along an urban-to-rural gradient. *Urban Ecosystems* 1:21–36.

National Geographic. 1999. *Field Guide to the Birds of North America*. Washington, DC: National Geographic.

Norberg, J. 1999. Linking nature's services to ecosystems: An ecological perspective. *Ecological Economics* 29:183–202.

Odum, E.P. and G.W. Barrett. 2004. *Fundamentals of Ecology,* 5th edition. Belmont, CA: Thomson Brooks/Cole.

Ostfeld, R.S. and F. Keesing. 2000. Biodiversity and disease risk: The case of Lyme disease. *Conservation Biology* 14:722–728.

Ostfeld, R.S., F. Keesing, C.G. Jones, C.D. Canham, and G.M. Lovett. 1998. Integrative ecology and dynamics of species in oak forests. *Integrative Biology* 1:178–186.

Pickett, S.T.A., W.R. Burch, Jr., S.E. Dalton, T.W. Foresman, J.M. Grove, and R.A. Rowntree. 1997. A conceptual framework for the study of human ecosystems in urban areas. *Urban Ecosystems* 1:185–201.

Sullivan, R. 2004. *Rats: Observations on the History and Habitat of the City's Most Unwanted Inhabitants*. London: Bloomsbury.

Wright, R.T. 2004. *Environmental Science: Toward a Sustainable Future,* 9th ed. Upper Saddle River, NJ: Pearson Education.

Zipperer, W.C., J. Wu, R.V. Pouyat, and S.T.A. Pickett. 2000. The application of ecological principles to urban and urbanizing landscapes. *Ecological Applications* 10:685–688.

CHAPTER 4

Urban Soils

Soil is a dynamic, natural system synthesized in profile form at the earth's surface from a variable mixture of weathered minerals and decaying organic matter, under the influence of climatic and topographic conditions and living organisms; which supplies, when containing the proper amounts of air and water, mechanical support, and in part, sustenance for plants.

—**Phillip J. Craul, Professor of Soil Science**

KEY CONCEPTS

1. There are distinct differences between urban and natural soil ecosystems.
2. Natural and urban soil ecosystems are different in terms of structure, function, trophic levels, soil biota, and ecosystem services.
3. Impacts of urbanization are the primary causes of structural and functional changes in urban soil ecosystems.
4. There are many strategies for taking better care of urban soils.

INTRODUCTION

The two renewable resources that are most critical for supporting all life on earth are fertile soil and adequate water. With proper management, both can be maintained, used, and reused indefinitely. Without proper management, soil and water can be depleted or rendered useless. Indeed, there is much evidence to suggest that many ancient civilizations collapsed and became extinct as a result of depleting their soil and/or water resources. In this regard, it is important to note that from 1982 to 1992, 1.4 million acres per year of farmland were converted to urban development in the United States alone, and that figure rose to 2.2 million acres per year from 1992 to 1997 (Scheyer and Hipple 2005). This chapter and Chapter 5 provide an overview of the essential information one needs to know in order to understand the critical ecological interrelationships between soil and water.

The underlying support of any civilization is a stable agricultural production system. No society can maintain itself for long in a civilized state without an ample and reliable food supply for its citizens, and such a supply can only come from the artificial propagation of various plants and animals. More than 90 percent of the world food supply comes from land-based systems that rely on soil as the fundamental support matrix. Propagation of plant species is even more fundamental than that of animals, because plants are always at the beginning of any food chain.

The soil ecosystem has several unique characteristics that distinguish it from ecosystems that exist in air and water. These distinguishing characteristics include:

1. A slow rate of nutrient and energy transfer. Decomposition rates vary from a few months in tropical rain forests to several years in temperate forests.
2. Different textures (e.g., different combinations of sand, silt, and clay).
3. A near total reliance on decomposers (bacteria, fungi) to begin the process of nutrient and energy transfer; literally millions of bacteria can be seen and counted in one gram of soil (Wright 2004).
4. The exclusive use of detritus (dead animal and plant material or wastes) as the source of energy and nutrients.
5. An extreme susceptibility to disturbance, with very slow recovery times.

Different soil textures demand some very unique adaptations on the part of animals that live in the soil. In short, moving through a soil medium is much more difficult than moving through air or water. For example, the earthworm (*Lumbricus terrestris*) is an invertebrate that literally eats its way through topsoil. It is estimated that as much as fifteen tons per acre of soil pass through earthworms each year in the course of their feeding (Wright 2004; see Perspective Essay 4.1). Mammals that live underground (fossorial) have special adaptations that include rudimentary eyes, highly evolved senses of touch and smell, and appendages adapted to rapid digging.

SOIL FORMATION

The development of topsoil from subsoil or parent material can take hundreds of years, depending on region and climate. The process begins when natural abrasive or weathering actions (e.g., wind and water) break down the parent rock leaving small fissures and a rough surface. The surface collects dust and bacteria from the atmosphere. Erosion (leaching) will force the mineral nutrients from the rock. The dust accumulations on the bedrock and the leached nutrients provide a substrate for the first biotic colonizers, such as lichens and algae. These early colonizers release substances that further disintegrate the rock. Many, many iterations of dust accumulation, addition of organic material, and weathering results in enough soil formation to support the growth of vascular green plants and soil biota (discussed below). This process is an example of ecological succession, defined in Chapter 3. Understanding the process of soil formation may lead one to appreciate the potential, long-term, destructive effect of urban sprawl on the soil ecosystem in different biotic regions in the United States.

SOIL STRUCTURE

The soil ecosystem exists to accomplish two tasks: providing a supporting substrate; and providing nutrients, water, and oxygen to green plants. As such, soils are complex systems of solid matter, pore spaces filled with water and oxygen, and a rich assemblage of organisms. Soil is a mixture of four basic components:

1. Forty-five percent inorganic materials, including rock, clay, silt, and sand, that give structure to the soil. Soil texture refers to the relative proportions of each mineral type in a given soil.
2. Five percent organic matter, including living and decomposing organisms and plant parts, supplies nutrients and helps hold moisture in the soil.
3. Twenty-five percent air that moves through the pore spaces and provides oxygen to plant roots.
4. Twenty-five percent water and dissolved nutrients, important for a number of a plant's life processes, which also move through the pores (U.S. Department of Agriculture 2001).

An alternative analysis of soil structure that considers only the mineral (inorganic) and organic composition revealed characteristics unique to this ecosystem and related to soil function discussed later (Figure 4.1). Note that only 7 percent of a total soil sample is organic. Eighty-five percent of

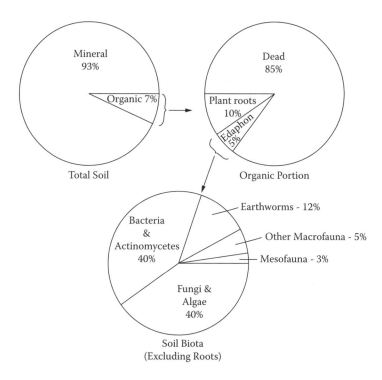

Figure 4.1   The living and nonliving components of meadow soil in terms of the dry weight. (From E.P. Odum. 1971. *Fundamentals of Ecology.* Philadelphia, PA: W.B. Saunders.)

the organic component is dead material. The remaining 15 percent is living organic material consisting of 10 percent root material and 5 percent of other life forms, called *edaphon*. The edaphon are classified by size. The bulk of the soil biota (80 percent) consists of bacteria, fungi, protozoans, viruses, and algae, which are considered *microbiota* because of their small size. *Mesobiota* or *mesofauna* include the smaller arthropods, nematodes, springtails, and smaller enchytraeids (pot worms). *Macrobiota* or *macrofauna* include all animals large enough to be seen by the naked eye, such as earthworms, larger arthropods, fossorial mammals, mollusks (slugs and snails), and larger enchytraeids (Harris 1991). In terms of sheer numbers, the meso- and macrobiota make up a small percentage of the total assemblage of soil biota (Figure 4.1) in spite of the fact that thousands of earthworms and pot worms can be found in a square meter of healthy soil (Table 4.1; Harris 1991). The dominant biota are those involved in decomposition and conversion of detritus into humus. These organisms can be considered to be specialists. The soil community is extremely unique given the complex adaptations required for existence in a soil matrix. As such, soil biota is highly susceptible to outside perturbations that might disrupt their ability to function or destroy them altogether. For example, because most soil biota live underground and never see the light of day, they are never exposed to ultraviolet light. Imagine what happens when soil is turned over by a shovel or plow exposing the soil biota to UV light rays for the first time—the result is instant death and the loss of the contributions these organisms made in the nutrient and energy transfer process.

SOIL HORIZONS

Soil consists of different layers called "horizons" (Figure 4.2). The arrangement and makeup of these horizons influence the amount and type of plant growth that can occur. The thickness of

Table 4.1   Abundance of Major Soil Animals

Group	Thousands per Square Meter
Nematodes	1.8×10^3–120×10^3
Mites	20–120
Collembola	10–40
Pot worms	0.5–20
Molluscs	0.5–9
Earthworms	0.1–2
Large myriapods	0.5–2
Beetles and larvae	0.5–1
Dipterous (fly) larvae	0.5–0.9
Spiders	0.2–0.9
Ants	0.2–0.5
Isopods	0.2–0.4

Source: J.A. Harris. 1991. In *Soils in the Urban Environment*, ed. P. Bullock and P.J. Gregory. Cambridge, MA: Blackwell Scientific. With permission.

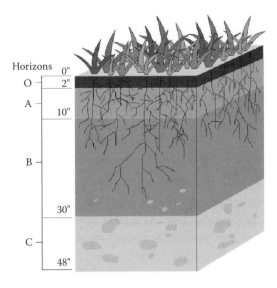

Figure 4.2   A typical soil profile showing the four horizons. (USDA photo)

each layer will vary depending on the climate and region and impacts of urbanization. Table 4.2 describes the four general horizons in a soil profile. *Loam* is a soil texture that contains roughly 40 percent sand, 40 percent silt, and 20 percent clay and is considered the optimal mineral composition for soil to facilitate healthy plant growth. An additional mixture of humus and other organic material from the O horizon increases soil porosity, enhancing the water-holding, nutrient-holding, and aeration qualities of soil. As earthworms eat their way through the O, A, and sometimes the B horizon, they form open channels that connect the horizons, increase water infiltration and oxygen transfer from the surface and, as a bonus, increase the humus content in the A horizon. Earthworms have an exceptional ability to displace large amounts of soil and play a major role in topsoil formation (see Perspective Essay 4.1). They can completely mix the top six inches of a humid grassland soil in ten to twenty years (U.S. Department of Agriculture 2001). Some of the same benefits are provided by burrowing mammals such as moles (five different genera) and pocket gophers.

Table 4.2 Description of Soil Profiles

Horizon	Description
O	The top layer or organic horizon consisting of decaying plant material called "humus."
A	The most productive layer of soil, commonly referred to as "topsoil." Most plant roots are found here. The A horizon is composed primarily of a mixture of mineral and humus material and is where most soil functions occur.
B	A subsoil layer formed by materials leached or moved from the A horizon, including clay, iron, aluminum, and some organic material.
C	The lowest layer of soil composed of disintegrated parent material and other minerals.

SOIL FUNCTIONS

Healthy soil ecosystems will support healthy plant growth above ground if they contain ample supplies of oxygen, water, and nutrients. In order to make these three key ingredients available, soils perform six critical functions simultaneously, regardless of region or climate.

1. Soils act like sponges, soaking up rainwater and limiting runoff. Soils also impact groundwater recharge and flood-control potentials in urban areas.
2. Soils act like faucets, storing and releasing water and air for plants and animals to use.
3. Soils act like supermarkets, providing valuable nutrients and air and water to plants and animals. Soils also store carbon and prevent its loss into the atmosphere.
4. Soils act like strainers or filters, filtering and purifying water and air that flow through them.
5. Soils buffer, degrade, immobilize, detoxify, and trap pollutants, such as oil, pesticides, herbicides, and heavy metals, and keep them from entering groundwater supplies.
6. Soils also store nutrients for future use by plants and animals above ground and by microbes within the soils (Scheyer and Hipple 2005).

These services are the result of topsoil formation, which is a function of the three constantly interacting ecological processes, that is, detritus decomposition, mineral leaching, and a detritus food web (Figure 4.3). No single soil function listed above can be assigned to any one of three ecological processes alone because each function is the result of the simultaneous completion of each process. They act in concert rather than as isolated entities. Imagine what might happen in terms of soil's ability to provide the above six functions if detritus were removed from the soil ecosystem—a common result of urbanization. This would shut down the decomposer food chain, resulting in the loss of soil nutrients—or the supermarket function. Above-ground plants could not survive, and without their root systems to hold the soil particles in place, wind and water erosion will eventually lead to soil loss.

SOIL BIOTA AND THEIR FUNCTIONS

"Tons of soil biota can live in an acre of soil, and are more diverse than the community of plants and animals above ground" (U.S. Department of Agriculture 2001). One can't help but wonder about the first agronomist (soil scientist) to examine the incredible diversity of organisms that can be found in the litter or top few inches of soil. One way to examine the diversity, or lack thereof, of soil biota in various types of urban green spaces (see Chapter 7) is to use the Berlese funnel sampling technique (Figure 4.4). A small soil sample is placed on the screen on top of the funnel. A bright light is placed above the sample. Soil biota are negatively phototrophic (they move away from light) and thus move down the funnel into the waiting bottle of alcohol which kills and preserves them. Next, the sample is poured into a Petri dish and examined against a black background using a compound microscope. What is usually present in a healthy soil ecosystem is a diverse assemblage

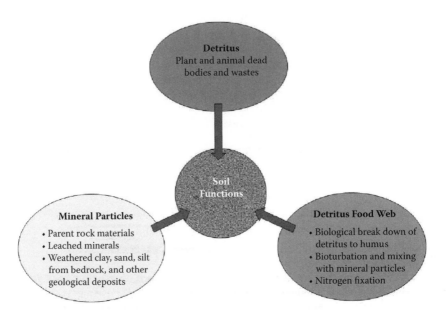

Figure 4.3 The three requirements in topsoil formation resulting in the primary soil functions. (Concept by Clark E. Adams, diagram by Linda Causey)

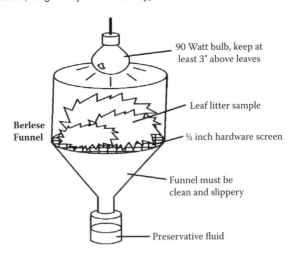

Figure 4.4 The Berlese funnel sampling array for capturing a sample of invertebrate soil biota. (Diagram by Linda Causey)

of soil arthropods, as illustrated in Figure 4.5. All of them will appear translucent and white against the black background. Soil biota that cannot be captured using the Berlese funnel are the larger mesobiota and macrobiota, but these organisms or their sign (e.g., burrows) can be observed directly with the naked eye. Other techniques such as gel cultures in Petri dishes will reveal the microbiota present in a soil sample.

Let's examine the cast of characters and their specific roles in the soil food web (a glossary is provided in Table 4.3) and match the type of soil organism to its major functions using Table 4.4. Soil biota perform three critical functions, including: residue decomposition; nutrient storage and release; and water storage, infiltration, and resistance to erosion. These functions are accomplished by a series of specific tasks carried out by organisms occupying different trophic levels or result from

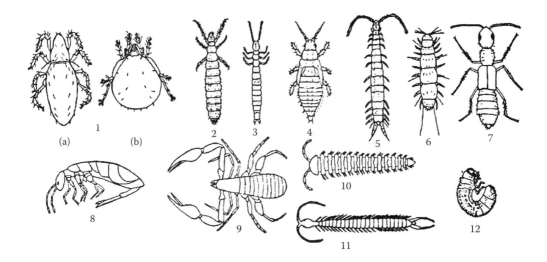

Figure 4.5   Representative soil arthropods found in soil litter using the Berlese funnel sample method. (1) Two oribatid mites; (2) a proturan; (3) a japygid; (4) a thrip; (5) a symphylan; (6) a pruropod; (7) a rove beetle; (8) a springtail or collembolan; (9) a pseudoscorpion; (10) a millipede (or diplopod); (11) a centipede (or chilopod); (12) a scarabaeid beetle larvae or grub. (From E.P. Odum. 1971. *Fundamentals of Ecology.* Philadelphia, PA: W.B. Saunders.)

Table 4.3   A Soil Food Web Glossary

Arthropods	Invertebrate animals with jointed legs. They include insects, crustaceans, sowbugs, arachnids (spiders), and others.
Bacteria	Microscopic, single-celled organisms that are mostly nonphotosynthetic. They include the photosynthetic cyanobacteria (formerly called blue-green algae), and actinomycetes (filamentous bacteria that give healthy soil its characteristic smell).
Fungi	Multicelled, nonphotosynthetic organisms that are neither plants nor animals. Fungal cells form long chains called hyphae and may form fruiting bodies such as mold or mushrooms to disperse spores. Some fungi, such as yeast are single-celled.
	• Saprophytic fungi: Fungi that decompose dead organic matter.
	• Mycorrhizal fungi: Fungi that form associations with plant roots. These fungi get energy from the plant and help supply nutrients to the plant.
Grazers	Organisms such as protozoa, nematodes, and microarthropods that feed on bacteria and fungi.
Microbes	An imprecise term referring to any microscopic organism. Generally, "microbes" includes bacteria, fungi, and sometimes protozoa.
Mutualists	Two organisms living in an association that is beneficial to both, such as the association of roots with mycorrhizal fungi or with nitrogen-fixing bacteria.
Nematodes	Tiny, usually microscopic, unsegmented worms. Most live free in the soil. Some are parasites of animals or plants.
Protozoa	Tiny, single-celled animals, including amoebas, ciliates, and flagellates.
Trophic levels	Levels of the food chain. The first trophic level includes photosynthesizers that get energy from the sun. Organisms that eat photosynthesizers make up the second trophic level. Third trophic level organisms eat those in the second level and so on. It is a simplified way of thinking of the food web. In reality, some organisms eat members of several trophic levels.

Source: Soil and Water Conservation Society. 2000. *Soil Biology Primer,* revised edition. Ankeny, IA: Soil and Water Conservation Society. With permission.

Table 4.4   Functions of Soil Organisms

Type of Soil Organism	Major Functions
Photosynthesizers • Plants • Algae • Bacteria	Capture energy • Use solar energy to fix CO_2. • Add organic matter to soil (biomass such as dead cells, plant litter, and secondary metabolites).
Decomposers • Bacteria • Fungi	Break down residue • Immobilize (retain) nutrients in their biomass. • Create new organic compounds (cell constituents, waste products) that are sources of energy and nutrients for other organisms. • Produce compounds that help bind soil into aggregates. • Bind soil aggregates with fungal hyphae. • Nitrifying and denitrifying bacteria convert forms of nitrogen. • Compete with or inhibit disease-causing organisms.
Mutualists • Bacteria • Fungi	Enhance plant growth • Protect plant roots from disease-causing organisms. • Some bacteria fix N_2. • Some fungi form mycorrhizal associations with roots and deliver nutrients (such as P) and water to the plant.
Pathogens • Bacteria • Fungi	Promote disease • Consume roots and other plant parts, causing disease. • Parasitize nematodes or insects, including disease-causing organisms.
Parasites • Nematodes • Microarthropods	
Root feeders • Nematodes • Macroarthropods (e.g., cutworm, weevil larvae, and symphylans)	Consume plant roots • Potentially cause significant crop yield losses.
Bacterial feeders • Protozoa • Nematodes	Graze • Release plant-available nitrogen (NH_4^+) and other nutrients when feeding on bacteria. • Control many root-feeding or disease-causing pests. • Stimulate and control the activity of bacterial populations.
Fungal feeders • Nematodes • Microarthropods	Graze • Release plant-available nitrogen (NH_4^+) and other nutrients when feeding on fungi. • Control many root-feeding or disease-causing pests. • Stimulate and control the activity of fungi populations.
Shredders • Earthworms • Macroarthropods	Break down residue and enhance soil structure • Shred plant litter as they feed on bacteria and fungi. • Provide habitat for bacteria in their guts and fecal pellets. • Enhance soil structure as they produce fecal pellets and burrow through soil.
Higher-level predators • Nematode-feeding nematodes • Larger arthropods, mice, voles, shrews, birds, other above-ground animals	Control populations • Control the populations of lower trophic level predators. • Larger organisms improve soil structure by burrowing and by passing soil through their guts. • Larger organisms carry smaller organisms long distances.

Source: Soil and Water Conservation Society. 2000. *Soil Biology Primer,* revised edition. Ankeny, IA: Soil and Water Conservation Society. With permission.

the complex interrelationships between trophic levels throughout the soil food web (Figure 4.6). In other words, soil organisms interact with one another, with plant roots, and with their environment to accomplish the following tasks:

1. Decomposition of organic matter;
2. Cycling of minerals and nutrients;
3. Reservoirs of minerals and nutrients;

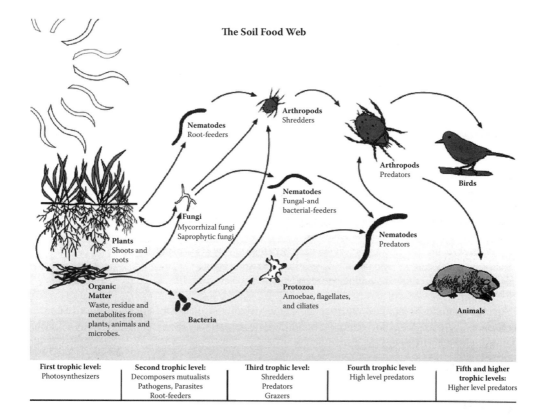

The Soil Food Web

Figure 4.6   The soil food web. Earthworms would be found in the third trophic level because their function is shredding. (From the Soil and Water Conservation Society. 2000. *Soil Biology Primer,* revised edition. Ankeny, IA: Soil and Water Conservation Society. With permission.)

4. Redistribution of minerals and nutrients;
5. Sequestration of carbon;
6. Degradation of pollutants, pesticides;
7. Modification of soil structure;
8. Community self-regulation;
9. Biological regulation of pest species.

These tasks and the soil functions mentioned earlier are the result of a "top down/bottom up" feedback loop that exists between above- and below-ground organisms. Note that the below-ground food web could not exist if above-ground photosynthesis did not take place, that is, the source of organic matter. In turn, the organic matter could not be reutilized by above-ground plants without the interrelated functions of the organism in the soil food web. One might consider this an ecosystem-level demonstration of a mutualistic relationship (see Chapter 3).

By now the reader should be impressed with the complexity and fragility of the soil ecosystem. It should probably come as no surprise that the soil ecosystem is the first to go in the process of urbanization and urban sprawl.

IMPACTS OF URBANIZATION ON SOIL STRUCTURE AND FUNCTION

Originally, the soils of urban systems were the same as those of the natural systems that exist in the area, but in developed areas the soils may have been hauled in from far away. So the important thing is to examine what happens to soil during the process of urbanization (Davis 2003; Schueler 2000; DNREC 2004). Urban soils have a definition all their own:

A soil material having a non-agricultural, manmade surface layer more than 50 cm thick, that has been produced by missing, filling, or by contamination of land surfaces in urban and suburban areas. (Bockheim 1974, as cited by Craul 1992)

The general characteristics of urban soil are in stark contrast to natural soil ecosystems. These include:

1. Great vertical and spatial variability;
2. Modified soil structure leading to compaction;
3. Presence of surface crust on bare soil that is usually hydrophobic (i.e., repels water);
4. Modified soil reaction, usually elevated;
5. Restricted aeration and water drainage;
6. Interrupted nutrient cycling and a modified soil organism population and activity;
7. Presence of anthropogenic materials and other contaminants;
8. Highly modified soil temperature regimes (Craul 1985; New York City Soil Survey Staff 2005).

Again, it should be pointed out that the above urban soil characteristics are interrelated; one condition leads to another. For example, condition 2 can lead to conditions 3 and 5. Some of these urban soil characteristics are discussed below.

Vertical and Spatial Variability

Bioturbation is the physical rearrangement of the soil profile by soil life, including insects, pot worms, earthworms, moles, and pocket gophers. This process mixes the organic matter in the O horizon with the A horizon thus "feeding" the decomposing bacteria and fungi, and forms channels and pores linking the O and A horizons. This linkage greatly enhances aeration and water infiltration. The impact of urbanization on the soil horizons is to greatly increase the O horizon through over-mulching (additions), and to lose or greatly diminish the A horizon through construction grading and scraping or erosion.

Grading and scraping are the most visible effects that urbanization has on the local soil (Figure 4.7). As new development begins, often the first step is to bring in heavy equipment to begin "sculpting" the land to meet the needs of the project. At the low-impact end of things, this may involve moving soil around and mixing up horizons. At the high-impact end, it may involve completely removing the topsoil and depositing it elsewhere (translocation). Removing topsoil eliminates the organic matter and its associated microbes, and destroys the soil community.

Structure Modification: Compaction and Surface Crusting

The most damaging process that happens to urban soil is compaction. Healthy soil is loose and contains many voids or pockets filled with air or water. Compaction occurs when anything applies sufficient pressure to the soil to compress these air pockets. Foot traffic compacts soil to some degree. Riding lawn mowers compact the soil to a larger degree. Automobiles compact the soil even more. Bulldozers, road graders, and other heavy construction equipment cause so much soil compaction that healthy full-grown trees in the area may die within a year of completing an urban development. Plants rely on air pockets in the soil to provide oxygen to their roots, and to the microorganisms that make up the soil community. Without oxygen in the soil, all life forms that depend on soil oxygen soon perish. Furthermore, roots often have a difficult time penetrating compacted soil to obtain nutrients.

Figure 4.7 Soil grading and scraping preparing urban soil for development. (Courtesy John M. Davis)

When soils are compacted there is decreased water absorption and increased surface water runoff. Urban soils have a large number of contaminants that ultimately end up in urban streams, rivers, and lakes. Compounds such as fertilizers, pesticides, herbicides, gasoline, oils, and other pollutants accumulate on these "created" impervious surfaces during dry weather conditions. These pollutants form a concentrated first "flush" to water bodies following a storm event. Soil compaction is so widespread in urbanized landscapes that impervious surfaces are the second highest source of pollutant concentration; only piped sources were higher. Furthermore, homeowners sometimes discover that one chemical or another was spilled on their property during construction and is the cause of soil problems later.

Modified Soil Reaction

Due to construction, urban soil is often left exposed to the elements for extended periods. Wind, rain, and ultraviolet radiation all take their toll on the soil. Wind and rain erode the soil, which removes topsoil and, as mentioned earlier, UV radiation kills soil bacteria that are essential in maintaining nutrient cycling.

In most cases soil pH values are higher (slightly alkaline) in urban areas. A pH of 7 is neutral, a pH above 7 is alkaline or basic, and a pH below 7 is acidic, as illustrated below (U.S. Department of Agriculture 2001).

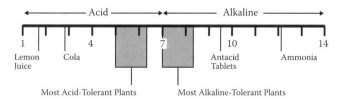

Factors that contribute to slightly alkaline pH levels in urban soils are:

1. Application of calcium or sodium chloride deicing salts in northern cities;
2. Irrigation of urban soils with calcium-enriched water;
3. Atmospheric pollution, for example, calcium-enriched ash fallout;
4. Release of calcium from weathering of construction rubble;
5. Liming of soil to correct suspected deficiencies (Craul 1992).

Finally, urban homeowners use prodigious quantities of salt-based, rather than organic, fertilizers. The use of so much salt-based fertilizer on the property causes salination or "salty soil" which sucks the water out of plants rather than allowing water to go from the soil into the plant.

Anthropogenic Materials

Often urban homeowners have problems growing things in certain areas of their yard. A common reason is that the contractor who built their home had buried all of the scrap mortar, bricks, concrete block, and other materials that altered the soil pH. The list of anthropogenic materials added to urban soils includes the list of contaminants above, plus construction debris, organic wastes (manures, industrial organic wastes, sewage sludge), heavy metals (cadmium, lead, mercury), radionuclides, materials dredged from waterways, and atmospheric acid deposition (Craul 1992; Scheyer and Hipple 2005).

Increased Soil Temperatures

The "heat loading" of urban soil is greater than that of rural or wildland soils (Craul 1992). Studies by the New York City Soil Survey have shown higher than average soil temperatures in landfills and other soils. There is also a wider range in soil temperatures in bare soil areas such as playgrounds compared to similar soils in wooded vegetation (New York Soil Survey Staff 2005). A vegetative canopy is missing in built-up areas, and the soil lacks the insulating property of an O horizon or litter layer, so the amount of radiation reaching the soil surface is great. Soil temperature is important because it controls the growth environment of roots and soil organisms, and drives rates of reactions (e.g., soil weathering) and biological processes (e.g., rate of decomposition of organic material; Craul 1992).

The following factors have a profound effect on the productivity of urban soils:

1. Little or no addition of organic material;
2. Artifacts that disrupt water movement;
3. Elevated salt content;
4. Interrupted nutrient cycling and modified activity of micro-organisms;
5. High soil temperatures that increase rate of chemical reactions;
6. Generally higher pH values resulting from additions of cement, plaster, and road salts;
7. Lateral (sideways) subsurface water flow resulting from compacted layers (Sheyer and Hipple 2005).

TAKING BETTER CARE OF URBAN SOIL

It is important to understand the value of a healthy soil ecosystem in sustaining a high quality of life for all organisms. The process of urbanization has seriously altered or even removed some of the services provided by the soil ecosystem. However, there are ways to restore and sustain the physical and biological properties of soil that benefit humans and all other organisms (U.S. Department of Agriculture 2001). The golden rules of soil conservation are:

- Cover the soil.
- Use minimal or zero tillage.
- Provide mulch for nutrients.
- Maximize biomass production.
- Maximize biodiversity.

Following the golden rules of soil conservation will produce productive soil with:

- A good supply of nutrients and nutrient-holding capacity;
- Good infiltration and water-holding capacity, and resistance to evaporative water loss;
- Porous structure for aeration;
- Near-neutral pH;
- Low salt content.

There are many sources of free information that provide strategies to restore soil fertility even in the most devastated urban soils. This information can be obtained from state extension services, garden centers, agricultural supply businesses, and web-based sources (use the search term "urban soils"). Homeowners who restore natural soil conditions reap the benefits of less watering, less fertilizer, healthier plant growth, and more plant and animal diversity. One easy way to increase soil organic content on a small scale is to develop a home composter (Perspective Essay 4.3). We have found the home composter to be a remarkably effective way to recycle kitchen wastes in a small space.

URBAN WILDLIFE MANAGEMENT IMPLICATIONS

Urban green spaces that receive the greatest attention (e.g., time and money) are residential lawns (see Perspective Essay 4.2). In fact, residential lawns, for the most part, represent the antithesis of nearly every aspect of healthy soil structure and function. For example, the requirements for developing and maintaining a typical residential lawn include:

1. Planting a monoculture exotic grass species, for example, blue grass in Nebraska or Saint Augustine grass in Texas.
2. Rodenticide treatments that eliminate moles (Talpidae) and pocket gophers (Geomyidae), and vermicides to kill earthworms.
3. Pesticide treatments that eliminate the natural predators in the soil ecosystem, such as scorpions, centipedes (Chilopoda), spiders (Araneae), mites (Arachnida), insects (Insecta), ants (Formicidae), and beetles (Coleoptera).
4. Heavy irrigation using the community's potable water supply. The pH and salination effects of over-irrigation are discussed above.
5. Over- fertilization to offset the soil's nutrient deficiency caused by the removal of most soil biota.
6. Herbicide treatments to eliminate plant diversity.

Generally, the idyllic residential lawn is nothing short of Astroturf™! Multi-million-dollar industries exist to offer urban residents the services required to sustain the "perfect" lawn. Even though there are many sources of information on alternative landscape designs that include a functioning healthy soil ecosystem (Arendt 1996; Craul 1992, 1999; Scheyer and Hipple 2005), few urban residents will adopt these alternative procedures. There are many barriers that need to be negotiated before urban residents will adopt alternative landscape designs (see Perspective Essay 4.2). Some of these barriers are personal preferences, cultural acceptance, neighborhood deed restrictions, knowledge, city governance, and retail industry support, among others. However, there are cases when urban residents have abandoned the Astroturf mentality for a more natural and sustainable landscape design. As might be expected, these lawns will appear strikingly different from the norm.

PERSPECTIVE ESSAY 4.1   Darwin's Earthworms

Charles Darwin (Figure 4.8) began and ended his career with "worms." His 1881 treatise on worms, *The Formation of Vegetable Mould through the Action of Worms with Observations on Their Habits*, was published in the 72nd year of his life and only six months before his death. The earthworm book is perhaps the third best-known among the general public of Darwin's books following the 1872 publication of *The Origin of Species* and *The Descent of Man*.

Darwin spent thirty-six years observing earthworms, meticulously recording their behavior, measuring the amount of soil moved through their activities, and evaluating their influence on soil fertility (Brown et al. 2003). To most people earthworms are unpleasant, slimy, ugly, blind, deaf, and senseless animals, of little use except for fish bait, and a general nuisance, particularly because of their unsightly surface castings (Feller et al. 2003). However, the resurgence of interest in organic farming and "biological agriculture" (in which earthworms play a more important role influencing soil fertility) in recent years has brought renewed recognition of Darwin's book and earthworms (Brown et al. 2003).

There has also be renewed interest in raising earthworms, called "vermicomposting," which is a fairly simple procedure, requires no more space than a wash tub, is inexpensive, and is an excellent technique for recycling food waste in the apartment as well as composting yard wastes in the backyard. The resultant "vermicompost" contains five to eleven times more nitrogen,

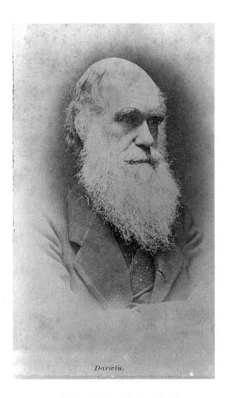

Darwin.

Figure 4.8   Charles Darwin.

phosphorus, and potassium than the surrounding soil and makes an excellent mulch and soil conditioner for potting plants and the home garden.

Night crawler hunting is nearly a lost form of recreation among urban residents. A generous supply of earthworms can be captured at night as they emerge to feed on surface litter and to mate. The best conditions for hunting night crawlers is just after a spring rain, at dusk or early darkness, with a flashlight covered with red cellophane. The best places to hunt are the yards of homes in the older areas of town, cemeteries, school yards, or other areas where the soil has not been disturbed by recent construction. The night crawlers will be lying on the surface with a portion of their body still in their hole. They are quick to retreat into their hole when disturbed! Holding the flashlight with one hand, use two fingers on the other hand to quickly pinch down on the crawler, clamping it gently but firmly between your fingers. The crawler will put up a struggle trying to retreat back into its hole, but a slow and gentle pulling action will cause the crawler to release its grip on the sides of its hole. Pull too hard and only a half of a night crawler will be collected. On a good night, an entire gallon can of night crawlers can be captured using these procedures.

Clark E. Adams

PERSPECTIVE ESSAY 4.2   For the Love of Lawns

Why does our society seem to be in love with the monotonous, sterile lawn? To understand this affinity for turf grass, it can help to examine our culture's history. According to Warren Schultz's book, *A Man's Turf: The Perfect Lawn* (1999), small strips of lawn began to show up in formal gardens in Europe during the 17th and 18th centuries. It was also during this time that vast expanses of short grass began to show up in the European countryside. Originally, these proto-lawns were kept grazed by flocks of sheep. The larger the expanse, the more sheep the landowner had, thereby demonstrating his wealth to the neighbors. Vast expanses of short grass became a symbol of wealth and power—yet another example of the human tendency to one-up the Joneses.

Many of the early U.S. colonists came from this cultural influence; George Washington was one of the first to install a lawn around his home (Schultz 1999). However, the lawn as we know it today didn't become popular until the late 19th century. The development of "improved" turf grass varieties later became a goal of the golf industry, and the invention of the lawn mower made an expanse of grass available to the common homeowner.

Today, the lawn is a symbol of territory, a "green moat" around our homes. Schultz (1999) says that a well kept lawn has come to indicate that the man of the house is powerful, in control of nature, and is taking care of business at home. He also states that a well kept lawn announces to our neighbors that we are abiding by the rules of society.

If you doubt this is the case, think about the typical urban landscape. It's really quite predictable. Just about every home or business will have a row of evergreen hedges around the perimeter of the building, called "foundation plantings." Unwanted views usually are "screened" with a row of evergreen hedges as well. With the exception of newly constructed developments, there will be large shade trees and some kind of groundcover, turf grass being the most common of these. Often you will see a flower bed of some sort at the base of trees and along walkways, usually containing non-native annuals and perennials to add color. The landscape is dominated by vast expanses of mowed turf grass. If traditional landscaping were ecologically sensitive, or even neutral, these practices would be a harmless cultural quirk, but the typical suburban lawn and garden creates a myriad of ecological problems.

As mentioned earlier, exotic plant species often are used in these settings. Because they are not native to the area they tend to require more intensive maintenance to survive (ironically, often these same plants are advertised as "easy care"). Those species that do well can escape the confines of the yard and compete with endemic species. As you'll remember from earlier in this chapter, a region's plants determine the animal species that live there. As native plants are lost, food and shelter resources are lost as well; this can have an adverse affect on resident wildlife.

Maintaining the perfect lawn is ecologically expensive. Exotic plants generally require more moisture than native landscapes, which puts more pressure on community water supplies. Turf grass needs pesticides, herbicides, and fertilizers to achieve the preferred deep green color and lush texture, all of which significantly decrease water quality. According to Schultz (1999), 70 million pounds of chemicals are applied to lawns in the United States each year! The majority of these chemicals are washed into local streams and reservoirs by rain and sprinkler systems. The deleterious effects of these chemicals on the population dynamics of aquatic and terrestrial wildlife are discussed in Chapters 4 and 7. A "healthy" lawn endangers more than our water supply. Running a lawn mower for one hour releases as much hydrocarbon into the air as driving a car for 11.5 hours (Schultz 1999).

From John M. Davis, 2003. Used with permission.

PERSPECTIVE ESSAY 4.3   Home Composting on a Small Scale

If a residential area contains some yard space, a simple composting apparatus for food scraps left over from meal preparation and nonconsumption can be utilized for home composting (Figure 4.9). The system only requires the purchase of a 30-gallon plastic garbage can. Holes are punched in the bottom of the can to allow liquid to escape. The can is buried in the

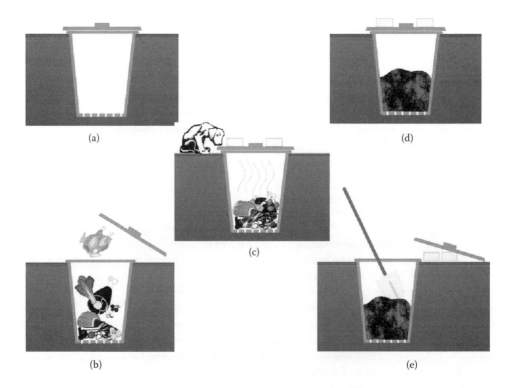

Figure 4.9   Home composter. (Courtesy Clark E. Adams and Linda Causey)

ground allowing enough room to replace the lid (Figure 4.9a). Nearly every kind of kitchen scrap can be thrown into the garbage can composters, including plant and animal material, and egg shells, but it is not designed for grass clippings and leaves (Figure 4.9b). One author used such a device, and added anything organic that would have gone down the garbage disposal to the composter. Some may not want to add animal material because of fear of disease and/or local ordinance restrictions. The organic material will begin to decompose on its own (Figure 4.9c). In addition, flies will lay their eggs on the organic material, facilitating a more rapid decomposition process as the larvae convert the detritus into humus. Odors are produced during the process of decomposition, which will attract pets and other animals; weights on the top of the lid will prevent their access. Over time, depending on climate, the organic material in the composter will turn into a black humus (Figure 4.9d). This humus can be retrieved from the composter and used as a mulching material on gardens, flower beds, and other areas that require a rich organic base (Figure 4.9e).

Clark E. Adams

SPECIES PROFILE: MOLES (*TALPIDAE* SPP.)

In shallow holes moles make fools of dragons.

Proverb

Moles swim through soil. Say what you will about these cursed and vilified creatures, but you can't deny the ability to breaststroke underground using enormous, outward-turned, long-clawed feet at a foot per minute is impressive. Some will swim in water, too. You wouldn't think a mole would pose a threat to anything but earthworms, slugs, grubs, and other subterranean invertebrates. They're relatively small, with streamlined, torpedo-shaped bodies and velvety fur that allows them to move forward and backwards in a narrow tunnel with equal ease. The front claws are intimidating, but their vision is poor. And yet homeowners across the continent have declared all out war on

Figure 4.10 Eastern or common mole. (Courtesy Bob Upcavage)

the mole, complete with lethal devices worthy of a horror movie torture chamber … all for the sake of a lawn (Whitaker 1998).

There are seven species of North American mole north of Mexico: shrew-moles (*Neurotrichus gibbsii*), broad-footed moles (*Scapanus latimanus*), coast mole (*Scapanus orarius*), Eastern mole (also known as the common mole, *Scalopus aquaticus*), hairy-tailed moles (*Parascalops breweri*), star-nosed mole (*Condylura cristata*), and the Townsend's mole (*Scapanus townsendii*). Most are solitary outside of breeding season. They have a relatively low reproduction rate, producing one litter per year of from two to six young that are raised in a nest chamber (Whitaker 1998).

Homeowners often refer to moles as rodents, but they are actually insectivores. Members of this scientific order are voracious eaters, consuming 60 to 100 percent of their body weight daily. It takes a lot of calories to tunnel in search of this food and moles remain active day and night all year long. One mole can dig up to 150 feet of new tunnels a day. They don't consume much plant material so they destroy very few plants by feeding, but tunneling can take a toll on turf grass, flower beds, and other landscaping. Commercial bulb growers and row crops producers can sustain significant economic loss when plants are dislodged, or when harvesting equipment coming into contact with the tunnel mounds is damaged (Parkhurst 1999).

The bulk of research literature related to urban moles is, no surprise, heavily weighted toward human-mole conflict management. One such paper, by Ferraro and Hygnstrom (1993), examined the perceptions of Omaha, Nebraska, residents related to wildlife encounters and found 44 percent of households surveyed experienced at least one conflict event in the past two years. Of these, 10 percent had encountered either common moles and/or pocket gophers. Suburban residents reported mole/gopher encounters more often (52 percent) than residents in all other residential (30 percent) or rural (17 percent) locations. Moles and gophers were grouped together to simplify the survey and, in part, because the researchers believed the public often had trouble distinguishing moles from gophers. Encounters with moles, as well as with mice, rats, bats (Chiroptera), and snakes, were most likely to invoke action on the part of residents.

Homeowners and wildlife managers have a large arsenal at their disposal for killing moles, including fumigants, repellents, toxicants, barriers, and trapping. Additionally, there are any number of "high tech" and "home remedies," most of which have either no proven effect and may even be illegal (Parkhurst 1999). Simply killing the offending animals, rather than addressing the cause of the problem, means that the solution is usually short-lived. The same grubs, snails, and slugs (Gastropoda) that attract one mole will attract another as soon as the territory has been vacated. Additionally, poisons and traps can cause bigger problems than they solve, particularly in urban landscapes, because nontarget animals, including pets or even humans, may be at risk. Insecticides may be recommended for reducing the number of beetle grubs—a major food source for moles— but moles feed on other invertebrates so they are likely to simply switch to another prey. What can we say? Not all human-wildlife conflict problems have simple solutions.

Besides, moles aren't all bad. They do consume the larvae and adult pest insects, including Japanese beetles, that affect garden, landscape, and flowering plants. Tunneling loosens the soil, improves aeration, and mixes deeper soils with surface organic material, all of which enhance soil quality (Parkhurst 1999). If only lumpy lawns were in fashion moles might be heroes instead of heels.

CHAPTER ACTIVITIES

1. Take a core sample of the soil in your backyard and determine how it fits the soil profile in Figure 4.2.
2. Compare the soil biota found in your backyard with other green spaces in your community using the Berlese funnel technique.

3. Tour the areas of your community that are under "development" and determine how soil is being treated.

4. Find those residences in your community that have abandoned the "Astroturf" model of landscape development, and describe what they have done as an alternative.

LITERATURE CITED

Arendt, R.G. 1996. *Conservation Design for Subdivisions*. Washington, DC: Island Press.

Bockheim, J.G. 1974. Nature and properties of highly disturbed urban soils. Paper presented before Division S-5, Soil Scientific Society of America, Annual Meeting, Chicago, Illinois.

Brown, G.G., C. Feller, E. Blanchart, P. Deleporte, and S.S. Chernyanskii. 2003. With Darwin, earthworms turn intelligent and become human friends. *Pedobiologia* 47:924–933.

Craul, P.J. 1985. A description of urban soils and their desired characteristics. *Journal of Arboriculture* 11:330–339.

Craul, P.J. 1992. *Urban Soil in Landscape Design*. New York: John Wiley and Sons.

Craul, P.J. 1999. *Urban Soils: Applications and Practices*. New York: John Wiley and Sons.

Davis, J.M. 2003. Urban Systems. In *Texas Master Naturalist Statewide Curriculum,* 1st edition, ed. M.M. Haggerty. College Station, TX: Texas Parks and Wildlife Department.

DNREC (Delaware Department of Natural Resources and Environmental Control, Water Resources Division). 2004. Influence of common landscaping and grading practices in the creation of impervious surfaces. *Tributary Times* 3(1):10–12.

Feller, C., G.G. Brown, E. Blanchart, P. Deleporte, and S.S. Chernyanskii. 2003. Charles Darwin, earthworms and the natural sciences: Various lessons from past to future. *Agriculture, Ecosystems and Environment* 99:29–49.

Ferraro, D.M. and S.E. Hygnstrom. 1993. Public perceptions of wildlife encounters in the Omaha, Nebraska, metropolitan area. *11th Great Plains Wildlife Damage Control Workshop Proceedings*, Kansas City, MO: Kansas State University.

Harris, J.A. 1991. The biology of soil in urban areas. In *Soils in the Urban Environment*, ed. P. Bullock and P.J. Gregory. Cambridge, MA: Blackwell Scientific.

New York City Soil Survey Staff. 2005. *New York City Reconnaissance Soil Survey*. Staten Island, NY: United States Department of Agriculture, Natural Resources Conservation Service.

Odum, E.P. 1971. *Fundamentals of Ecology*. Philadelphia, PA: W.B. Saunders.

Parkhurst, J. 1999. *Managing Wildlife Damage: Moles*. Publication Number 420-201. Blacksburg, VA: Virginia Cooperative Extension.

Scheyer, J.M. and K.W. Hipple. 2005. *Urban Soil Primer*. Lincoln, NE: United States Department of Agriculture, Natural Resources Conservation Service, National Soil Survey Center.

Schueler, T. 2000. The compaction of urban soils? Technical Note No. 107. *Watershed Protection Techniques* 3(2): 661–665.

Schultz, W. 1999. *A Man's Turf: The Perfect Lawn*. New York: Clarkson Potter.

Soil and Water Conservation Society (SWCS). 2000. *Soil Biology Primer,* revised edition. Ankeny, IA: Soil and Water Conservation Society.

U.S. Department of Agriculture. 2001. *Urban Soils*. Lincoln, NE: Natural Resource Conservations Service, National Soil Survey Center.

Whitaker, J.O., Jr. 1998. *National Audubon Society Field Guide to North American Mammals*. New York: Alfred A. Knopf.

Wright, R. 2004. *Environmental Science: Toward a Sustainable Future*, 9th edition. Upper Saddle River, NJ: Pearson Education.

CHAPTER 5

Urban Waters

When the well is dry, we know the worth of water.

—**Benjamin Franklin,** *Poor Richard's Almanac*

KEY CONCEPTS

1. The ways in which urban residents obtain, use, and discharge water from their communities have a definite impact on wildlife.
2. The water cycle consists of three major filters.
3. There are many ways the water cycle can be protected in urbanized environments.
4. The primary abiotic effects of urbanization on stream structure and function are:
    a. Impervious surface covers (ISC);
    b. Pollutants;
    c. Stream channelization.
5. The primary biotic effects of urbanization on stream ecology are:
    a. Eutrophication;
    b. Changes in the aquatic food chain;
    c. Changes in dominant fish assemblages, that is, fish diversity and abundance;
    d. Introductions of nonnative species.
6. Fish populations provide fundamental and demand-driven ecosystem services.
7. Urban wetlands consist of riparian streams and corridors and constructed impoundments.

INTRODUCTION

We have a special chapter on urban waters because there are definite impacts on wildlife in terms of how urban residents obtain, use, and discharge water from their communities. Today, as in the past, it is quite common for humans to settle near some type of water, whether it be coastal zones, rivers, streams, or lakes. People like water, and they are willing to pay premium prices for real estate next to or in close proximity to water. In some cases, where it is not possible to be near natural waterways, constructed wetlands serve as an acceptable alternative. There is a profound public ignorance of what water is, how nature provides a continual supply of filtered water, community sources of water, uses in urban communities, water pollution, water conservation and recycling, and wastewater treatment.

Freshwater habitats are among the most altered ecosystems on earth because they are the ultimate catchments (low areas that receive drainage) for watershed pollutants. Water is increasingly diverted for human use through dams and reservoirs that alter flow regimes and fragment drainages.

Freshwater habitats are the focus of most human activity, both in large cities and agricultural regions, that degrades water quality, and their biota is subject to intense exploitation (Rahel 2002).

The effects of urbanization on the water cycle are (1) changing much of the earth's surface from permeable to impermeable, (2) pollution, and (3) excessive groundwater withdrawal rates (Paul and Meyer 2001). We will examine each of these, as well as a few other factors, in terms of the abiotic and biotic effects of urbanization on riparian corridors (streams and rivers) and standing (wetlands) bodies of water in urban communities.

THE FLOW OF WATER THROUGH AN URBAN COMMUNITY

Sources

One of the most extensive and devastating impacts of urbanization on wildlife and their habitats is the development of the urban "water works," the system by which urban communities obtain, use, and discharge water (Figure 5.1). Urban communities obtain water from surface impoundments (e.g., lakes, rivers, or reservoirs) or from underground aquifers. Of particular importance are the abiotic and biotic alterations of the natural ecosystem resulting from human-constructed reservoirs, for example, dams on rivers or streams. There are many reasons for building dams, including:

1. Hydropower (there are more than 2000 major hydropower dams in the United States);
2. Recreational purposes (creation of lakes);
3. Supply of water for human consumption;
4. Irrigation of agricultural lands;
5. Flood control;
6. Political pork and public works projects;
7. Navigation;
8. Tourism (e.g., Hoover Dam near Las Vegas).

Regardless of the reason for building a dam, the alterations of the natural ecosystem and impacts on wildlife can be found behind the dam, at the dam, and below the dam (Perry and Vanderklein 1996).

Behind the dam there is flooding of terrestrial habitats, and replacement of the riparian habitat with a large surface reservoir. Riparian plants and animals are lost. Sediment build-up leads to accumulations of nutrients, heavy metals, and organochlorines. Increased nutrient retention time leads to eutrophication (Figure 5.2). Eutrophic waterways are defined as "nutrient rich" (high concentrations of phosphorus and nitrogen) caused by large inputs of fertilizer from storm-water drainage, wastewater treatment plants, and combined sewer outflows that cause rapid and extensive algal growth called "blooms." Eutrophic water can look like green pea soup because of the explosive growth of phytoplankton (e.g., algae of various species). The nutrient-rich water could also be covered with floating aquatic plants such as water hyacinth (e.g., *Eichhornia* sp.), or duck weed (e.g., *Lemna* sp.). Excessive growth of hydrilla (e.g., *Hydrilla* sp.), a submerged macrophyte, is also indicative of eutrophic waterways. In any case, eutrophication is a classic example of the abiotic × biotic interrelationship discussed in Chapter 3. In this case a dramatic change in nutrient levels causes an equally dramatic change in the life forms that occupy freshwater ecosystems.

Eutrophic water ways are characterized by low oxygen concentrations, high biochemical oxygen demand (BOD), low light penetration, low biotic diversity, high phosphate and/or nitrate content, elevated water temperatures, high sediment content, and high algae (e.g., green or blue-green) densities. As more plants and animals die in eutrophic water, more decomposition is required to remove them from the system. Decomposers need oxygen to accomplish the recycling process, which results in less oxygen for other living organisms, which in turn die and require decomposition,

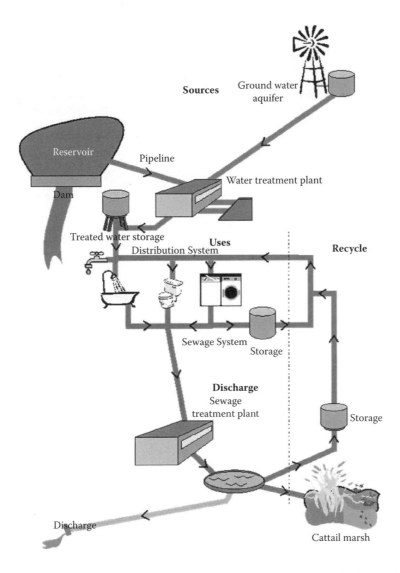

Figure 5.1 The flow of water through an urban community from sources, to uses, to discharge with suggested recycling options. (Concept by Clark E. Adams, diagram by Linda Causey)

which further reduces oxygen levels. Only a few plant and animal species that have a wide oxygen tolerance range can survive in highly eutrophic waterways.

At the dam the normal flow of the river is interrupted. Water chemistry, temperatures, flow dynamics, and sediment transport are dramatically modified. The dam fragments the river corridor and obstructs fish runs. Dams are one of the primary causes of aquatic species endangerment because they prevent migratory fish [e.g., salmon (*Oncorhynchus* sp.), long nose or alligator gar (*Lepisosteus osseus*), American paddlefish (*Polyodon spathula*), and northern pike (*Esox lucius*)] from reaching their spawning grounds in the upper streams and tributaries.

Below the dam the river is subject to episodic flow rates based on the timing of reservoir discharges rather than the more predictable seasonal variations. Reservoir water dumped into the stream below the dam alters normal temperature, chemistry, and pollutant levels, which affects species diversity—only species that are highly tolerant to these conditions survive. The loss of native fish assemblages is discussed later in terms of biotic homogenization.

Figure 5.2 *(A color version of this figure follows page 158.)* Example of eutrophication in an inland fresh-water pond. Note how the surface of the water is covered with a macrophyte, in this case a plant called giant salvina (*Salvinia molesta*). (Courtesy Michael Masser)

Regardless of source, before water can be used by urban residents, it needs to be "purified" in a water treatment plant. The purification process may involve removal of solids, disease vectors, and pollutants, as well as pH or temperature adjustments. Treated water is then transferred to a storage tower for community distribution.

Uses

A typical household in the United States consumes an average of 100 gallons per person per day. Indirect uses, such as irrigation, recreation, and car washes bump this figure up to 1300 gallons per person per day. However, it is not only the amount, per se, that affects wildlife. The impacts on wildlife also result from what is put into the water to facilitate hygiene needs, food preparation, cleaning, irrigation, and waste disposal. In fact, whatever urban residents put into household water to expedite any of the aforementioned uses will become a problem for aquatic, semiaquatic, and terrestrial animals. These problems are discussed in detail later in this chapter.

Discharge

The most common method of "treating" wastewater from urban communities is to run it through a sewage treatment plant or wastewater treatment facility (WTF). Raw wastewater is about 1000 parts water for every part waste, or 99.9 percent water to 0.1 percent waste. The wastes (pollutants) consist of the following categories of materials.

1. Debris and grit: rags, plastic bags, course sand, gravel, sticks of wood, paper clips, rubber balls, dead animals, tampons, and condoms. These items are removed during the *Preliminary Treatment Process* at the bar screen (Figure 5.3) and the grit tank (Figure 5.4).
2. Particulate organic material: fecal matter, food wastes from garbage disposal units, toilet paper, lettuce leaves, and other materials that settle out in still water are removed during the *Primary Treatment Process* (Figure 5.5).
3. Colloidal and dissolved organic material: very fine particulate organic material, bacteria, urine, detergents, soaps, and other cleaning agents removed during the *Secondary Treatment Process*, also called *Biological Nutrient Removal* (Figure 5.6).

Figure 5.3 *(A color version of this figure follows page 158.)* Material removed during the first step in preliminary wastewater treatment at the bar screen. (Courtesy Clark E. Adams)

Figure 5.4 Material removed during the second step in the preliminary treatment of wastewater at the grit-settling tank. (Courtesy Clark E. Adams)

Figure 5.5 The primary treatment settling tanks, called "primary clarifiers," at the WFT where large organic material settles out and is collected as "raw sludge" at the bottom of the tank. The raw sludge is transported to hay fields and used as an organic fertilizer. (Courtesy Clark E. Adams)

Figure 5.6 Secondary treatment consisting of an activated sludge system wherein natural decomposers and detritus feeders feed on the colloidal and dissolved organic material and break these materials down into their elemental parts. The agitation of the water is due to a forced aeration process that keeps the living organisms alive and functioning. (Courtesy Clark E. Adams)

Figure 5.7 *(A color version of this figure follows page 158.)* Cattail Marsh, a reconstructed wetland built at the wastewater treatment facility in Beaumont, Texas. (Courtesy Clark E. Adams)

4. Dissolved inorganic materials: nitrogen and phosphorus, and other nutrients from excretory wastes and detergents cannot be removed in a conventional WTF. However, they can be removed by "polishing" the water in a reconstructed wetland (Figure 5.7). This procedure is often called *Tertiary Treatment* or *Biological Removal*.
5. Toxic materials: pesticides, heavy metals, and prescription drugs appear in wastewater because people pour unused portions of these materials down sinks, tubs, or toilet drains. These materials cannot be removed in a conventional WTF (Wright 2006:448).

Imagine the environmental abuse on our streams, rivers, bays, and estuaries and their biota when none of the above categories of pollutants were removed. At one time in U.S. history, raw sewage was dumped directly into streams and rivers without treatment of any kind. The Clean Water Act of 1972 mandated the production by WTFs in order to stop or abate the rampant water pollution caused by raw wastewater. Even today, WTF technology is designed to remove only pollutant categories 1, 2, and some of 3 (listed above). For the most part, WTFs exist to remove solids only. A higher-order and more expensive technology is required to remove pollutant categories 4 and 5, or they are not removed at all and discharged directly into the receiving body of water used by the WTF. The lives and life cycles of many animals, particularly aquatic and semiaquatic, are compromised by the existence of these pollutants in urban waters. However, it is difficult to conduct field studies that provide definitive cause and effect associations between these pollutants and abnormalities in resident biota.

Most urban residents have no idea what happens to the material they flush down the toilet, or who is responsible for each flush. However, urban wildlife biologists need to be informed about how the urban community WTF functions in order to make the correct associations between these functions and the presence or absence of wildlife. In fact, a useful exercise would be to visit a WTF in order to appreciate its value and the lack thereof in cleansing the wastewater from urban communities.

Recycle

The right-hand side of Figure 5.1 illustrates three recycling opportunities for the millions of gallons of water that pass through WTFs on a daily basis. The first begins at the residence where used or "gray" water (all household water except from the toilet) is stored on the residence to be used for other functions, e.g., watering lawns, flushing toilets, or washing cars. The other two methods of recycling gray water originate at the WTF. One is to store the water coming out of the WFT (adjusted for pH and disinfected) in a reconstructed wetland, such as Cattail Marsh in Beaumont, Texas (Figure 5.7). Another possible method of recycling WTF water is to develop a separate water distribution system that pumps the water from the WFT back into the community for watering lawns, washing cars, flushing toilets, or other functions that do not need to use the potable water supply to the community, such as the $480 million Groundwater Replacement System in Orange County, California. There are of course economic, cultural, political, and ecological considerations in the design, development, and implementation of any one of the above WTF water recycling options. However, a cost-benefit analysis may demonstrate a positive and long-term influence on water use and conservation in urban communities if these recycling options could be implemented.

THE WATER CYCLE—NATURE'S FILTER

Not many people know the source of their water supplies and even fewer know how their water supply is tied to the water cycle. The cycle does not produce new water; it simply filters out impurities and other pollutants in existing water. It is important to examine the water cycle if one wants to begin to comprehend the drawbacks of poor water management practices on urban waters.

The water cycle can be explained in terms of three major filters: (1) the transpiration loop, (2) the groundwater loop, and (3) the evaporation loop. Given the opportunity, the three loops illustrated in Figure 5.8 provide a continuous supply of fresh water for all living things on earth.

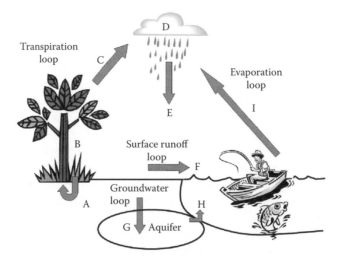

Figure 5.8 Diagram of the water cycle and filters. (A) Infiltration, (B) plant xylem tissue, (C). transpiration (evaporation through plant leaves), (D) condensation (cloud formation), (E) precipitation (rain, snow, ice), (F) surface runoff, (G) gravitational or capillary water, (H) seeps or artesian flow from the aquifer, (I) air evaporation. (Diagram by Clark E. Adams)

The Transpiration Loop

In the transpiration loop, represented by letters D-E-A-B-C in Figure 5.8, water infiltrates into the soil, but only to a level that allows the plant roots to capture it. Water is filtered as it passes through the plant tissues and out through the leaves, returning to the atmosphere in a process called *evapotranspiration*. The evaporated water condenses into clouds, which eventually produce some form of filtered precipitation, such as rain, sleet, hail, or snow. The human impact on this filter, particularly in urban and suburban areas, is the replacement of vegetation with impervious surface covers (ISC, discussed later), thus significantly reducing the effect of this filter.

The Groundwater Loop

In the groundwater loop (D-E-G-H-I in Figure 5.8) water falls to the ground as precipitation, but instead of being completely captured by plant roots, some of the water passes through the soil into an underground water supply called an *aquifer*. In this loop soil acts as the filter. The aquifer water returns to the surface in the form of seeps or artesian flow, or it can be pumped out of the ground by wells. If the water ends up in a surface water impoundment, such as a lake, the water evaporates and reenters the water cycle.

The process of urbanization through ISCs and compaction reduces or nullifies infiltration of water through the soil. This may lead to reduced rates of recharge of underground aquifers or ground table water reserves. As mentioned in the previous chapter, soil pollution is common in urban areas. There are many cases of pollutants moving through the soil and contaminating the aquifers that communities depend on for their water supply (see Figure 5.1). For example, in the early 1980s, it was discovered that the extremely heavy use of nitrate fertilizers in the cornfields of Nebraska for several decades ultimately led to groundwater contamination (nitrates) throughout large portions of the state (C.E. Adams, personal observation). High nitrate levels in drinking water pose a risk to infants because when nitrate is converted to nitrite it may cause a blood disorder called methemologlobinemia, a condition known as "blue baby." Nitrite in the blood combines with hemoglobin to form methemoglobin, which reduces the capability of the blood to carry oxygen to all parts of the body and the baby's skin takes on a bluish cast (Illinois Department of Public Health 1999).

An additional human impact of urbanization on the groundwater loop is using more aquifer water than can be replaced through the natural recharge rate. A drop in the water table occurs, which is particularly noticeable when artesian seeps and springs cease to flow, and the streams and associated wildlife that depend on this artesian system cease to exist [e.g., the Edwards Aquifer Habitat Conservation Plan (http://www.edwardsaquifer.org)].

The Evaporation Loop

In the evaporation loop (D-E-F-I in Figure 5.8) the atmosphere acts as the filter, removing the impurities as water evaporates from streams, rivers, ponds, lakes, and the oceans. A simple demonstration of how this filter works would be to allow a spoonful of saltwater to evaporate. Once all of the water is gone all that is left behind is salt. This process is similar to that seen in the transpiration filter, but without plants playing an active filtration role.

Human activities, such as manufacturing and burning fossil fuels, release pollutants into the atmosphere that are unaffected by the filtering effect of evaporation. These pollutants include carbon dioxide (CO_2), as well as sulfur oxide (SO) and nitrous oxide (NO) radicals. The latter two pollutants are responsible for the formation of acid rains.

CARING FOR THE WATER CYCLE

The process of urbanization has led to the concentration of many people into a relatively small area, which has had profound negative effects on the water cycle and its filters. The most immediate effect is the rate at which water is withdrawn compared to nature's capability to replace it—called the recharge rate. Urbanites waste water! They use far more water than is required to accomplish the task at hand. For example, the green golf courses in the middle of the xeric Southwest (e.g., Tucson, Arizona) exemplify how urbanites attempt to manipulate the habitat to produce plant and animal communities that could never survive in that region without heavy water subsidies.

What really captures urban residents' attention regarding the water cycle are those times when a catastrophic natural event (e.g., hurricanes and tornados) turns the water off altogether. Deprivation seems to make people disciples of water quality standards, and they begin to focus their attention on the gravity of degrading the benefits of the water cycle. Since water usually appears to be abundant and seemingly everlasting in urban communities, and catastrophic events are rare, an educational process has to be implemented on the benefits of preserving the integrity of the water cycle. Education should always be a proactive rather than reactive management approach. Educational programs could focus on:

1. The origin of the community water supply and what processes are involved in providing a continuous supply of potable water to residents;
2. Water consumption rates for various resident activities, including business, recreational, and household uses;
3. Opportunities for water conservation at the community, for example, industrial, residential, and personal levels;
4. Multiple water use strategies that reduce withdrawal rates from community water sources (e.g., use of "gray water");
5. The technologies required to replicate nature's water filters;
6. Adoption of alternative water use lifestyles such as landscaping with endemic vegetation that is adapted to the climate of the area.

There are two well-developed and successful community and public education programs about water. One is *Project WET* (Water Education for Teachers). The mission of Project WET is to reach children, parents, educators, and communities of the world with water education. "We invite you to join us in educating children about the most precious resource on the planet—water" (http://www.projectwet.org/). The other, *Project WILD Aquatic K-12 Curriculum and Activity Guide* emphasizes aquatic wildlife and aquatic ecosystems (http://www.projectwild.org/index.htm). State coordinators for *Project WET* can be found at their URL given above. *Project WILD Aquatic* is administered through the communications divisions of state fish and wildlife management agencies offices (http://www.fws.gov/offices/statelinks.html).

RIPARIAN CORRIDORS: STREAMS AND RIVERS

Riparian corridors are the lush green areas of vegetation adjacent to streams and rivers (Figure 5.9). The key to maintaining the ecological integrity of the riparian corridor is the buffering effect of zones 1, 2, and 3. The ecological services and materials provided by the buffer zone vegetation are listed below.

Zone 1: This zone hosts species found along the water's edge. Common species include sedges and rushes that are water loving and capable of stabilizing stream banks with their deep roots. These species are critical for promoting water recharge and water table depth.

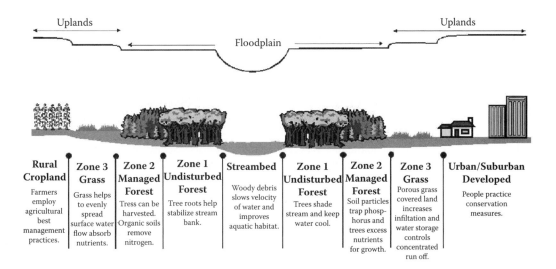

Figure 5.9   Diagram of a riparian corridor including zones, floodplain, and uplands. (Rendered from Tjaden and Weber 1998 by L. Causey)

Zone 2: This zone contains species that are found in wet ground and consists of shrubs, trees, moisture-loving grasses, and water-tolerant broad-leaved plants. These plants catch water, facilitate absorption of nutrients transported into the area by runoff and groundwater, and provide wildlife habitat and travel corridors for terrestrial animals.

Zone 3: This zone is located where the riparian area merges with the uplands and includes a mixture of riparian and upland species. The area is also host to many terrestrial animals, including early successional, edge-loving species. This zone is critical in the control of sediment, nutrient, pesticide or non-point sources of pollution, as is the case in urban and agricultural situations (South Dakota Department of Agriculture website). Riparian buffer zone vegetation provides many important ecological services and materials that support aquatic biodiversity.

These services and materials include:

1. Stabilizing stream banks, controlling sediment input and bank erosion.
2. Shading river channels, thereby maintaining lower water temperatures and regulating dissolved oxygen levels.
3. Contributing substantial quantities of large woody debris to the aquatic system, providing in-stream complexity essential for the success of insects, amphibians, reptiles, and fish.
4. Depositing large amounts of leaf litter, insects, and nutrients that are crucial for the viability of aquatic food webs.
5. Reducing damage caused during flood events by preventing large woody debris from entering agricultural lands, roads, and properties during flood peaks.
6. Discouraging geese congregations (see Chapter 14).
7. Providing a visually appealing greenbelt and recreational activities.
8. Sequestering pollutants and excess nutrients entering streams from surrounding landscapes.

Riparian buffer zone vegetation provides so many important ecosystem services and materials that regulatory agencies are increasingly recognizing that effective management of aquatic ecosystems requires effective management of the terrestrial plant communities that border these environments (http://www.theforum.org/Davis/riparian.htm and http://www.aspenpitkin.com/depts/33/riparian.cfm). The latter source also provides a plant list for riparian buffers that includes twenty-seven different species of trees and understory plants.

Note that the riparian corridor or *floodplain* consists of zones 1 and 2. It is often the case that urban development would proceed in zones 1 and 2 rather than farther back in the upland area, as illustrated in Figure 5.9. In fact, the western expansion of civilization across the United States took place, in large part, by using the major rivers and their tributaries as a primary means of transport and settlement. Many major cities and municipalities in the United States were built starting at the river's edge, for example, St. Louis, Missouri, and Minneapolis, Minnesota, among others.

Riparian zones are areas of transition between aquatic and upland ecosystems and they offer numerous, yet often overlooked, benefits to wildlife and people. The high value of riparian areas as wildlife habitats is due to the abundance of water combined with the convergence of many species along the edges and ecological transition zones between aquatic/wetland, aquatic/upland, wetland/upland, and river channel/backwaters habitats. This dynamic equilibrium between the water and land creates a corresponding dynamic equilibrium of life within a river system (Cohen 1997). To put it bluntly, the plant and animal diversity in the riparian corridor (e.g., bottomland hardwood forests) is impressive, including mammals, birds, reptiles, amphibians, migratory species, and rare and endangered species. A survey by the U.S. Fish and Wildlife Service recorded 273 species of birds, 45 mammals, 54 reptiles, 31 amphibians, 116 species of fish, and countless invertebrates. Furthermore, an estimated 80 percent of all vertebrate species in the desert Southwest depend on riparian areas for at least some part of their life cycle (Wagner 2004). A useful exercise to illustrate the potential animal diversity in the riparian corridor would be to review the species in any taxonomic guide of mammals, birds, reptiles, and amphibians, and develop a checklist of those species that would be found there.

It has been estimated that as much as 70 percent of riparian ecosystems present at the time of European colonization in the United States have been destroyed (President's Council on Environmental Quality 1978). The changes that have stressed flowing water systems have impaired their value for both human use and environmental services. Stresses arise from: (1) water quantity or flow mistiming, (2) morphological modifications of the channel and riparian zone, (3) excessive erosion and sedimentation, (4) deterioration of substrate quality, (5) deterioration of water quality, (6) decline of native species, and (7) introduction of alien species. The locus of the problem can be in the watershed, along the riparian or floodplain zone, or in the channels and pools (National Research Council 1992). Over 130,000 km (80,730 miles) of streams and rivers in the United States are impaired by urbanization (Paul and Meyer 2001). The quality of any given stream is negatively correlated with the amount of urbanization in its surrounding watershed (Miltner, White, and Yoder 2004).

ABIOTIC EFFECTS OF URBANIZATION ON RIPARIAN ECOLOGY

There is very little published information on the abiotic effects of urbanization on stream ecology (Table 5.1; Paul and Meyer 2001), although there is more data on water chemistry in urban streams than any other aspect of their ecology.

Pollutants

Water chemistry is affected by the extent and type of urbanization, amount of WTF effluent and/or combined sewer outflows (CSOs), and the extent of storm-water drainage area facilitating non-point surface (NPS) runoff. In addition, illicit discharges, leaking sewer systems, and faulty septic systems are large and persistent contributors of pollutants to urban waters. These organic wastes are of particular importance from a human health aspect given the potential for diseases associated with pathogens such as bacteria, viruses, protozoa, and parasitic worms (Mason 2000). A fairly detailed analysis concerning the extent, environmental impacts, and control measures associated with organic pollutants is provided later in this chapter.

Table 5.1   Deleterious Effects of Varying Levels of ISC on
             Urban Streams

Effect	Cause
Stream hydrology	Increased surface runoff
	Sedimentation
	Floods and discharge rates
	Decreased lag times
Stream geomorphology	Changing channel width
	Changing channel depth
Stream temperature	Runoff over heated surfaces
	Changes in fish and invertebrate diversity

Source: Data from M.J. Paul and J.L. Meyer. 2001. *Annual Review of Ecology and Systematics* 32:333–365.

Table 5.2   Types of Chemicals Found in Elevated Concentrations in Urban Streams and Their
             Sources

Chemical	Source
Phosphorus	Wastewater and fertilizer runoff from lawns
Nitrate and ammonium	Wastewater and fertilizer runoff from lawns
Metals: Pb, Zn, Cr, Cu, Mn, Ni, Cd, Hg, As, Fe, B, Co, Ag, Sr, Sb, Sc, Mo, Li, Sn	Industrial discharge, brake linings, metal alloys, and accumulations on roads and parking lots
Pesticides and herbicides	Urban use around homes and work places and lawn and golf course management
Polychlorinated biphenyls (PCBs), polycyclic aromatic hydrocarbons (PAHs), petroleum-based aliphatic hydrocarbons	Storm-water runoff from industrial point sources, episodic spills, oil and gasoline spills
Pharmaceuticals: antibiotics, chemotherapeutic drugs, analgesics, narcotics, psychotherapeutics	Hospital effluent and drugs flushed down the toilet

Source: Data from M.J. Paul and J.L. Meyer. 2001. *Annual Review of Ecology and Systematics* 32:333–365.

Nearly every chemical used directly or indirectly by urban residents ends up in urban waterways and lakes. Paul and Meyer (2001) provide a detailed analysis of chemical type and source, summarized in Table 5.2. Non-point sources of chemical pollutants are more extensive than point sources due to the increase in runoff from impervious surface covers (ISC) in urban areas.

Other types of pollutants entering streams and rivers flowing through urban communities include fecal wastes from companion animals, leaves and grass, fossil fuels such as oil and gasoline, road salts, sediments (i.e., sand, gravel, and clay), paper products, and pharmaceuticals. The majority of these pollutants enter aquatic ecosystems through storm-water runoff. This water contains whatever potentially polluting material was deposited on hard surfaces prior to the rain storm.

A dominant feature of urbanization is a high level of ISC—parking lots, paved roads, and buildings—that prevents water from infiltrating into the soil. ISC has become an accurate predictor of urban impacts on streams because thresholds of degradation in streams are associated with ISC. For example, as ISC increases, the surface runoff increases proportionately. The rate and flow volume of storm-water runoff leads to decreased groundwater recharge, and alteration of natural drainage patterns through changes in stream bed and bank geomorphology. The results include an increase in localized flooding, surface runoff, erosion, and channelization.

Stream Channelization

Urban storm-water management systems are designed to collect and discharge large volumes of water from the community as quickly as possible through channelization (Figure 5.10). Channelization

Figure 5.10 *(A color version of this figure follows page 158.)* An example of stream channelization. (Courtesy John M. Davis)

is the straightening and/or deepening of a watercourse for purposes of storm runoff control, or ease of navigation on urban waterways. Channelization is necessary because, as mentioned above, urban communities consist of large expanses of ISCs (concrete or asphalt) that prevent infiltration, increasing surface runoff rates and the potential for flooding. During heavy rainstorms, huge amounts of water run off from residential lawns, streets and roads, parking lots, rooftops, and other hard surfaces (nearly 80 percent of the surface cover at the urban core). In order to deal with increased surface runoff, many cities have channelized their streams. Channelization usually involves removing trees and other vegetation from the stream bank, then sculpting the bank into a smooth, straight channel. This channel may then be seeded with grass or, worse yet, lined with concrete.

Converting natural streams into concrete channels: (1) conveys more water more quickly than does a natural stream, and (2) separates the water from the soil, thus reducing erosion at a particular site. This may sound like a good thing, but it isn't. Channelizing a stream fixes one problem, but creates many other problems.

Stream channelization removes the native streamside (riparian) vegetation. This has several negative effects. First, removal of riparian vegetation eliminates habitat for a vast assemblage of amphibians, reptiles, birds, and mammals that depend on that vegetation. Second, replacing the vegetation along urban streams with concrete eliminates the natural beauty of the stream. Since people prefer to live adjacent to natural areas (Lane 1991; Adams, Dove, and Leedy 1984) property values tend to be higher where vegetation is present. In Wichita, Kansas, for example, lots bordering a wetland sold for fifty percent more than comparable lots away from the natural area (Ferguson 1998). Natural areas provide natural beauty. Once that natural beauty is removed, property values are adversely affected.

A third effect of removing the native vegetation along urban streams eliminates the water cleansing ability of the vegetation. As discussed earlier, aquatic vegetation locks up excess nutrients and contaminants found in the water. In fact, wetlands planted with aquatic vegetation are currently being used to clean up or polish wastewater (APWA 1981; Hammer 1997; see Figure 5.7).

A final effect of removing the vegetation along urban streams is the loss of the cooling effect that shade trees have on the water. Parking lots, rooftops, and streets all become very hot during the summer. Naturally, as runoff water travels across these surfaces it heats up. This heated water enters urban streams, raising the average temperature of the stream. Water that flows down an unshaded concrete channel does not have a chance to cool down; the result is low dissolved oxygen levels and a reduced capacity for supporting life. The entire biotic community in a stream can be affected by a change of only a few degrees in the water temperature. There are many other specific effects caused by stream channelization, including:

1. Elimination of pools and riffles. Natural streams have a normal pattern of deep, slow-moving pools of water and shallow, faster-moving riffles. These habitats support their own unique assemblage of plant and animal species, which are lost.
2. Reduced organic material. Most stream systems rely on organic material that is carried in from outside the system to form the basis of the aquatic food web.
3. Increased water velocity. The smooth concrete sides of the stream channel increase the velocity of the water in the channel.
4. Reduced or eliminated infiltration. The concrete lining of the channel doesn't allow water to infil- trate the soil. Therefore, the amount of water moving through the channel is actually higher than would be found in a natural stream because none is allowed to seep into the ground.
5. Increased downstream flooding and erosion. Channelized water has to be discharged at some point. When this happens, the area downstream from the channel gets flooded and scoured.
6. Increased danger for urban residents. Swiftly moving water rushes down these smooth-walled chan- nels. It is not uncommon to hear of people being swept away and killed in these raging torrents (Davis 2003; Riley 1998).

BIOTIC EFFECTS OF URBANIZATION ON STREAM ECOLOGY

The biological and ecological effects of urbanization on urban streams have received even less research attention than the abiotic effects. However, Paul and Meyer (2001) provided an overview of studies conducted on microbes, algae, macrophytes, invertebrates, and fish.

There are several deleterious biological and ecological effects of chemical and organic pollut- ants reaching natural waterways. Stream biota can be lost to pesticide or heavy metal poisoning, osmotic imbalances from excess road salt, suffocation due to sediment clogging their respiratory mechanisms, or loss of oxygen from high biological oxygen demand (BOD) by decomposers. High water temperatures and low saturation of dissolved oxygen are correlated with high relative abun- dance of introduced species and fish with parasites and diseases.

Fecal coliform and other bacterial pathogens appear in urban waterways from a variety of sources, including WTFs, companion animals, and high concentrations of urban wildlife (e.g., resi- dent Canada geese, discussed in Chapter 14). Bacterial pathogen levels rise in urban streams dur- ing and after storms. Some species of these bacteria demonstrated increased antibiotic resistance attributed to genetic transfer and level of metal toxicity in the stream. Nitrifying bacteria densities increase in urban streams due to WTF outflows. Algae species diversity decreases in urban streams, largely due to changes in water chemistry. Bed sediment changes, nutrient enrichment, turbidity, and alien species introductions result in reduced native macrophyte diversity in urban streams.

THE AQUATIC FOOD CHAIN

The aquatic food chain (diagrammed below) begins with phytoplankton, algae attached to hard surfaces and covering the stream bed, and aquatic macrophytes as the primary producers. The pri- mary consumers are zooplankton and benthic invertebrates that feed on the phytoplankton, attached algae, detritus, and macrophytes, as well as each other. Fish larvae feed on the zooplankton, and small fish also eat small benthic invertebrates. The smaller fish are fed on by large invertebrates and larger fish that are fed on in turn by even larger fish.

Aquatic food chain = phytoplankton → zooplankton → fish larvae → small fish → large fish → fish-eating birds

Additionally, fish and invertebrates are fed on by reptiles, birds, and mammals, forming a link in energy and nutrient transfer between aquatic and terrestrial ecosystems. One can sometimes predict the ecological consequences if any one of the trophic levels is stressed. However, this depends on the strength of both direct and indirect interactions among organisms across trophic levels. For example, if some environmental pollutant prevented production of fish larvae then zooplankton would increase and phytoplankton would decrease, thus removing a primary producer from the system.

Environmental pollutants are accumulated at each trophic level, resulting in very high levels of toxic materials accumulated at the top of the food chain. The consequences of this accumulation are discussed in Chapter 6 as they relate to causes of the near extinction of the peregrine falcon, a fish-eating bird.

All aspects of aquatic invertebrate habitat are altered by urbanization, and directly correlated with the extent of urban land use. Invertebrate presence declined significantly with increasing ISC. General invertebrate responses were decreased diversity due to toxins, temperature changes, siltation and organic nutrients; decreased abundance due to toxins and siltation; and increased abundance due to inorganic and organic nutrients. Decreases were especially evident for sensitive orders of mayflies (*Ephemeroptera*), stoneflies (*Plecoptera*), and caddisflies (*Tricoptera*). In contrast, the abundance of bloodworms (*Chironomids*), segmented worms (e.g., sludge worms, known as oligochaetes), fly larvae (*Diptera*), and tolerant gastropods (snails) increased. Dominant indicator invertebrates of polluted water are Chironomidae (Diptera) larvae and oligochaete annelids.

FISH AS INDICATOR SPECIES

Fish communities are indicators of the environmental health of natural waterways. For example, polluted urban streams may: (1) contain fish communities that lack intolerant species, (2) include significant numbers of nonnative or introduced fishes, and (3) show low abundance and diversity of the native species. These conditions indicate poor fish habitat and significant ecological disturbances (Paul and Meyer 2001).

The biological and ecological effects of urbanization on fish assemblages (species present and relative abundance) are given special attention because:

1. Fish are good indicators of long-term (several years) effects and broad habitat conditions because they are relatively long-lived and mobile.
2. Fish communities generally include a range of species that represent a variety of trophic levels (omnivores, herbivores, insectivores, detritovores, planktivores, and piscivores). They tend to integrate effects of lower trophic levels; thus, fish assemblage structure is reflective of integrated environmental health.
3. Fish are at the top of the aquatic food web and are consumed by humans, making them important for assessing contamination.
4. Fish are relatively easy to collect and identify to the species level. More common species can be sorted and identified in the field by experienced fisheries professionals, and subsequently released unharmed.
5. Environmental requirements of more common species are comparatively well known. Life history information is extensive for many game and forage species, and information on fish distributions is readily available.

6. Aquatic life uses (e.g., water quality standards) are typically characterized in terms of fisheries (cold-water, cool water, warm water, sport, forage). Monitoring fish provides direct evaluation of "fish-ability" and "fish propagation," which emphasizes the importance of fish to anglers and commercial fishermen.
7. Fish account for nearly half of the endangered vertebrate species and subspecies in the United States (Barbour et al. 1999).
8. Fish provide many ecosystem services (Table 5.3; Holmlund and Hammer 1999).

Increasing urbanization is related to declines in fish diversity and abundance, and a rise in the relative abundance of tolerant taxa increases with increasing urbanization. Invasive species increase in the more urbanized reaches of streams. Extensive fish kills are more common in urban streams after storms, given the extensive deposition of pollutants from impervious surface cover runoff. The extirpation of fish species is not uncommon in urban river systems. WTF effluent selects for fish species that have wide tolerances for changes in dissolved oxygen concentration, temperature, and siltation.

Barbour et al. (1999) summarized a study where several physical and chemical measures of degradation were used to test changes in fish assemblages for Maryland streams. The study found that out of 270 fish species found in Maryland streams and tested, 8 percent were tolerant, 62 percent moderately tolerant, and 30 percent intolerant of a wide range of chemical and physical changes.

It is helpful to know which species were tolerant to the chemical and physical change because they can act as predictors of deteriorated stream quality in urban areas (Table 5.4; Barbour et al. 1999). For example, the red shiner was found in fifty of sixty-five sample sites by Marsh-Matthews and Matthews (2000) and accounted for 90 percent of the fish community in the Chattahoochee River in Atlanta, Georgia (Devivo 1995, cited in Paul and Meyer 2001). The red shiner (*Cyrpinella*

Table 5.3   Major Fundamental and Demand-Derived Ecosystem Services Generated by Fish Populations

Regulating Services	Linking Services
Fundamental Ecosystem Services	
Regulation of food web dynamics	Linkage within aquatic ecosystems
Recycling of nutrients	Linkage between aquatic and terrestrial ecosystems
Regulation of ecosystem resilience	Transport of nutrients, carbon and minerals
Redistribution of bottom substrates	Transport of energy + matter
Regulation of carbon fluxes from water to atmosphere	Acting as ecological memory: migratory fish
Maintenance of sediment processes; beneficial stream bed alterations (+ Stream Bed)	
Maintenance of genetic, species, ecosystem diversity	

Cultural Services	Information Services
Demand-Derived Ecosystem Services	
Production of food	Assessment of ecosystem stress: fish assemblage changes
Aquaculture production	Assessment of ecosystem resilience: fish assemblage changes
Production of medicine: neurotoxin	Revealing evolutionary tracks: co-evolution of fish and other species
Control of hazardous diseases: malaria	Provision of historical information: climatic changes
Control of algae and macrophytes	Provision of scientific and educational information
Reduction of waste: scavenger fish	
Supply of aesthetic values: aquarium industry	
Supply of recreational activities: fishing	

Source: C.M. Holmlund and M. Hammer. 1999. *Ecological Economics* 29:253–268. With permission.

Table 5.4   Fish Species in Urban Stream Communities Tolerant of Wide Ranges in Chemical and Physical Changes

Common Name	Scientific Name	Trophic Designation[a]
1. Banded killifish	*Fundulus diaphanous*	I
2. Black bullhead	*Ameiurus melas*	O
3. Blacknose dace	*Rhinichthys atratulus*	G
4. Bluegill	*Lepomis macrochirus*	I
5. Bluntnose minnow	*Pimephales notatus*	O
6. Brown bullhead	*Ameiurus nebulosus*	I
7. Catfish	*Ictalurus* spp.	G
8. Central mudminnow	*Umbra limi*	I
9. Comely shiner	*Notropis amoenus*	I
10. Common carp	*Cyprinus carpio*	O
11. Creek chub	*Semotilus atromaculatus*	G
12. Eastern mudminnow	*Umbra pygmaea*	G
13. Fathead minnow	*Pimephales promelas*	O
14. Golden shiner	*Notemigonus crysoleucas*	O
15. Goldfish	*Carassius auratus*	O
16. Green sunfish	*Lepomis cyanellus*	I
17. Largescale sucker	*Catostomus macrocheilus*	O
18. Northern squawfish	*Ptychocheilus oregonensis*	P
19. Red shiner	*Cyrpinella lutrensis*	O
20. Redear sunfish	*Lepomis microlophus*	O
21. Reticulate sculpin	*Cottus perplexus*	I
22. Rudd	*Scardinius erythrophthalmus*	O
23. Silver carp	*Hypophthalmichthys molitrix*	O
24. Spotfin chub	*Cyprinella monacha*	I
25. Western mosquito fish	*Gambusia affinis*	O
26. White sucker	*Catostomus commersoni*	O
27. Yellow bullhead	*Ameiurus natalis*	I

Source: M.T. Barbour et al. 1999. *Rapid Bioassessment Protocols for Use in Streams and Wadeable Rivers: Periphyton, Benthic Macroinvertebrates and Fish,* 2nd edition. EPA 841-B-99-002. Washington, DC: U.S. Environmental Protection Agency, Office of Water.

[a] Trophic designations: G = generalist, plant and animal material; I = insectivore, opportunistic predator on aquatic insects; O = omnivore, bottom feeder; P = piscivore, opportunistic predator on other fish.

lutrensis) is a popular baitfish and has been introduced into streams and rivers through accidental or intentional releases. The common carp, goldfish, and bullhead catfish (*Ameiurus* spp.) are listed as pollution-tolerant species by Rahel (2002). The majority of species listed in Table 5.4 have a trophic classification denoting a broad range of feeding habits characteristic of tolerant and potentially invasive species.

The primary source of invasive exotic fish is the global aquarium industry, which generates $7 billion annually (Holmlund and Hammer 1999). Each year, more than 2000 nonnative fish species, representing nearly 150 million exotic freshwater and marine fishes, are imported into the United States for use in the aquarium trade. Dumping aquaria fish into the nearest body of water when they are no longer wanted creates a problem for the native fish species and for ecosystems in general (Phipps 2001).

Each of the twenty-seven species listed in Table 5.4 is also on the freshwater invasive species list provided by the U.S. Geological Survey (http://nas.er.usgs.gov). The report, titled *Nonindigenous Aquatic Species*, includes a list of 672 species of which 35 percent are exotics. The remaining 65

Table 5.5 Causes of Nonnative Species Introductions and Suggested Responses

Causes of Species Introduction	Strategies for Reducing Homogenization
Deliberate transfers (e.g., stocking by sport fishing industry and state agencies)	Stop deliberate transfers and by-product introductions
By-product introductions (e.g., canal building, ballast water discharge, aquaculture operations)	Minimize habitat alterations, e.g., restore natural flow regimes
Human introductions in urban landscape designs (e.g., garden ponds)	Increase public education concerning the magnitude of the problem and ecological ramifications
	Use sterile hybrids
	Remove naturalized populations of nonnative species prior to reestablishing native species

percent on the U.S. Geological Survey (USGS) list are native species to the United States that were transplanted outside their native range. The *Field Guide to Freshwater Fishes* by Page and Brooks (1991) lists 790 freshwater species native to the United States. The USGS list suggests that 435 of the 790 native species (55 percent) have become invasive species! How did this happen and what are some of the ecological impacts?

The occurrence and ecological impacts of invasive fish species in terms of biotic homogenization was discussed by Rahel (2002). Biotic homogenization is the increased similarity of biota over time caused by the replacement of native species with nonindigenous species, usually as a result of introductions by humans. In human-created habitats, endemic species typically are replaced by cosmopolitan species, with the result that entire ecosystems resembling each other now occur in disparate parts of the country. The interacting mechanisms that result in homogenization are introductions of nonnative species (the key factor homogenizing fish faunas), extirpations of native species, and habitat alterations (e.g., abiotics, stream flow patterns, sedimentation) that facilitate the first two processes. The nonnative species are usually pollutant tolerant and have a wide range of tolerance to a host of abiotic conditions at the extreme end of the spectrum (e.g., low dissolved oxygen concentrations, high temperatures; see Figure 6.1). Native endemic species are usually highly specialized having adapted to a narrow set of abiotic conditions. A list of fifty-three fish species, native to Ohio rivers and streams, that are highly sensitive to either habitat degradation, pollution, or both was provided by Miltner, White, and Yoder (2004). They found significant declines in biological integrity detectable when the amount of impervious surface cover exceeded 13.8 percent. Table 5.5 illustrates both the causes of nonnative species introductions, and suggested strategies to reduce the rate of biotic homogenization.

RESTORATION OF RIPARIAN HABITATS

Restoring the ecology and natural history of streams and rivers destroyed by urbanizations is not simple and definitely not cheap. For example, the Kissimmee River in Florida was channelized in the 1960s by the U.S. Army Corp of Engineers. A 103-mile meandering river was straightened into a 56-mile canal that drained the surrounding wetlands and heavily impacted wildlife and other natural resources throughout the Everglades. The restoration plan (National Research Council 1992) was estimated to take fifteen years and cost $350 million (Campbell and Ogden 1999). The restoration tasks outlined below represent a reactive management process. The proactive management process would have protected the physical and biological integrity of the stream or river during the urban development process. In fact, "Protect the Wetlands" should be at the top of every page of every urban development blueprint. Unfortunately, this is not the case and urban planners are forced to rely on their available technologies and engineering skills to reconstruct what nature had provided at the outset.

Table 5.6   Realized Benefits from Urban Stream Restoration

Cultural	Preserves or restores a historic or cultural resource
	Upgrades the quality of life in urban and neighborhood environments
	Restores a regional or local identity
	Creates interesting educational opportunities for schools
Ecological	Encourages the return of birds and wildlife in urban neighborhoods
	Improves water quality
Economic	Revives a decaying downtown and depressed commercial economy
	Creates meaningful jobs and provides job training
	Increases property values
	Returns public life and commerce to urban waterfronts
Hydrological	Reduces flood damage
	Reduces damages from stream bank erosion
	Corrects performance problems and reverses the damages of large or small engineering projects, which include:
	• Sediment filtering
	• Bank stabilization
	• Water storage and release
	• Aquifer recharge
Recreational	Develops pedestrian and bicycle trails
	Provides greenbelts, open spaces, and parks
	Creates boating and other in-stream recreational opportunities
	Returns or improves recreational and commercial fishing
	Provides a safe food source for family fishers

Source: Riley (1998); Groffman et al. (2003).

It is not our intent to go into a detailed discussion of stream and river restoration strategies. Extensive information on stream and river restoration can be found in Riley (1998) and Mason (2000), in addition to many agricultural extension publications. We recommend two comprehensive documents on stream corridor restoration; the first is a book with the same title (*Stream Corridor Restoration*) written by the Federal Interagency Stream Restoration Working Group (1998), and the second is *Restoration of Aquatic Ecosystems Science, Technology, and Public Policy* (National Research Council 1992). There are also many national, state, and community river restoration organizations throughout the United States (http://riverrestoration.org/). It appears that as we begin the twenty-first century, stream and river restoration has reached the front burner in the urban planning process in many cities and municipalities, and government agencies. For example, in 2007 the Washington State Legislature approved $100 million in funding for riparian ecosystem restoration projects through the Washington Wildlife and Recreation Program.

There are many different incentives for stream restoration classified as hydrological, cultural, recreational, economic, and ecological benefits (Table 5.6). The goals of any riparian restoration program should be to convert an unhealthy system into a healthy one that would have some or all of the characteristics and functions listed earlier in this chapter.

Natural resources management in urban environments must always begin with identification and involvement of all stakeholders, and getting organized (see Chapter 9). Ensuring the involvement of all partners and securing their commitment to the project is a central aspect of getting organized. Getting organized involves:

1. Setting boundaries: establishing the geographic scale of the restoration process.
2. Forming an advisory group—a collection of key participants, including private citizens, public interest groups, economic interests, public officials, and others who are interested in or might be affected by the restoration initiative.

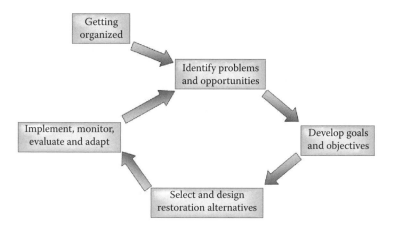

Figure 5.11   The model for the stream restoration development process. (From FISRWG 1998.)

3. Establishing technical teams that represent the knowledge and skills of different agencies, organizations, and individuals.
4. Identifying local, state, and federal funding sources, for example, the Wildlife Habitat Incentives Program (WHIP) administered by the U.S. Department of Agriculture's Natural Resources Conservation Service.
5. Establishing points of contact and a decision structure—networking with others and identifying a group that coordinates the work and information derived from all other groups.
6. Facilitating involvement and information sharing among participants—communicating the status of information from and to all restoration participants.
7. Documenting the process—print and electronic media, websites, newsletters, town hall meetings, and other mass media communication strategies (authors and FISRWG 1998).

The model, called the stream corridor restoration plan development process (Figure 5.11) includes a cycle of events.

1. Identifying problems and opportunities: data analysis of resource conditions in the stream corridor.
2. Developing goals and objectives: guides the development and implementation of restoration efforts and establishes the means of measuring progress and evaluating success.
3. Selecting and designing restoration alternatives: provides the range of management alternatives that fit the data analysis, budget, and desired outcomes.
4. Implementing, monitoring, evaluating, and adapting: activates the restoration plan, matching outcomes with expectations and revising the plan as needed (authors and FISRWG 1998).

The above list only briefly summarizes the complex process of stream corridor restoration. A more detailed analysis of the process can be found in the sources referenced at the end of this chapter.

URBAN WETLANDS

Our examination of the literature on the definition of "wetlands" made it difficult to define and give examples of wetlands in the context of the urban environment. A definitive textbook on *Wetlands,* by Mitsch and Gosselink (2000) provides an explanation: at the outset, there are no definitions of wetlands that can be accepted by all interested parties (e.g., wetland scientists, managers, and regulators). They listed seven extensive yet thematically different definitions of wetlands that depended on the objectives and field of interest of the users. The fields of interest of users included geology, agronomy, hydrology, biology, ecology, sociology, political science, public health, and legal. We found it more useful to discuss urban wetlands in the context of Campbell and Ogden's (1999) *Constructed Wetlands in the Sustainable Landscape,* rather than Mitsch and Gosselink (2000). This being the case, we developed a definition that drew from those features common to all seven definitions and emphasized urban wildlife considerations:

> Urban wetlands consist of land that is transitional between aquatic and terrestrial ecosystems and is covered with standing water for at least part of the year.

Wetland vegetation is adapted to saturated soil conditions and can be distinguished in form and function from upland (dry-land) vegetation. Urban wetlands select for aquatic and semiaquatic invertebrate and vertebrate species. Examples are residential water gardens, ponds, and lakes, and constructed wetlands for storm-water or WTF effluent retention.

As is the case with riparian buffer zones, wetlands provide a variety of valuable ecosystem services, including water purification, filtration, retention of nutrients, flood control, groundwater recharge, as well as habitat for a highly diverse biotic community. Wetlands are also valued for recreational and aesthetic reasons. Urbanization directly impacts these remaining wetlands by changing their hydrology, increasing runoff of nutrients and pollution, increasing exposure to introduced species, and increasing fragmentation. From 1980 to 1990, 30 percent of natural wetland losses were to urban development, while 26 percent of the losses were to agriculture (Boyer and Polasky 2004). Major cities in the United States, including Houston, Texas, and many Florida communities, were built directly in natural wetland habitats.

Urban wetlands occur throughout urban communities because they are a habitat amendment that is highly valued by urban residents. In fact, developers will often increase population densities around constructed wetlands, for example, lakes, by reducing lot size in order to pack more people around the lake. High human population densities living on or in close proximity to the lake shore cause abiotic and biotic changes, just as dramatic as those discussed above for streams and rivers (see Sidebar 5.1).

Water Gardens

Water gardens (Figure 5.12) have become a popular part of landscape architecture in the United States (Masser 1999). In fact, water gardens have become a multi-billion-dollar industry (http://www.pondscapesinc.com/). They offer the urban resident an opportunity to connect to the natural world, similar to putting up bird feeders. Unlike bird feeders, water gardens range in cost from a few hundred to tens of thousands of dollars to build, which does not include maintenance once constructed. The designs (size, construction materials, plants, and animals) of water gardens are as diverse as the personal preferences of the urban residents who build them. Water gardens are included in this chapter on urban waters because they are a wonderful way to enjoy the natural beauty of aquatic plants and animals, and gain a better understanding of the complexities of aquatic ecosystems (Masser 1999).

One needs to do some extensive homework on water gardens before beginning construction. There are several considerations before, during, and after the construction of a water garden in an urban community. These considerations include selection of location, size, and building materials; review of related city ordinances; understanding water quality issues and filtering requirements; selection of plants and animals and their maintenance; and recurring management problems. Recurring management problems include algae control; controlling fish reproduction, diseases, and predators; and preventing the accidental release of exotic plants and animals into the natural waterways (Masser 1999).

Now, where is the urban wildlife management connection in water gardens given their provision of food, water, shelter, and an appropriate space to complete life's functions? Consider what

Figure 5.12 *(A color version of this figure follows page 158.)* An urban water garden. (Courtesy Michael Masser)

might attract the neighborhood wildlife to a water garden. The water alone will draw any thirsty animal to it. It could become the neighborhood bath house for some animals. Semiaquatic animals (invertebrates and vertebrates) could consider it a good breeding ground. Fish-eating animals will soon realize it contains an accessible food supply. So, another recurring and challenging problem is determining what types of management strategies will be required to invite or exclude neighborhood wildlife populations from water gardens.

Constructed Wetlands

The common constructed wetlands in urban communities are ponds, lakes, and storm- and wastewater catchment basins (see Figure 5.7). By definition, any body of water less than 500 surface acres is considered a pond, whereas any body of water greater in size is considered a lake (M. Masser, personal communication). Of course, the size of a standing body of water makes all the difference in terms of what ecological, economic, and recreation functions it can provide. It needs to be understood that constructed wetlands are the results of human manipulation of the landscape and, as such, these wetland habitats are artificial and can be transitory. Ecological succession (see Chapter 3) is a force to be reckoned with. This is particularly the case for smaller bodies of water that can be filled in with sediment, lost entirely because of a sustained drought, or become so nutrient rich from surface runoff of urban lawn fertilizers and so chocked with macrophytes (see Figure 5.2) that its original intended purpose is lost. Constructed wetlands in urban communities need to be continually monitored and cared for in order to sustain the human purposes for their existence.

Earlier in this chapter we addressed the types of materials found in wastewater, and what can and cannot be removed in a traditional WTF. It may be difficult to believe, but storm water appears to contain many more toxic materials and in higher concentrations than wastewater (Burton and Pitt 2001). Storm-water runoff comes from many different sources, including all those that drain into the local watershed. These sources can include agriculture, silviculture, resource extraction, modification of natural waterways, urban, and land disposal. Examples of each of these sources are given in Burton

Figure 5.13 *(A color version of this figure follows page 158.)* A storm-water catchment area consisting of about 10 acres of surface water located between a municipal golf course and urban neighborhood. Note the white signpost in the distance. (Courtesy Clark E. Adams)

and Pitt (2001). They also listed twenty-three examples of chemical toxicants and their sources, such as automobiles, pesticides, and industry, found in storm-water runoff (Burton and Pitt 2001). It might be an overstatement, but in some geographical regions, storm-water runoff can be a "hell's kitchen" of abiotic substances unknown to the biota of the region. So the message here, in terms of urban wildlife management and human safety, is to be aware of the chemistry of storm-water runoff if the urban planning design calls for the development of storm-water catchment basins.

There may be local examples of highly polluted storm-water catchment basins in your neighborhood. One example, shown in Figure 5.13, is a five-acre storm-water reservoir (Finfeather Lake) located between a municipal golf course and an urban neighborhood. Beginning in the early 1930s and for two decades thereafter, there was a high level of arsenic in the surface runoff into the reservoir. The source was the Cotton Poisons, Inc. company, located just across the street from the reservoir. In the old days, trains carrying arsenic would off-load on to a sloppy conveyor system that carried the arsenic into the plant, but left a lot of material on the ground as well. The empty railroad cars were washed out on the tracks. Storm-water runoff from the plant area carried the arsenic contained in the spillage and wash-down water from unlined retention ponds into the reservoir. Control measures were implemented in the late 1960s, but arsenic pollution was extensive. At one time the arsenic levels in Finfeather Lake (College Station, Texas) were fifty times higher than the EPA's maximum for dissolved arsenic concentrations in water (Seale 1992). As recently as 2008, the lake was still declared a danger zone (Figure 5.14). Nevertheless, the reservoir contains fish (species unknown), native and domesticated waterfowl, nutria (*Myocastor coypus*), herons (several species), and probably many other animals known to frequent standing bodies of water in urban areas.

Urban (Community) Fisheries Programs

Fishing is an activity that cuts across generation gaps, social classes, and all levels of ability to unite people in a common respect for the environment and appreciation of the outdoors. To provide fishing opportunities in urban areas, many state and federal fisheries management agencies have developed urban, also called community, fisheries programs. The primary goal of community

Figure 5.14 What the white signpost in Figure 5.13 tells the urban public. (Courtesy Clark E. Adams)

fisheries programs is to make greater use of aquatic resources by enhancing recreational fishing opportunities and facilitating conservation education programs. The primary guiding principle in planning, developing, implementing, and sustaining a community fisheries program is "community." The term community implies the vested interests of multiple stakeholder groups working together, which include anglers, the regulatory agency, city government, fisheries managers and biologists, natural resource educators, local schools, community civic groups, and recreation centers, among others.

Community fisheries locations include ponds or lakes within public parks, streams or rivers through areas with public access, waterfronts along reservoirs or coastal areas, or artificial sites such as swimming pools or hatchery ponds. Community fisheries programs have been offered using several venues such as put-and-take, put-grow-and-take, and self-sustaining fisheries. The American Fisheries Society, in conjunction with their state chapters, offers guides for developing community fisheries programs (e.g., Texas Chapter of the American Fisheries Society, *Guide to Developing Community Fisheries*, published in 1999).

Community fishing resources provide recreational fishing opportunities to a diversity of users, including individuals who lack convenient access to more distant (and costly) fishing opportunities, lower mobility anglers, individuals who lack the time necessary to fish elsewhere, and those being introduced to fishing by family members (many with extensive fishing experience). There are many different examples of urban waters that can be used for community fisheries, including streams, rivers, ponds, lakes, and constructed impoundments such as swimming pools (see Figure 2.7). If stocking is required, the selection of which fish may be based on tolerance levels to water temperature and oxygen concentrations, supplementary feeding requirements, and growth rates. Any guides to developing community fisheries will cover the basic considerations on getting started and sustaining the program.

SUMMARY

The difficulty in writing this chapter was our attempt to tell the whole story to the whole community of readers. The story about urban waters is central in importance in any consideration of urban wildlife management because of the ultimate dependency of all life forms on this precious resource. Two repeated observations about the urban waters story are: (1) how perfectly aquatic ecosystems functioned before human intervention, and (2) the draconian efforts required to restore this perfect function after human intervention.

SIDEBAR 5.1   Case Study of an Exotic Animal Controlling an Exotic Plant

A classic example of what can happen to a constructed impoundment surrounded by a high human population density is the case of Lake Conroe in Conroe, Texas (Klussman et al. 1988). Lake Conroe is a 20,000 acre impoundment, the result of a dam across the West Fork of the San Jacinto River. It was built in 1973 principally to serve as a water supply for the city of Houston, but it also became a site for prime real estate development and water recreation. Invasive exotic macrophytes (e.g., hydrilla [*Hydrilla verticillata*] and water hyacinth [*Eichhornia* spp.]) began to appear by 1975, taking advantage of high levels of dissolved nutrients (non-point sources from agricultural fields and lake shore residences, and point sources from septic tank drainage fields), and the lack of herbivores. The most probable cause for macrophyte presence was human introductions. There is an irony in this story given the fact that humans introduced both the plant and the abiotic conditions it needed to live and reproduce abundantly.

In 1979, hydrilla covered over 4500 acres or 23 percent of the lake's surface area. By 1981, almost 50 percent of the lake was covered with aquatic plants. Color infrared remote sensing of the lake showed only a small spot of open water in the deepest part of the lake. There were measurable changes in the abiotic and biotic conditions of the lake. For example, macrophytes replaced phytoplankton, the open-water aquatic food chain was disrupted, fish assemblages changed in terms of dominant species, and there were measurable changes in the availability of dissolved nutrients and depth of light penetration. The impacts on residents and recreationists included the loss of boating, shoreline fishing, swimming, water skiing, and most other aquatic recreational opportunities. These losses galvanized the Lake Conroe Association and the Texas Parks and Wildlife Department (TPWD) into an aggressive management plan to eradicate the "water weeds."

None of the usual mechanical and chemical methods of controlling the expansion of or eradicating the hydrilla invasion were effective. So in 1979, a permit was obtained to introduce the natural herbivore consumer of hydrilla into the lake, which is the grass carp (*Ctenopharyngodon idella*). Since grass carp feed exclusively on plant material, that is, hydrilla, in their natural habitat in Asia, they represented the logical biological control option here in the United States. The use of a nonnative exotic fish species to control the hydrilla raised concerns among wildlife and fisheries biologists about the potential for accidental introduction into other aquatic systems, where they might do more harm than good. This concern was somewhat alleviated by the assurance that the grass carp were in an enclosed area, which made it difficult for them to establish other populations in the river. However, the possibility of grass carp escaping into the San Jacinto River was monitored.

From September 1981 to 1982, 270,000 advanced, fingerling, diploid grass carp were released at twenty-nine sites distributed throughout the reservoir. Within two years following their release, they had removed virtually all submersed aquatic macrophytes from the reservoir. The grass carp grew rapidly (fourfold increase in weight) during the first year of introduction into what might be called a bountiful food supply. The grass carp literally ate themselves out of house and home. Once they had eaten all of the hydrilla, they had to resort to feeding on alternative forms of vegetation that occurred in the reservoir, including leaves, sticks, algae, and detritus. Eventually, all of the grass carp completed their life cycle and died. However, their legacy was that they had changed Lake Conroe from a macrophyte to a phytoplankton dominant lake, restored recreational and real estate opportunities, and gave some wildlife and fisheries biologists a sense of satisfaction in a job well done.

This case study is a classic example of the dilemma faced by wildlife management programs that can only address the symptoms rather than the root cause. After all the grass carp from the 1980 stocking died, the hydrilla started to come back. Around 2005, another grass carp stocking program took place in Lake Conroe to curtail hydrilla growth and expansion. It did not work initially, so more grass carp were added. Then, after 119,301 grass carp were added to the reservoir, the hydrilla disappeared. Now the problem is what to do with all the grass carp that do not have a sufficient food supply to survive. There is a legitimate concern that they will forage on native plants. In 2008, the Lake Conroe Association, the TPWD, wildlife and fisheries biologists, and other stakeholder groups remained aware of the fact that they may have won some management battles, but certainly not the war.

SPECIES PROFILE: AMERICAN BEAVER (*CASTOR CANADENSIS*)

... Being body'd like a boat, with such a mighty tail
As served him for a bridge, a helm, or for a sail.

—Horace T. Martin, Author

Talk about a comeback story! Beavers were abundant throughout New England and beyond when Europeans first arrived in North America (DeGraaf and Yamasaki 2001), and became the focus of a thriving fur trade. Beaver pelts were the engine that drove the exploration of large tracts of what would become the United States and Canada. The fur was used for robes, coats, and top hats. Trapping was unregulated, well into the twentieth century in some areas, and the American beaver disappeared from much of its original range (Whitaker 1998). But this is an adaptable animal, as is typical of species that thrive in urban landscapes, and given some legal protection it has returned to former stomping grounds with a vengeance.

There are actually two North American species that go by the name beaver—the American beaver, which is the animal most people connect to the name, and the mountain beaver (*Aplodontia*

Figure 5.15   American beaver. (Courtesy Steve Hersey)

rufa). Both are rodents, but that's about as far as the kinship goes. The mountain beaver bears more of a resemblance to a woodchuck. It lacks the signature paddle tail and is much smaller—adults weigh an average of one to three pounds (0.5 to 1.4 kg)—than the American beaver, our largest rodent at forty-five to sixty pounds (20 to 27 kg; Whitaker 1998).

The beaver is built for an aquatic life; its habitat consists of ponds, rivers, streams, marshes, and wetlands. Trees are used as both shelter and food source, with an emphasis on the bark of beech, maple, willow, and aspen, along with a variety of aquatic vegetation, buds, and roots. Branches trimmed to a convenient size may be stored underwater for later use by poking the ends into the mud bottom of the pond or stream (National Zoo website; Whitaker 1998).

Mating occurs in winter and after a gestation period of about 105 days the female gives birth to a litter of up to eight young, although the average litter size is between two and four. Beavers are social creatures, living in colonies of four to eight family members. Communication occurs through posture, scent marking, vocalizations, and slapping of the broad tail on water (a sound that can be heard up to a mile away).

Dam maintenance is constant, and the sound of running water stimulates repair activity (National Zoo website). It should come as no surprise, then, that beavers are commonly involved in human-wildlife conflicts. People may initially thrill to the idea that a family of beavers has moved into the neighborhood … that is, until those same beavers decide to do a little landscaping of their own, appropriating backyard trees and flooding local roads. Suddenly, it's neighbor versus neighbor, both intraspecies and interspecies.

There aren't many cases in which changes in human attitudes can be tracked with dates and wildlife population numbers, but the beaver is one exception (DeStefano and Deblinger 2005). By the 1750s the species had been extirpated in Massachusetts. In 1928, the species was observed in the state once more. Populations were low so beavers were protected, even though conflicts with humans did occur. The population grew to approximately 300 animals by 1946, aided by protection and translocation, and by 1952 Massachusetts once again established a fur-trapping season. A modest fur harvest of about 200 to 2000 pelts were taken annually over the next several decades, and conflicts between beaver and property owners were low. By 1994, Massachusetts' beaver population was estimated at 18,500 and the trapping season was lengthened (DeStefano and Deblinger 2005).

In 1996, a ballot initiative made foothold and other body-gripping traps illegal in the state (Deblinger, Woytek, and Zwick 1999). Fur trapping was still allowed, but only using box or suitcase-type traps, and harvest declined dramatically and beaver populations jumped to 48,000 by 1999. Human-beaver conflicts grew in tandem with rising numbers of beaver. When the population reached between 60,000 and 70,000 in 2000, special legislation was passed to allow limited use of conibear traps under permit for situations related to human health and safety. Reported conflicts and beaver populations dropped modestly, but the beaver is now viewed as a pest by many, including some individuals who voted for the ban on body-gripping traps (DeStefano and Deblinger 2005).

In the past, wildlife management relied almost exclusively on ecological information (Organ et al. 1998). Now, and especially in urban and suburban areas, managers must consider input from various stakeholders as well. Managers and stakeholders need to understand the relationships between wildlife populations and human perceptions and attitudes, and the consequences of the resulting actions (DeStefano and Deblinger 2005). Individuals who want to limit or exclude consumptive use and/or lethal management practices are no longer on the fringe; they are the dominant social movement and they influence management programs whether they are invited to do so by wildlife professionals or not (Minnis 1998).

CHAPTER ACTIVITIES

1. Sample the fish populations of streams and ponds in your community and determine the degree to which they match the list in Table 5.4.
2. Look for examples and the extent of stream channelization in your community.
3. Find alternative construction materials that would decrease the amount of impervious surface cover on sidewalks, driveways, and parking lots.
4. Find the source of your city water supply, determine how people use water in your community, and then visit your wastewater treatment facility.
5. Construct a species list of wildlife that visit water gardens.
6. Make a list of wetlands in your community.

LITERATURE CITED

Adams, L.W., L.E. Dove, and D.L. Leedy. 1984. Public attitudes toward urban wetlands for stormwater control and wildlife enhancement. *Wildlife Society Bulletin* 12:299–303.
APWA. 1981. *Urban Stormwater Management.* Special Report No. 49. Chicago, IL: American Public Works Association.
Barbour, M.T., J. Gerritsen, B.D. Snyder, and J.B. Stribling. 1999. *Rapid Bioassessment Protocols for Use in Streams and Wadeable Rivers: Periphyton, Benthic Macroinvertebrates and Fish,* 2nd edition. EPA 841-B-99-002. Washington, DC: U.S. Environmental Protection Agency, Office of Water.
Boyer, T. and S. Polasky. 2004. Valuing urban wetlands: A review of non-market valuation studies. *Wetlands* 24:744–755.
Burton, G.A., Jr. and R.E. Pitt. 2001. *Storm Water Effects Handbook: A Tool Book for Watershed Managers, Scientists, and Engineers.* Boca Raton, FL: CRC Press.
Campbell, C.S. and M.H. Ogden. 1999. *Constructed Wetlands in the Sustainable Landscape.* New York: John Wiley and Sons.
Cohen, R. 1997. *Fact Sheet No. 3: Functions of Riparian Areas for Wildlife Habitat.* Massachusetts Department of Fish and Game.

Davis, J.M. 2003. Urban systems. In *Texas Master Naturalist Statewide Curriculum*, 1st edition, ed. M.M. Haggerty. College Station, TX: Texas Parks and Wildlife Department.

Deblinger, R. D., W. A. Woytek, and R. R. Zwick. 1999. Demographics of voting on the 1996 Massachusetts ballot referendum. *Human Dimensions of Wildlife* 4:40–55.

Deblinger, R.D., R. Field, J.T. Finn, and D.K. Loomis. 2004. A conceptual model of suburban wildlife management: A case study of beaver in Massachusetts. In *Proceedings of the 4th International Urban Wildlife Symposium*, ed. W.W. Shaw, L.K. Harris, and L. VanDruff. Tucson: University of Arizona.

DeGraaf, R.M. and M. Yamasaki. 2001. *New England Wildlife*. Hanover, NH: University of New England Press.

DeStefano, S. and R.D. Deblinger. 2005. Wildlife as valuable natural resources vs. intolerable pests: A suburban wildlife management model. *Urban Ecosystems* 8:179–190.

DeVivo, J.C. 1995. Impact of introduced red shiners (*Cyprinella lutrensis*) on stream fishes near Atlanta, Georgia. In *Proceedings of the 1995 Georgia Water Resources Conference*, K. Hatcher, ed. Athens, GA: University of Georgia Press.

Federal Interagency Stream Restoration Working Group (FISRWG). 1998. *Stream Corridor Restoration: Principles, Processes, and Practices*. Federal Interagency Stream Restoration Working Group (FISRWG). GPO Item No. 0120-A; SuDocs No. A 57.6/2:EN 3/PT.653. Washington, DC: Government Printing Office.

Ferguson, B.K. 1998. *Introduction to Stormwater*. New York: John Wiley and Sons.

Groffman, P.M., D.J. Bain, L.E. Band, K.T. Belt, G.S. Brush, J.M. Grove, R.V. Pouyat, I.C. Yesilonis, and W.C. Zipperer. 2003. Down by the riverside: Urban riparian ecology. *Frontiers in Ecology and the Environment* 1:315–321.

Hammer, D.A. 1997. *Creating Freshwater Wetlands*, 2nd edition. Boca Raton, FL: CRC Press.

Holmlund, C.M. and M. Hammer. 1999. Ecosystem services generated by fish populations. *Ecological Economics* 29:253–268.

Illinois Department of Public Health. 1999. *Nitrates in Drinking Water*. Environmental Health Fact Sheet. http://www.idph.state.il.us/envhealth/factsheets/NitrateFS.htm.

Klussmann, W.G., R.L. Noble, R.D. Martyn, W.J. Clark, R.K. Betsill, P.W. Bettoli, M.F. Cichra, and J.M. Campbell. 1988. *Control of aquatic macrophytes by grass carp in Lake Conroe, Texas, and the effects on the reservoir ecosystem*. College Station, TX: Texas Agricultural Experiment Station MP-1664.

Lane, K.F. 1991. Landscape Planning and Wildlife: Methods and Motives in Wildlife Conservation in Metropolitan Environments. In *Wildlife Conservation in Metropolitan Environments*, ed. L.W. Adams and D.L. Leedy. Columbia, MD: National Institute for Urban Wildlife.

Marsh-Matthews, E. and W.J. Matthews. 2000. Geographic, terrestrial and aquatic factors: Which most influence the structure of stream fish assemblages in the Midwestern United States? *Ecology of Freshwater Fish* 9: 9–21.

Mason, C. 2000. *Biology of Freshwater Pollution*, 4th edition. Essex, UK: Pearson Education Limited.

Masser, M.P. 1999. *Water Gardens*. Southern Regional Aquaculture Center Publication No. 435, Stoneville, MS.

Minnis, D.L. 1998. Wildlife policy-making by the electorate: An overview of citizen-sponsored ballot measures on hunting and trapping. *Wildlife Society Bulletin* 26:75–83.

Miltner, R.J., D. White, and C. Yoder. 2004. The biotic integrity of streams in urban and suburbanizing landscapes. *Landscape and Urban Planning* 69:87–100.

Mitsch, W.J., and J.G. Gosselink. 2000. *Wetlands*. New York: John Wiley & Sons.

National Research Council (NRC). (1992). *Restoration of Aquatic Ecosystems: Science, Technology and Public Policy*. Committee on Restoration of Aquatic Systems. National Academy Press, Washington, DC, USA.

Organ, J.F., R. Gotie, T.A. Decker, and G.R. Batcheller. 1998. A case study in the sustained use of wildlife: The management of beaver in the Northeast United States. In: H.A. van der Linde and M.H. Danskins, eds. *Enhancing Sustainability: Resources for Our Future*. World Conservation Union, Gland, Switzerland.

Page, M. and B.M. Brooks. 1991. *A Field Guide to Freshwater Fishes*. New York: Houghton Mifflin.

Paul, M.J. and J.L. Meyer. 2001. Streams in the urban landscape. *Annual Review of Ecology and Systematics* 32:333–365.

Perry, J. and E. Vanderklein. 1996. *Water Quality: Management of a Natural Resource*. Cambridge, MA: Blackwell Press.

Phipps, R. 2001. *Got Fish? Already Tired of that Holiday Gift Aquarium? Think Before You Dump and Create an Even Bigger Problem*. U.S. Geological Survey News Release.

President's Council on Environmental Quality. 1978. *Environmental Quality*. The 9th Annual Report of the
    Council on Environmental Quality, Washington, DC.

Rahel, F.J. 2002. Homogenization of freshwater faunas. *Annual Review of Ecology and Systematics*
    33:291–315.

Riley, A. 1998. *Restoring Streams in Cities: A Guide for Planners, Policymakers, and Citizens*. Washington,
    DC: Island Press.

Seale, A. 1992. Arsenic and old lakes. *Insite Magazine*, February, 11–14.

Texas Chapter of the American Fisheries Society. 1999. *Guide to Developing Community Fisheries*. Austin:
    Texas Parks and Wildlife Department.

Tjaden, R.L. and G.M. Weber. 1998. *Riparian Buffer Management: Grasses for Riparian Buffers and Wildlife
    Habitat Improvement*. FS728. College Park: Maryland Cooperative Extension, University of Maryland.

Wagner, M. 2004. *Managing Riparian Habitats for Wildlife*. Austin: Texas Parks and Wildlife Department.

Whitaker, J.O., Jr. 1998. *National Audubon Society Field Guide to North American Mammals*. New York:
    Alfred. A. Knopf.

Wright, R.T. 2006. *Environmental Science: Toward a Sustainable Future,* 10th edition. Upper Saddle River,
    NJ: Pearson Prentice Hall.

CHAPTER 6

Principles of Population Dynamics

If any population was miraculously allowed to grow exponentially and indefinitely, it would eventually expand outward at the speed of light and come to weight as much as the visible universe.

—Author Unknown

KEY CONCEPTS

1. Definition of *population dynamics* with examples of the kinds of questions that are addressed in a study of animal populations.
2. Definition of *survival* with explanations of how various abiotic and biotic conditions determine the level of survivability of animal species known as "generalists" and "specialists."
3. Adaptations, natural selection, and survival of the fittest defined as these terms pertain to the study of population dynamics.
4. Five unique adaptations that promote the abundance of some urban species are listed.
5. How populations grow and two common population growth patterns are discussed and illustrated.
6. *Population density* is defined and discussed in terms of the effects of urban habitats.
7. The positive and negative effects of habitat fragmentation, supplemental feeding, animal damage control, and environmental pollutants on population dynamics are discussed.

INTRODUCTION

Population dynamics is the study of those concepts that predict or explain an animal's presence, abundance, and concentration in a particular habitat. This chapter focuses on those principles of population dynamics that moderate species presence, abundance, and concentration in urban communities. When studying a wildlife population, scientists may ask the following broad questions:

1. Can individuals of the population survive in the particular habitat (i.e., urban ecosystems)?
2. What adaptations does the species have that enhance its survival in the habitat?
3. What is the average density of the population in this type of habitat?
4. What abiotic and/or biotic factors maintain and/or precipitate changes in the average population density over time?

The final section of this chapter examines the impacts of human development on the population dynamics of urban wildlife species. These impacts are discussed in terms of habitat fragmentation (i.e., habitat patches), including the influence of corridors (issues not discussed in Chapter 7), supplemental feeding, animal damage control, and introduction of toxic elements into the urban environment.

SURVIVAL

When a wildlife population is found in an urban habitat (or any habitat, for that matter), two definitive statements can be made. First, at least some of the individuals that make up the population will be able to tolerate the conditions found in the habitat. Second, some individuals within the population will be able to acquire enough resources to survive and reproduce. However, reduced predation and supplemental feeding can alter the natural sequence of events listed (see the section on supplemental feeding below). Before we discuss these statements further, let's review the definitions of *conditions* and *resources* in ecological terms (see Chapter 3). Conditions, as you no doubt remember, are those abiotic factors that vary in space and time; temperature and wind speed are examples of conditions. Resources are those things that are consumed or used by the organisms, such as food and nesting sites (Begon, Harper, and Townsend 1996).

Few if any vertebrate species are equipped to survive the entire range of conditions found on earth. Consequently, the geographic distributions of species are limited to particular areas. However, many wildlife species that inhabit urban areas usually have large geographic distributions, sometimes close to worldwide distributions (e.g., Norway rat, *Rattus norvegicus*). Earlier, in Chapter 1, we discussed the distribution of the raccoon (*Procyon lotor*) throughout the Western Hemisphere except northern Canada. This pattern indicates that rats and raccoons, among others, have a wide range of tolerance for various abiotic conditions. Ranges of tolerance for different conditions will vary between species. For example, a specific temperature range will predict whether brook trout (*Salvelinus fontinalis*) or the common carp (*Cyprinus carpio*) are likely to be found in an aquatic ecosystem. Additionally, water temperature will predict the amount of dissolved oxygen in the aquatic ecosystem—the lower the temperature, the greater the oxygen concentration.

Trout have a narrow range of tolerance to both water temperature and oxygen concentration; they need cold water and high levels of dissolved oxygen. On the other hand, the common carp has a broad range of tolerance to both temperature and dissolved oxygen. These relations are illustrated in Figures 6.1a and 6.1b. What might be some other examples of plants or animals that exhibit the abiotic tolerance ranges similar to those of the brook trout (*Salvelinus fontinalis*) and carp (Cypriniformes)?

One can predict the loss of the trout from a stream if the temperature increases or oxygen decreases beyond the trout's narrow range of tolerance for these two variables. A carp, on the other hand, having a broad range of tolerance for extreme changes in temperature and oxygen

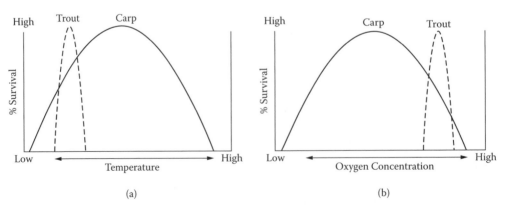

Figure 6.1   (a) Temperature tolerance curve. (b) Oxygen tolerance curve.

concentration, is far less affected than the trout and would likely continue to exist in the changed environment. In general, those species that survive well in urban ecosystems are like the common carp in their ability to survive wide ranges and shifts in abiotic conditions.

If an organism can tolerate the abiotic conditions, then it must still be able to obtain enough resources to survive. Further, in order for the population to persist in the habitat, at least some of its individuals must access enough resources to devote energy to reproduction. Several species of urban wildlife are masters at accessing and using the resources. For example, chimney swifts (*Chaetura pelagica*) are so named because they commonly use chimneys as nesting sites during the summer and as roosting sites during their spring and fall migrations. Populations of these birds likely initially increased as a result of European settlement in North America because they were able to both tolerate the conditions of urban areas and readily use the available resources.

ADAPTATIONS

Whether an organism lives or dies is based on the types of genetic adaptations it has to enhance its survival and provide priority access to resources in order to reproduce. The types of adaptations an organism has are reflected partly in how it looks and how it behaves. For example, the teeth of a coyote and a beaver are adaptations to carnivore and herbivore diets, respectively (Figures 6.2a and 6.2b). The canine and carnassial teeth of the coyote (*Canis latrans*) are carnivore adaptations for grabbing and holding prey and then biting off large chunks of meat. The large incisors and molars of the beaver (*Castor canadensis*) are herbivore adaptations for chiseling off and then grinding up woody plant material. The stalking ability of a coyote and the innate nervous twitch of a cottontail rabbit (*Sylvilagus floridanus*) are behavioral adaptations for catching prey and avoiding capture, respectively.

The process of natural selection is illustrated by any animal that survives in its environment and produces viable offspring. In the urban ecosystem, the process of natural selection usually works against species that were present prior to the urbanization process. Recall that the process of urbanization results in a destabilization and simplification of natural ecosystems. Many endemic species are selected against because they are specialists (require special conditions for survival, e.g., food, habitats, plant and animal community interrelationships) that are usually lost during urbanization.

The environment an organism lives in "tests" its genetic adaptations, determining whether it is fit for that particular environment. Once the minimum criteria for survival have been met the next challenge is to obtain sufficient resources to reproduce. Reproduction is the process through which genes are introduced into the ecosystem to be tested. Successful reproduction allows the individual the opportunity to pass along its adaptive genetic code to a new generation. In time, the adapted species, by sheer numbers, may force the less adapted species out of the habitat. This scenario is played out constantly and consistently in natural as well as urban ecosystems.

There are several general adaptations that wildlife must have to survive and reproduce in urban environments. Two of these adaptations include the following:

1. Generalists or specialists—urban wildlife may be generalist, able to use a wide assemblage of food sources and shelters, and recognize alternative opportunities for survival in the structure of urban ecosystems. On the other hand, urban wildlife can be fortunate specialists that need and have access to a resource readily available in urban environments, for example, artificial nectar for hummingbirds.
2. Tolerance of human presence or being active when humans are not—animals that are successful in the urban ecosystem must become habituated to human activity; in other words, humans are not seen as a threat. In fact, some urban coyotes have grown so familiar with the presence of humans that they will prey on their pets in broad daylight and in the presence of pet owner(s). Urban raccoons

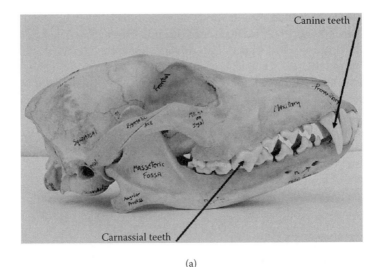

Canine teeth

Carnassial teeth

(a)

Incisor teeth

Molar teeth

(b)

Figure 6.2   (a) Coyote skull showing canine and carnassial teeth adaptations for meat-eating. (b) Beaver skull showing large incisors and molars adapted for eating woody plant material. (Courtesy of Clark E. Adams)

(*Procyon lotor*) are notorious for launching night-time raids on pet food dishes or garbage cans left unprotected. There are even incidences of cougars (*Puma concolor*) stalking and sometimes attempting to prey on urban joggers. White-tailed deer (*Odocoileus virginianus*) will forage on lawn shrubs and gardens while urban residents are mowing their lawns. It is this tolerance of human presence that allows people to enjoy wildlife viewing in urban and suburban environments.

Urban wildlife has been categorized based on the ability of various species to adapt to urban habitats. A taxonomy developed by Blair (1996) was based on the average daily densities of all bird species in a business district, office park, residential area, golf course, open-space recreation area, and natural preserve. Avian species were grouped into "urban exploiters" (three species including pigeons [*Columba livia*], swifts [Apodidae], and house sparrows [*Passer domesticus*])—these species are adept at exploiting the ecosystem changes caused by urban sprawl. This group reaches its highest densities in developed sites and usually represents a very small subset of the world's species.

Another category was the "urban avoiders," which are particularly sensitive to human-induced changes in the landscape. In part, urban avoiders include forest-interior (especially insectivorous) birds that disappear quickly in the initial stages of suburban encroachment. This group reaches their highest densities in natural areas.

A final category was those birds that are "suburban adaptable." This group is able to exploit the additional resources (e.g., ornamental vegetation) that accompany moderate levels of development. This category contained thirty avian species that were found in all or several sites. Suburban adapters—mammals and birds that are mainly adapted to forest edges and open areas—flourish in suburban habitats, especially older subdivisions where ecological succession has produced extensive vegetation. One of the most important traits that separates the three categories is the extent to which species depend on human-subsidized resources to exist in an area (Blair 1996; McKinney 2002).

Finally, there are several characteristics that distinguish urban from rural wildlife communities. A community is defined as all the populations of wildlife living in a particular area. For example, when compared to rural communities, urban wildlife communities tend to:

1. Demonstrate reduced species diversity.
2. Consist of species without a common evolutionary past that have not been found elsewhere in such a combination.
3. Be represented by more exotic or introduced species than native species.
4. Have many generalist species rather than specialists in utilizing the resources in urban habitats.
5. Demonstrate changed behavioral patterns between species and within species groups (Adams 1994; Alberti et al. 2003; Blair and Launer 1997; Blair 1996; Kloor 1999; Rebele 1994).

DENSITY

In general, species that live in both rural and urban habitats (e.g., raccoons) typically reach higher densities in urban areas. However, urban wildlife population densities follow some general patterns seen in other ecosystems. First, populations of species that feed lower on the food chain (e.g., herbivores) are higher in density than populations that feed higher on the food chain (e.g., carnivores). This occurs because energy transfer between the different trophic levels is usually between 2 and 24 percent efficient (Pauly and Christensen 1995). Consequently, less energy for reproduction is available to the populations that feed at higher trophic levels. However, an urban ecosystem can support higher densities of mice and hawks because of the available levels of energy provided to each population from supplemental and other anthropogenic sources as discussed below.

A second pattern of population densities that is similar between urban and natural ecosystems is the dispersion pattern of individuals in the population. Urban populations exhibit what is referred to as clumped dispersion patterns, meaning the individuals making up the population are in closer proximity to each other than would be expected at random. The main factor in determining this dispersion pattern in urban areas is the distribution of resources. As stated before, food sources in urban areas are often clumped, resulting in the attraction of several individuals to a common location. This phenomenon has important consequences for both humans and wildlife. For example, urban residents commonly place multiple hummingbird feeders in close proximity to each other to maximize observations of the delicate hummers. This dispersion of food sources for the hummingbirds is more clumped, richer, and consistent than would be found in natural habitats. An unfortunate consequence of this phenomenon is increased fighting between the male hummingbirds for access to these food sources, which can result in serious injuries or even death (K.J. Lindsey, personal observation).

However, it is important to note that urban ecosystems are significantly different from natural ecosystems in at least two respects: the amount of anthropogenic garbage and intentional supplemental feeding that serves as a food source for some wildlife populations. As a result of these artificial food sources, omnivore populations (e.g., opossums [*Didelphis virginiana*], rats [*Rattus* spp.], coyotes [*Canis latrans*]) and many seed-eating birds probably maintain higher densities in urban areas than in natural ecosystems, although this hypothesis has not been definitively tested.

Factors Affecting Population Densities

The actual number of any species of plant or animal within ecosystems is controlled by the opposing factors that either increase or decrease the numbers of individuals in the population. Populations will increase in density by reproduction or via immigration (moving in) of individuals into the population. Populations will decrease in density by mortality or emigration (dispersal or moving out) of individuals from the population. Each species has its own reproductive potential for increase and potential for longevity. As described before, natural selection has shaped different strategies of reproduction and survival for various species. For example, small mammals tend to reproduce at very young ages and produce relatively high numbers of offspring at each reproductive episode. However, this high reproductive potential is offset by relatively short life spans and high mortality rates of the young. In contrast, larger mammals are usually older at sexual maturity, and produce fewer offspring at each reproductive episode, but tend to have longer life spans and lower mortality rates of the young. (There are, of course, exceptions to this general rule.)

But what factors control how much reproduction and/or mortality occurs at a given time in each population? The most common examples of factors that limit growth of populations by changing reproductive and/or mortality rates are food supply, available nesting or denning sites, predators, diseases, and climate.

In general, urban habitats greatly reduce limiting factors for some species. What draws some wildlife species to urban areas is an abundance of food (a free lunch) and water, alternative shelters, and lack of natural predators. A variety of virtually limitless food sources are presented in the form of garbage, pet foods, bird feeders, pets, fish and amphibians in backyard ponds, gardens, orchards, and landscape plants. During 2006, 55.5 million U.S. residents reported feeding wildlife around their homes (U.S. Census Bureau 2006). As mentioned earlier, urban food resources are consistent, rich, and clumped, which leads to year-round urban wildlife populations. As a result, the numbers of some urban wildlife species have increased at nearly exponential rates (e.g., resident Canada geese, urban white-tailed deer, gulls [Laridae], pigeons, blackbirds [Icteridae], starlings [*Sturnus vulgaris*], rats, and mice [*Mus* spp.]). For example, in the early 1960s it was estimated there were only 50,000 giant Canada geese left in all of North America. In 1998, the figure stood at over two million in the eastern United States alone (see Chapter 14).

Sometimes food is so abundant in urban areas that further supplementing the food sources would not change the numbers of animals in any significant way. For example, Calhoon and Haspel (1989) demonstrated that increasing supplemental food of urban cats in Brooklyn, New York, did not affect the numbers of cats. This finding indicated that above a certain level, food was not a limiting factor to an increase in the population. However, other limiting factors might have existed (e.g., climate changes, or shortages of food, water, shelter, and mates).

How Populations Grow

The formula to calculate population growth is

$$N_{(t+1)} = N_t + (B - D + I - E)$$

In words, the formula means population change over one unit of time equals the population at the beginning unit of time plus (the difference in births and deaths plus the difference in immigrants and emigrants). The population size increases as a result of births and immigrants and decreases as a result of deaths and emigrants. Consider the following application of this formula where immigration or emigration are not considered.

> The rate of birth, death, or population change is expressed as a percent which is the number per 100 members of the population, over a unit of time. For example, a birth rate of 30% per year (meaning that, on average, 30 individuals are born each year for every 100 members of the population) and a death rate of 10% per year (meaning that, on average, 10 individuals die each year for every 100 members of the population), which would result in a population growth rate of 20% per year (meaning that the population increases by 20 individuals each year for every 100 members of the population). (Brower et al. 1989) The population growth rate is also referred to as the *intrinsic rate of increase* or *biotic potential* in the population ecology literature.

Even though immigration was left out in the above example, the movement of animals into the urban environment is an extremely important contributor to population growth in urban environments. Immigration becomes particularly evident when members of specific urban wildlife populations are removed by various animal control methods. Examples were given in the discussions on urban coyotes and raccoons in other chapters. Conversely, those species that are not urban adaptable die in place or move out of the area entirely—pointing out the need to consider emigration in the population growth rate formula.

Population Growth Rate Patterns

The two population growth patterns (how densities change over time) are illustrated in Figure 6.3. The J-curve demonstrates exponential population growth under optimal conditions. The S-curve shows a population at equilibrium constrained by factors that limit additional growth. Some species of urban wildlife are increasing exponentially with no end in sight, for example, the J-curve in Figure 6.3. However, no population can grow exponentially for an indefinite period of time. Limiting factors place constraints on population growth, preventing sustained exponential growth.

The J-type of growth curve, characteristic of some urban wildlife, represents a population explosion; members of a population have produced or will produce more offspring than can be adequately supported by the urban ecosystem, also known as the *carrying capacity*. For example, urban white-tailed deer have a propensity to produce more twins than single offspring in urban compared to rural ecosystems (video production titled *Suburban Deer Management*, Cornell University). This results in deer populations that cannot find adequate food resources to meet the nutritional needs of all herd

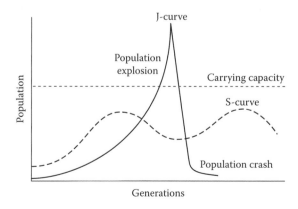

Figure 6.3 Population growth curves. (Diagram by Clark E. Adams)

members. This results in overgrazing on all plant resources, very small deer from under-nourishment or starvation, highly localized deer densities (indicated by high numbers of road kills), and the possibility of a massive population crash as illustrated in Figure 6.3. However, barring disease, a serious crash is unlikely to happen in the near future given the steady input of nutrients, and the protectionist attitudes and behavior by some human residents of urban ecosystems. Disease can have a much more devastating effect on a highly concentrated population within an urban environment.

There are few natural predators, except the automobile, for some of the most densely populated species of urban wildlife. As pointed out earlier in this book, the few predators that exist may not engage in a normal predator-prey relationship (e.g., urban cats are less likely to prey on urban rats). Coyotes and cougars are the largest predators in urban ecosystems, but they often opt for prey species that are naïve or ill-adapted to win the predatory encounter, for example, pets.

Any population of animals with the above characteristics can potentially become so numerous in urban communities that they can qualify as a "nuisance" species. A nuisance species is defined as "any animal that has a negative impact on human health or economics." For example, the top three bird nuisance species in urban areas were pigeons, blackbirds, and starlings (Fitzwater 1988). Raccoons, coyotes, and skunks (Mephitidae) were the most common nuisance mammals (see Table 2.1). Chapter 14 analyzes the complexity of the pest species management issues associated with resident Canada geese and urban white-tailed deer.

The most significant population controls on urban wildlife are the wildlife damage management programs conducted by state, federal, and private organizations. The task of confronting wildlife damage problems in urban communities is an overwhelming obligation resulting in a scenario of winning a few battles, but certainly not the war. As pointed out in Chapter 2, state wildlife agencies have neither the budget nor the personnel, nor do they have a tradition of engagement in the daunting task of urban wildlife management. Most urban wildlife species have been given the opportunity to grow and prosper, and they are!

EFFECTS OF HABITAT FRAGMENTATION ON POPULATION DYNAMICS

One of the most profound affects of urban development on the population dynamics of urban wildlife is habitat fragmentation, which creates islands of habitat or "patches" in urban areas (discussed in detail in Chapter 7). Habitat patches (or islands) vary in size, edge circumference, shape, vegetative makeup, connectivity to other patches (i.e., corridors), and surrounding land use patterns. The selective pressures imposed by habitat patch characteristics results in variations in the presence, abundance, and densities of animals within urbanized areas. In fact, habitat patches affect animal behavior, reproductive patterns, survivability, immigration and emigration or dispersal capabilities, and foraging activity.

From the standpoint of population dynamics, it is reasonable to expect repetitive "boom and bust" (the J-shaped growth pattern, Figure 6.3) cycles of population growth by those species occupying habitat patches if:

1. There are fluctuations in the availability of food, space, and cover, for example, climatic changes and urban development.
2. Competition within and between species groups for these resources increases, particularly when population numbers and densities increase beyond the carrying capacity of the patch.
3. The species is small and has a high reproductive rate under any circumstances.
4. The availability of a specific prey species becomes a limiting factor for predators.
5. There are no dispersal corridors at all or unsafe (e.g., roads and highways) dispersal corridors between patches.

This "boom and bust" growth cycle has been documented by Barko et al. (2003) for white-footed mice (*Peromyscus leucopus*). Ostfeld et al. (1998) used the source-sink model to explain the occurrence of boom and bust growth cycles in habitat patches. They divided patches into those in which birth rates exceed death rates (sources) and those in which death rates exceed birth rates (sinks). In other words, populations that become isolated either decline to extinction or increase rapidly and overshoot the carrying capacity of the habitat patch.

Another aspect of habitat fragmentation in urban areas that influences certain aspects of the population dynamics of resident wildlife was how they used corridors between habitat patches for dispersal and/or foraging activity. Of some note is the relative frequency of corridor use by large predators, including cougars (*Puma concolor*), bobcats (*Lynx rufus*), foxes (*Vulpes vulpes*), and coyotes (*Canis latrans*). Bobcats and coyotes used corridors more as habitat rather than for travel, and tended to cross over a road rather than use safer underpasses such as culverts (Tigas, Van Vuren, and Sauvajot 2002). They found that both species can persist in urban environments because they can adjust behaviorally to habitat fragmentation and human activities, in part through temporal and spatial avoidance. Juvenile cougars use urban habitat corridors for dispersal (Beier 1995). Mammalian carnivores generally prosper in the dissected habitats within urban developments. They prosper because they become resource generalists benefiting from the supplemental resources (i.e., garden fruits and vegetables, garbage, direct feeding by humans), and a variety of landscape elements (i.e., forest and grassland patches and corridors), associated with residential development (Crooks 2002; Atwood, Weeks, and Gehring 2004). One only has to begin paying attention to newspaper and magazine articles that report human encounters with cougars, coyotes, bears (*Ursus* spp.), and other predatory animals to become convinced of how often they do live among and interact with us. More often than not, it comes as a total shock to an urban or suburban dweller to learn that they have lost a valued pet to the predatory activities of coyotes. Cougar attacks on hikers, joggers, and cyclists are becoming more frequent and publicized in the print and electronic media. Once predators lose their inherent fear of humans, the rules of the cohabitation game change dramatically.

Tree-lined streets can function as habitat corridors for some species. Birds and tree squirrels are frequent users of tree-lined streets. Blackbirds (*Turdus merula*), wood pigeons (*Columba palumbus*), magpies (*Pica pica*), and starlings (*Sturnus vulgaris*) can feed and reproduce within the tree-lined streets. Fernandez-Juricic (2000) found that out of twenty-four species that inhabited a local park, fourteen were also observed in tree-lined streets. Use of tree-lined streets provides a way for birds to travel from patch to patch while avoiding the inhospitable urban matrix. Squirrels have been observed using power lines and the canopy of tree-lined streets for the same purpose. In urban habitats with little ground cover, high dog and cat populations, as well as auto traffic, make movement along the ground especially risky for squirrels (Williamson 1983).

Rights-of-way are managed vegetated areas that line railroad tracks, highways, and power lines. Although usually highly managed by periodic mowing, these greenways consist of herbaceous cover, shrubs, and occasional trees. Small mammals, such as rabbits, shrews, and mice, can feed and reproduce in these strips of land. Mammals as large as foxes can survive in rights-of-way, as can many bird species (Transportation Research Board 2002). In fact, these areas can be managed for the explicit purpose of inviting wildlife.

Constructed corridors between habitat fragments were found to cause increased movement and associated changes in the demography (e.g., age and sex ratios) of meadow voles (*Microtus pennsylvanicaus*). However, there was no clear pattern of detectable differences in vole population size or recruitment in unfragmented, isolated or nonlinked fragments, and corridor-linked fragments (Coffman, Nicols, and Pollock 2001).

EFFECTS OF SUPPLEMENTAL FEEDING ON POPULATION DYNAMICS

While people enjoy viewing wildlife up close and personal, there are potential deleterious effects of providing supplemental food (i.e., human-derived subsidies; Fedriani 2001) to urban wildlife.

These effects on game animals were summarized in 2000 by the Wildlife Management Institute in a booklet titled: "Feeding Wildlife...Just say NO!" by Scot Williamson. In straightforward language and with the use of illustrations, it carefully explains why such feeding programs are invariably costly and rarely beneficial to wildlife in the shorter or longer run. Fedriani (2001) said the impacts of human-derived subsidies on the population dynamics of urban wildlife need to be examined in the contexts of the effect on consumer populations, resource availability, and food web and community dynamics. Any urban animal that is offered an alternative food supply that is free (does not have to hunt for it), abundant or in some situations "everlasting," and satisfies its daily caloric requirements is going to be affected in several ways. The urban animal may:

1. Switch its foraging activity to concentrate on the subsidized resources (perhaps exclusively).
2. Improve its nutritional condition and gain weight faster because of the lower expenditure in foraging energy required to access the subsidized resource (i.e., reduce the size of its home range).
3. Begin reproductive activity earlier in life and have larger numbers of offspring more often.
4. Increase its population density in proximity to the location of the subsidized resource.
5. Pick more territorial battles with its own species and other urban animals (including domestic pets) to compete for priority access to the subsidized resource.
6. Survive longer and during climatic conditions that would normally limit survival (e.g., winter).
7. Run a higher risk of predation and disease transmission due to increased species densities at or around the subsidized resource.
8. Become accustomed to the presence of humans—even entering their homes, more often than not, as an invited guest.
9. Rapidly colonize areas where the subsidized resource is available.

Additionally, supplemental feeding affects other ecosystem components. Martinson and Flaspohler (2003) found that aggregations of bark-foraging birds near a feeder resulted in increased predation on nearby bark-dwelling arthropods. The management implication of this study was that birds may increase tree health by consuming leaf-chewing arthropods over-wintering in tree bark. So increasing the density of bark-foraging and excavating birds through supplemental feeding may increase predation on certain pest species, thus minimizing the need for chemical alternatives. The use of natural predators for pest control has also been demonstrated by large bat colonies occupying bridges (see Chapter 8, Bridges and Bats).

EFFECTS OF ANIMAL DAMAGE CONTROL
ACTIVITIES ON POPULATION DYNAMICS

Animal damage control (ADC) activities become necessary as a reaction to urban wildlife management policies and practices that have facilitated rather than prevented pest problems in the first place. ADC efforts target those species that have "crossed the line" in terms of having (1) a negative impact on human health or economics, (2) worn out their welcome, or (3) become a cause of fear. Species fitting the first category and their impacts on human health and economics were covered in Chapter 2. Species that have worn out their welcome could be urban deer, that is, a few hanging around in the neighborhood is tolerable; a few hundred is not. Species that create fear in humans as a the result of urbanites not knowing what they are dealing with; for example, an opossum or very big rat, an assumption of danger (such as all snakes are poisonous—venomous being the correct term), or knowing that a real potential for danger exists (predatory animals). Regardless of the reason for the need for ADC activities, there are consequences of these activities that affect the population dynamics of targeted species.

In order to understand the effects of ADC activities on the population dynamics of selected species, it is necessary to reexamine the techniques and strategies used in ADC (Conover 2002). At

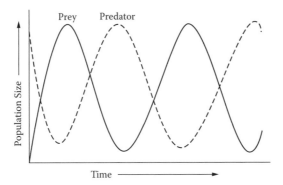

Figure 6.4 Idealized oscillating population growth patterns of predator and prey. (Diagram by Clark E. Adams)

the risk of becoming too simplistic, ADC activities are basically designed to remove the "offending" animals permanently or to another location (e.g., translocation) or to discourage the depredatory effect through aversive conditioning. The removal of any animal from its population can have several effects on the population dynamics of the species. For example, the population density decreases depending on the number of animals removed. Removal of some members of the population releases more resources for the remaining individuals, and may provide opportunity for accelerated immigration. The status of the removed animals within their population hierarchy may be an important factor if the removed individuals had priority access to mates and control of the reproductive activities within the population. This could lead to increased intraspecific competition for the abandoned hierarchical position. Furthermore, sex and age ratios within the population might be altered, particularly if these are variables that predict capture and removal. ADC activities may also have an effect on the predator and prey trophic interrelationships within the urban animal community. Figure 6.4 is an idealized representation of the oscillating cycle of the numbers of predators and prey species in a community, also known as the Lotka–Volterra model in the population ecology literature. The prey species numbers are high because the predator species is low in the community. However, since there is an abundance of prey species for the few predators, predation increases. The predator's numbers increase over time because of an abundant food supply, causing the prey numbers to decrease over time. Now the cycle shows predator numbers high and prey numbers low. The prey species will again increase because the predator is at a point in the cycle where few prey species exist, causing a decrease in predator numbers. The application of the predatory and prey cycle to ADC activities can be explained as follows. If predators are the primary targeted species group, and ADC activities are successful in the near elimination of predators, one can certainly expect the prey species to increase—perhaps dramatically.

Coyotes rank second in importance in the number of calls received by Wildlife Services for public assistance (see Table 2.1 and Chapter 9). Coyotes are one of the most "controlled" species across their range, often with high percentages of local populations removed annually. Although coyote population numbers seem to recover rapidly from these control efforts, the genetic consequences have not previously been explored (Williams et al. 2003). Speculation existed concerning the degree to which inbreeding by subdominant males (often offspring of removed dominant males) would affect the gene pool of the population. Williams et al. (2003) found that various aspects of coyote social structure and dispersal patterns adequately maintained genetic variation and promoted genetic homogeneity over relatively small geographic scales during periods of locally aggressive removal. In other words, control efforts tended to stabilize rather than disrupt various aspects of the population dynamics of this species. A more extensive treatment of the coyote as a species of special management concern is given in Chapter 9.

EFFECTS OF ENVIRONMENTAL POLLUTANTS ON
WILDLIFE POPULATION DYNAMICS

Chemicals have replaced bacteria and viruses as the main threat to health. The diseases we're beginning to see as the major causes of death in the latter part of this century and into the 21st century are
diseases of chemical origin.

—Dick Irwin, toxicologist at Texas A&M Universities

Some of the most insidious materials to humans and wildlife today are the witches' brew of
toxic chemicals that are present in the soil, air, and water. In fact, humans have probably produced
and released, directly or indirectly, every conceivable toxicant possible into the natural ecosystem.
There is an endless list of toxic chemicals, but they can be categorized into hazardous materials
(HAZMATS), pesticides, herbicides, rodenticides, heavy metals, industrial discharges, radioactive
wastes, and pharmaceutical products (see Table 5.2). All of these materials represent pollutants of
"type" which means that living organisms do not know what to do with them once they enter the ecosystem. They do not represent the seventeen essential nutrients for life (see Chapter 3) so organisms
cannot process them into normal bodily functions once they have been assimilated. The inability
to process "unnatural" chemicals, more often than not, leads to death (immediately or eventually),
cancer, abnormal thyroid function, decreased fertility, decreased hatching success, demasculiniza-
tion and feminization of males, and alteration of immune function, among others. Susceptibility to
environmental toxicants is more noticeable in fish and amphibians, which may be due to the highly
specialized genetic programs they need to survive in their natural surroundings. Frogs and other
amphibians have been vanishing worldwide over the last few decades (Cone 2005).

The literature on environmental toxicology is extensive and cannot be given a full treatment in
this text. Rather, our purpose is to raise the readers' level of awareness concerning the existence of
environmental toxicants, and alert them to what one might call the "hidden dangers" that now exist
in higher concentrations in urban rather than rural environments. Our treatment of the impacts of
environmental toxicants on the population dynamics of urban wildlife will focus on selected pesticides, lead, mercury, and pharmaceuticals.

Urban runoff (point and non-point) is a significant pollutant source. Point sources include industrial waste discharges into the air and water or on the soil, and municipal wastewater treatment plant
discharges into aquatic habitats. Non-point sources consist primarily of storm-water runoff from
impervious surface covers (e.g., concrete pavements). Some of the more urban-specific environmental toxicants are pharmaceutical, personal care, and common household products. Recall from
Chapter 5 how these toxicants enter the environment from hospitals, pharmaceutical manufacturing
plants, and household wastes contained in the wastewater treatment stream. Few, if any, wastewater
treatment facilities can extract these toxicants from the water and they are discharged into streams
and rivers. Manure used as fertilizer (e.g., applied to urban gardens) may contain veterinary pharmaceuticals that directly contaminate soils and runoff into streams and rivers after heavy rains
(Glaser 2004; Ternes, Joss, and Siegrist 2004).

Another review of toxicoses of urban wildlife from exposure to pesticides and lead poisoning
was provided by Burton and Doblar (2004). They found that in most cases of toxicoses the clinical evidence of pesticide poisoning begins with signs of restlessness and irritability and then progresses to hypersalivation, diarrhea, and respiratory difficulty with open mouth breathing. Death
usually results from central nervous system depression, respiratory failure, or drowning in the case
of waterfowl. They cited a study by Stone and Crandall (1985) that reported a large die-off of Brandt
geese (*Branta bernicla*) on a New York golf course after diazon was applied to the grass.

Lead poisoning has been an important cause of mortality in waterfowl for more than 100 years
(Paine 1996, as cited by Burton and Dolbar 2004). More recently, The Wildlife Society commissioned a comprehensive and technical review of sources and implications of lead ammunition

and fishing tackle on natural resources (Rattner et al. 2008). Waterfowl and other wildlife species can fall victim to lead poisoning as a result of exposure to lead shot (directly or ingesting shot-contaminated flesh), and ingesting lead sinkers. Lead inhibits the production of enzymes necessary for the synthesis of heme (a vital molecule used in the production of red blood cells). Clinical signs of lead poisoning include lethargy, depression, weakness, vomiting, green fluid diarrhea, convulsions, and coma ending in death. Lead has been found in urban wildlife. For example, Chandler, Strong, and Kaufman (2004) found greater levels of lead in urban house sparrows (*Passer domesticus*) than rural. They suggested that these high levels of lead in urban house sparrows could lead to adverse effects on predators such as small falconiformes due to bioaccumulation.

The effect of increasing mercury concentrations on alligator (*Alligator mississippiensis*) populations is a concern, given its importance as a keystone species in aquatic ecosystems. Alligators are particularly good environmental biomonitors of mercury because of their position in upper trophic levels, prey preferences, long life spans, and nonmigratory habits. The anthropogenic contributors of environmental mercury are municipal solid waste combustion, medical waste incineration, paint applications, and the electric utilities industry (Kahn and Tansel 2000). The most toxic form of mercury is methylmercury, capable of causing chromosome damage and nuclear changes. However, Kahn and Tansel (2000) were not able to establish a firm link of mercury toxicosis with reptiles, suggesting further long-term studies would be required.

Without question, environmental toxicants have significant negative impacts on the population dynamics of various species of wildlife. For example, with the advent of modern pesticides (i.e., DDT in 1943), came a marked decline in the populations of many birds. The decline of the American robin (*Turdus migratorius*) in the early 1950s in many Midwestern states was linked to DDT spraying for Dutch elm disease. For example, a recurring spring phenomenon in Nebraska during the 1960s was the delivery of dead or dying robins to my biology classroom by my students (personal observation, C.E. Adams). These robins had the clinical signs of pesticide poisoning of respiratory difficulty evidenced by open mouth breathing discussed above. DDT in the soil was incorporated into and accumulated in the tissue of earthworms (voracious detritus feeders), which is a mainstay in a robin's daily diet. The ingestion of one too many contaminated earthworms constituted a lethal DDT dose for the robin and its chicks. In this case, the impact of DDT on the population dynamic of urban wildlife was to adversely influence avian survival and reproductive success. The ability for DDT to bioaccumulate in the food chain put those organisms at the top of the food chain (e.g., fish-eating birds and mammals) at the highest risk of adverse effects on their reproductive capability (see Perspective Essay 8.1, The Peregrine Story).

Another impact on the population dynamics of urban and other wildlife caused by pharmaceutical toxicants is the development of hermaphroditic (intersexes: individuals with gonads containing both female and male reproductive tissue) members of a population, for example, fish, frogs, and alligators (Cone 2005; Gibson et al. 2005). It should not be a surprise to anyone that the hermaphroditic condition in organisms that were once discretely male or female will cause problems in reproduction and possibly a population crash, for example, it was for this reason that cricket frogs (*Acris crepitans*), once abundant, declined dramatically around Chicago (Cone 2005). Unfortunately, there do not seem to be any enforceable mechanisms in place at this time to control the release of pharmaceutical toxicants into natural ecosystems. However, media reports on the deleterious effects of these toxicants on various wildlife species is increasing, which may increase public awareness and action to alter existing conditions.

This chapter ends with an essay that describes the anthropogenic causes of bird mortality. It is an appropriate ending to a chapter on population dynamics because it seems that bird populations have always been subject to many more selective forces of human origin and design than other vertebrate species.

PERSPECTIVE ESSAY 6.1   Bird Kills

Birds are one group of terrestrial vertebrates likely to be noticed by urbanites, when they are overabundant or missing. To begin with, many birds are diurnal, so they are active during the same time of day when people are out and about. Also, it is well documented that many people have a special attraction to the avifauna around them and will travel many miles to establish contacts with birds outside their urban neighborhoods. The imagery of birds can be one of delicacy and fragility (e.g., the bee hummingbird, *Mellisuga helenae*) or a fully coordinated mass of raw power (e.g., the Harpy eagle, *Harpia harpyja*). Birds provide many forms of entertainment: songs, flight patterns, colors, and feeding and brood-rearing behavior. Even though some bird species can become pests, the overall attitudes of humans toward birds are positive, as evidenced by the time and money they spend to get close to birds. This perspective essay was developed to focus on a contradiction: birds contribute to our quality of life, but our activities, accidentally or by design, may keep us from having access to this group of animals that offers many benefits.

BIRDS = DIVERSITY

Contemporary representatives of Class Aves are the result of an evolutionary explosion beginning with *Archaeopteryx lithographica*, an archaic Jurassic bird (Romer 1966). An examination of the species level of taxonomic classification captures the complexity of form and function within this group of vertebrates. One early estimate by Mayr (1946) included 2600 different genera and 8616 different species discounting for subspecies and hybridization between species. A twenty-first century estimate is over 10,000 species of birds in the world. About 925 have been seen in the United States and Canada. About 1000 have been seen in Europe. By far the largest concentration of bird species is found in South America. Over 3200 species have been seen there. In Colombia, Bolivia, and Peru the species count for each country tops 1700. The species count for each continent includes:

1. South America, 3200
2. Asia, 2900
3. Africa, 2300
4. North America (from Panama north, and the Caribbean), 2000
5. Australia and surrounding islands, 1700
6. Europe, 1000
7. Antarctica, 65 (http://www.birding.com/species.asp)

The species level of classification denotes populations that are capable of interbreeding with one another and producing viable offspring with the same taxonomic characteristics. The taxonomic characteristics of birds are a staggering array of extreme variations in anatomical and physiological designs. These variations in bird anatomical designs are evident in sizes, colors, beaks, body shapes, wings, feathers, feet, syringes (sing = syrinx = voice box), and internal organs. Variations in bird anatomical designs are not only evident in structural form, but also in terms of behavioral attributes. Variations in behavioral attributes include feeding, breeding, singing, flying, swimming, migrating, nesting, flocking, escaping, running, rearing young, and defense mechanisms. It would be going beyond the intent of this essay to include examples of the variations in bird form and function listed above, but there are many fine ornithology texts and Internet websites that provide examples (e.g., smallest vs. largest bird, slowest vs. fastest, etc.). Physiological variations in avian taxonomic groups provide the adaptations required to live in virtually every habitat, for example, terrestrial vs. water (salt and fresh); climate, for example, tropical rainforest vs. desert; and geographic region of the world.

BIRD ECONOMICS

For millions of urbanites, birds are the centerpiece for an economic bonanza derived from bird watching, purchasing bird food and feeders, and hunting. A direct assessment showed that nearly $33 billion were spent in 2006 by U.S. residents to watch wildlife which, more often than not, means "bird" watching (U.S. Department of the Interior 2005). Of the total $33 billion spent, $3.4 billion were spent on bird feed, and $790 million were spent on nest boxes, bird houses, feeders, and baths. Total expenditures for migratory bird hunting were about $1.4 billion. So on the positive sides of the ledger, birds support a $34 billion industry. Additionally, one needs to consider the ecological services provided by birds to protect human health and economics. For example, there are economic benefits realized by the insect feeding behavior of night hawks and whippoorwills, rodent control by hawks and owls, birds as a human food source (meat and eggs), the pollination activity of hummingbirds, and even the danger warnings of flushing coveys of ground-nesting birds. Except for the poultry market (another billion dollar bird-related industry), there does not appear to be any direct or indirect assessments of the value of the ecological services performed by birds.

However, there are also negative economic consequences related to birds due to crop damage, disease transmission, collisions with aircraft, and environmental pollution. These costs (in the tens of millions of dollars) are primarily indirect assessments, and are discussed in subsequent chapters and in Conover (2002). Except in isolated cases, for example, the loss of human life, the positive side of the bird economics ledger is much larger than the negative.

BIRD LOSSES DUE TO URBANIZATION

Over increasingly large areas of the United States, spring now comes unheralded by the return of the birds, and the early mornings are strangely silent where once they were filled with the beauty of bird song.

—Rachel Carson (circa 1964)

The death toll on birds began many decades prior to the onset of urbanization. Two classic examples of the loss of once abundant bird populations were the passenger pigeon (*Ectopistes migratorius*) and the whooping crane (*Grus americana*). Passenger pigeons lived in enormous flocks and during migration it was possible to see flocks of them a mile wide and 300 miles long, taking several days to pass and containing up to a billion birds. Some estimate that there were as many as five billion passenger pigeons in the United States at the time Europeans arrived in North America. It did not take long after their arrival for commercial hunting, habitat loss (i.e., deforestation), and disease, to decimate the passenger pigeon to extinction. On September 1, 1914, Martha, the last known passenger pigeon, died in the Cincinnati Zoo, Cincinnati, Ohio.

The whooping crane population used to range in the millions throughout most of the eastern and central portions of North America, central Canada, and Mexico. The species range shrank rapidly in the second half of the nineteenth century. Habitat (wetlands) loss and conversion to farms due to westward expansion of settlers, unregulated hunting for sport and food, and the popularity of egg and specimen collecting at the turn of the century were the major factors leading to the species decline. In 1914 there were only sixteen left. The whooping crane was put on the endangered species list in March 1967. Recovery and restocking programs have increased the wild populations and those in captivity to around 486 surviving whooping cranes. Recovery programs are seriously compromised by the low numbers of breeding pairs of whooping cranes. Population numbers are at the point called the "threshold level," which is the lowest number possible for staging a return to population increases.

In 1962, Rachael Carson published her famous (some say infamous) book, *Silent Spring*. It was the first effort to bring together a large body of information on the effects of pesticides on birds, other wildlife, and humans. Her book finally galvanized the policy and decision makers into reactive management efforts to curb the deleterious environmental effects of DDT (dichloro-diphenyl-trichloroethane). DDT was a common, highly marketed, unregulated, long-lasting broad spectrum pesticide used indiscriminately by agriculturists, urban residents, and anyone else or any other government agency with a "bug" problem. DDT was banned in the United States in 1972 because it contributed to the near extinction of birds, including the bald eagle, brown pelican, and the peregrine falcon (see Perspective Essay 8.1). DDT is a persistent chemical that becomes concentrated in animal tissues, rising in concentration in animals that are higher in the food chain. While not immediately toxic to birds, DDT causes long-term reproductive problems by causing eggshells to weaken and crack, threatening the survival of many bird species. Because of its chemical nature, once DDT is applied in a field or other environment, it remains in an active form for decades (Environmental Defense Fund website). It is amazing how much Rachel Carson understood, even in the early 1960s, about biodiversity and ecosystems and the relationship between pesticides, the environment, and health. Rachel Carson's work still has the power to awaken a profound sense of connection between human beings and the rest of the natural world (Hawkins 1994). The irony behind the DDT story is that the Swiss chemist Paul Hermann Müller was awarded the Nobel Prize in Physiology or Medicine in 1948 "for his discovery of the high efficiency of DDT as a contact poison against several arthropods" (Nobel Foundation website).

Given all of the positive contributions of birds to human health and economics, it seems logical that conserving bird diversity would be a high priority in urban wildlife management. Unfortunately, this has not been the case, and the situation has not changed much in recent years. There are astounding bird losses due to anthropogenic causes, primarily involving collisions with human-made structures. One study estimated that from 500 million to possible over one billion birds are killed annually in the United States (Erickson, Johnson, and Young 2005). Our illustration of the bird cemetery (Figure 6.5) is a fitting illustration. Below, we summarize the lethal impacts of fifteen urban and exurban features on migratory birds. In numerical order, these features are:

1. Windows
2. Cats
3. Automobiles
4. Electric transmission line collisions
5. Agriculture
6. Urban development
7. Communication towers
8. Stock tank drowning
9. Oil and gas exploration
10. Logging and strip mining
11. Commercial fishing
12. Electrocution
13. Hunting
14. Wind towers
15. High-rise buildings

Some of these features and associated impacts are discussed in Chapter 8. Bird kills in categories 5, 6, 9, and 10 are probably associated with chemical pollution (the DDT story revisited) and habitat loss. Many bird deaths have been associated with categories 8 and 9, but there does not appear to be any field studies that provide actual numbers.

It needs to be made clear that the numbers given in Figure 6.5 are estimates, or extrapolations from direct counts as explained in the footnotes below Table 6.1. As such, there will be differences in the reported number of bird fatalities in the literature. Nevertheless, the loss of many millions of birds each year due to habitat alterations or human activities in natural and

Figure 6.5 Fifteen examples and levels of bird mortality caused by direct impacts with urban structures or associated indirect effects of urbanization. In numerical order the features are windows, cats, automobiles, electric transmission line collisions, agriculture, urban development, communication towers, stock tank drowning, gas and oil exploration, logging and strip mining, commercial fishing, electrocution, hunting, wind towers, and high-rise buildings. (Concept by Clark E. Adams, illustration by Linda Causey)

Table 6.1 Summary of Predicted Annual Avian Mortality

Mortality Source	Annual Mortality Estimate	% Composition
Buildings[a]	550 million	58.2
Power lines[b]	130 million	13.7
Cats[c]	100 million	10.6
Automobiles[d]	80 million	8.5
Pesticides[e]	67 million	7.1
Communications towers[f]	4.5 million	0.5
Wind turbines[g]	28.5 thousand	<0.01
Airplanes	25 thousand	<0.01
Oil spills, oil seeps, fishing by-catch	Not calculated	Not calculated

Source: W.P. Erickson et al. 2005. A Summary and Comparison of Bird Mortality from Anthropogenic Causes with an Emphasis on Collisions. USDA Forest Service Gen. Tech. Rep. PSW-GTR-191.

[a] Mid-range of fatality estimates reported from Klem (1990), 1 to 10 bird fatalities per house, extrapolated to 100 million residences.

[b] Based primarily on a study in the Netherlands (Koops 1987), extrapolated to 500,000 miles of bulk transmission line in the United States.

[c] One study in Wisconsin estimated 40 million (Coleman and Temple 1996); there are 60 million cats claimed as pets in the United States.

[d] Based primarily on one study in England (Hodson and Snow 1965; Banks 1979) that estimated 15.1 fatalities per mile of road each year, no searcher efficiency or bias adjustments in that study, updated based on increase in vehicle registrations.

[e] Conservative estimate using low range of empirical fatality rate (0.1 to 3.6 birds/acre), studies typically adjusted from search efficiency and scavenging.

[f] Estimates from models derived by Manville and Evans (M. Manville, personal comm.).

[g] Mid-range per turbine and per MW estimates derived from empirical data collected at several wind projects.

urban areas begs the question of how some species of birds survive as reproducing populations at all. Is it possible that natural selection mechanisms that facilitated the evolutionary explosion of one of the most diverse classes of terrestrial vertebrates on earth will again function to provide those adaptations necessary to survive the structural impediments of urbanization? No one can predict, but this is my hope!

Clark E. Adams

SPECIES PROFILE: TREE SQUIRRELS (*SCIURUS* SPP.)

Oh downy woodsman, saucey, rakish, brazen fellow!

—William Shakespeare

They have become as closely associated with cities as pigeons, and a suburban icon as well. Love them or hate them—and it's easy to find passionate members of both clubs—chances are you have an opinion. Squirrels are the bane of bird feeding devotees, who go to great lengths and great expense to control which of their wildlife neighbors will have access to the bountiful backyard buffet (usually with little success). Squirrel aficionados have their own array of serving options, including munch boxes, corn cob Ferris wheels, bungee jumps, and more, all designed to force the guests of honor to sing—make that swing—for their supper. The local wildlife boutique is a demilitarized zone of sorts, where both camps must honor a ceasefire for the sake of reprovisioning.

Tree squirrels are so familiar it feels a bit silly to include a description here, like trying to describe the sensation of breathing. We'll try to keep it brief. The usual suspects belong to one of two gangs: the grays and the foxes. Grays—well, the name pretty much says it all, but fox squirrels (Figure 6.6) also have a fair amount of gray. The difference is that while grays have silver-tipped hair and a white belly, foxes have orange- or yellow-tipped hairs with a rusty belly. This description doesn't take into account both albino (white) and melanistic (black) color phases. The eastern fox squirrel (*Sciurus niger*) is our largest tree squirrel and, on average, foxes are larger than grays. And,

Figure 6.6   Fox squirrel. (Courtesy John J. Mosesso/NBII)

of course, squirrels have a bushy, expressive tail, a curious nature, an energetic demeanor, and an ability to look incredibly cute while performing the most mundane task. The latter trait may be the ultimate secret to their success. As Carrie Bradshaw of *Sex and the City* fame points out, what is a squirrel but "a rat with a cuter outfit"? Clothes, it seems, make more than just the man.

Breeding takes place early in the year, with the first litter born sometime in late February to early March. Both grays and foxes build large leaf nests in trees; the nests are hard to spot in June but obvious in January. Other signs of nesting areas include food debris; apparently, humans aren't the only mammals who like to eat in bed. Depending on the region, squirrels may have two or even three litters annually (Whitaker 1998). Young squirrels grow up fast and are out on the street, or causing trouble at the bird feeder, by the time mom is busy with the next brood.

If only the trouble began and ended at the feeder, but there's more to human-squirrel conflict than sunflower seeds. Squirrels can damage buildings (e.g., siding, insulation) and cause safety hazards by gnawing electric wires and cables. In rural areas they compete with producers for nuts, fruit, and grain crops, and may damage trees grown for timber by stripping bark (Kenward 1983). They use power lines as travel routes and occasionally short out transformers (which ends the damage that particular squirrel can do, but the problem returns as soon as the transformer is repaired; Curtis and Sullivan 2001).

Squirrels are so prevalent that any management efforts focused on reducing populations are doomed to fail. Lethal management methods are not only ineffective long term, they are a public relations nightmare. Trapping (lethal or live-capture) has little effect on squirrel populations, although it can be useful for removing individual animals from specific locations if combined with preventative action. Repellants, including capsaicin (the active ingredient in hot peppers) has been shown to have some success in protecting seeds, plastic tubing, fittings, and the like, but it can be a time consuming to use, especially if the items are exposed to weather and require regular applications. Conflict prevention appears to be the most cost-effective approach. Educating homeowners about the importance of regular wildlife-proofing can go a long way toward heading off squirrel-caused damage to buildings. Fencing and other barriers can protect small gardens and specific trees, although it must be admitted that protecting an entire orchard or timber lot is generally unfeasible (Curtis and Sullivan 2001).

When it comes to keeping squirrels out of the bird seed we can offer two suggestions: (1) buy the best squirrel-proof feeder you can find—it's sure to be effective for as long as it takes the squirrels to outsmart the designer—and be prepared to spend more money every time the next "ultimate" squirrel-proof feeder comes to market; or (2) learn to like squirrels.

CHAPTER ACTIVITY

1. Develop a PowerPoint presentation consisting of 20 or more slides on an urban adaptable animal, which can be a fish, amphibian, reptile, or mammal, that includes some facts about:
   a. its natural history,
   b. characteristics that make the animal so well adapted to urban ecosystems,
   c. the animal's impacts on urban habitats and humans, and
   d. suggested control measures.

LITERATURE CITED

Adams, L.W. 1994. *Urban Wildlife Habitats: A Landscape Perspective*. Minneapolis: University of Minnesota Press.

Alberti, M., J.M. Marzluff, E. Shulenberger, G Bradley, C. Ryan, and C. Zumbrunnen. 2003. Integrating humans into ecology: Opportunities and challenges for studying urban ecosystems. *BioScience* 53:1169–1179.

Atwood, T.C., H.P. Weeks, and T.M. Gehring. 2004. Spatial ecology of coyotes along a suburban-to-rural gradient. *Journal of Wildlife Management* 68:1000–1009.

Banks, R.C. 1979. *Human Related Mortality of Birds in the United States*. Special Scientific Report, Wildlife No. 215. Washington, DC: Fish and Wildlife Service, U.S. Department of the Interior.

Barko, V.A., G.A. Feldhamer, M.C. Nicholson, and D.K. Davie. 2003. Urban habitat: A determinant of white-footed mouse (*Peromyscus leucopus*) abundance in southern Illinois. *Southeastern Naturalist* 2(3):369–376.

Beier, P. 1995. Dispersal of juvenile cougars in fragmented habitat. *Journal of Wildlife Management* 59:228–237.

Begon, M., J.L. Harper, and C.R. Townsend. 1996. *Ecology: Individuals, Populations and Communities*. Oxford, UK: Blackwell Science Ltd.

Blair, R.B. 1996. Land use and avian species diversity along an urban gradient. *Ecological Applications* 6:506–519.

Blair, R.B. and A.E. Launer. 1997. Butterfly diversity and human land use: Species assemblages along an urban gradient. *Biological Conservation* 80:113–125.

Brower, J.E., J.H. Zar, and C.N. von Ende. 1989. *Field and Laboratory Methods for General Ecology*. Dubuque, IA: William C. Brown Publishers.

Burton, D.L. and K.A. Doblar. 2004. Morbidity and mortality of urban wildlife in the Midwestern United States. In *Proceedings of the 4th International Symposium on Urban Wildlife Conservation,* ed. W.W. Shaw, L.K. Harris, and L. VanDruff. Tucson, AZ.

Calhoon, R.E. and C. Haspel. 1989. Urban cat populations compared by season, subhabitat, and supplemental feeding. *Journal of Animal Ecology* 58:321–328.

Carson, R. 1962. *Silent Spring*. New York: Houghton Mifflin.

Chandler, R.B., A.M. Strong, and C.C. Kaufman. 2004. Elevated lead levels in urban house sparrows: A threat to sharp-shinned hawks and merlins? *Journal of Raptor Research* 38:62–68.

Coffman, C.J., J.D. Nicols, and K.H. Pollock. 2001. Population dynamics of *Microtus pennsylvanicus* in corridor-linked patches. *Oikos* 93:3–21.

Coleman, J.S. and S.A. Temple. 1996. On the prowl. *Wisconsin Natural Resource* December.

Cone, M. 2005. Hermaphrodite frogs linked to pesticide use. *Los Angeles Times*, March 2.

Conover, M. 2002. *Resolving Human-Wildlife Conflicts: The Science of Wildlife Damage Management*. Boca Raton, FL: Lewis Publishers.

Curtis, P.D. and K.L. Sullivan. 2001. *Tree Squirrels*. Wildlife Damage Management Fact Sheet Series. Ithaca, NY: Cornell Cooperative Extension.

Crooks, K.R. 2002. Relative sensitivities of mammalian carnivores to habitat fragmentation. *Conservation Biology* 16(2):488–502.

Erickson, W.P., G.D. Johnson, and D.P. Young. 2005. *A Summary and Comparison of Bird Mortality from Anthropogenic Causes with an Emphasis on Collisions*. USDA Forest Service Gen. Tech. Rep. PSW-GTR-191.

Fedriani, J.M., T.K. Fuller, and R.M. Sauvajot. 2001. Does availability of anthropogenic food enhance densities of omnivorous mammals? An example with coyotes in southern California. *Ecography* 24:325–331.

Fernandez-Juricic, E. 2000. Avifaunal use of wooded streets in an urban landscape. *Conservation Biology* 14:513–521.

Fitzwater, W.D. 1988. Solutions to urban bird problems. In *Proceedings of the 13th Vertebrate Pest Conference,* ed. A.C. Crabb and R.E. Marsh. University of California, Davis.

Gibson, R., M.D. Smith, C.J. Spary, C.R. Tyler, and E.M. Hill. 2005. Mixtures of estrogenic contaminants in bile of fish exposed to wastewater treatment works effluents. *Environmental Science and Technology* 39:2461–2471.

Glaser, A. 2004. The ubiquitous Triclosan: A common antibacterial agent exposed. *Pesticides and You* 24:12–17.

Hawkins, T.R. 1994. Re-reading *Silent Spring*. *Environmental Health Perspectives* 102:536–538.

Hodson, N.L. and D.W. Snow. 1965. The road deaths enquiry, 1960–61. *Bird Study* 9:90–99.

Kahn, B. and B. Tansel. 2000. Mercury bioconcentration factors in American alligators (*Alligator mississippiensis*) in the Florida everglades. *Exotoxicology and Environmental Safety* 47:54–58.

Kenward, R.E. 1983. The causes of damage by red and grey squirrels. *Mammal Review* 13(2–4):159–166.

Klem, D., Jr. 1990. Collisions between birds and windows: Mortality and prevention. *Journal of Field Ornithology* 61(1):120–128.

Kloor, K. 1999. A surprising tale of life in the city. *Science* 286:663.

Koops, F.B.J. 1987. *Collision Victims of High-Tension Lines in the Netherlands and Effects of Marking*. KRMA Report 01282-MOB 86-3048.

Martinson, T.J. and D.J. Flaspohler. 2003. Winter bird feeding and localized predation on simulated bark-dwelling arthropods. *Wildlife Society Bulletin* 31:510–516.

Mayr, E. 1946. The number of species of birds. *Auk* 69:64–69.

McKinney, M.L. 2002. Urbanization, biodiversity, and conservation. *Bioscience* 52:883–890.

Ostfeld, R.S., F. Keesing, C.G. Jones, C.D. Canham, and G.M. Lovett. 1998. Integrative ecology and the dynamics of species in oak forests. *Integrative Biology* 1:178–186.

Pauly, D. and V. Christensen. 1995. Primary production required to sustain global fisheries. *Nature* 374:255–257.

Rattner, B. A., J.C. Franson, S.R. Sheffield, C.I. Goddard, N.J. Leonard, D. Stand, and P.J. Wingate. 2008. *Sources and Implications of Lead-Based Ammunition and Fishing Tackle to Natural Resources*. Wildlife Society Technical Review. Bethesda, MD: The Wildlife Society.

Rebele, F. 1994. Urban ecology and special features of urban ecosystems. *Global Ecology and Biogeography Letters* 4:173–187.

Romer, A.S. 1966. *Vertebrate Paleontology*. Chicago, IL: University of Chicago Press.

Stone, R.M. and T.B. Crandall. 1985. Recent poisoning of wild birds by diazinon and carbofuran. *Northeast Environmental Science* 4:160–164.

Ternes, T.H., A. Joss, and H. Siegrist. 2004. Scrutinizing pharmaceuticals and personal care products in waste-water treatment. *Environmental Science and Technology* October:393–399.

Tigas, L.A., D.H. Van Vuren, and R.M. Sauvajot. 2002. Behavioral responses of bobcats and coyotes to habitat fragmentation and corridors in an urban environment. *Biological Conservation* 108:299–306.

Transportation Research Board. 2002. *Interactions between Roadways and Wildlife Ecology: A Synthesis of Highway Practice*. Washington, DC: National Cooperative Highway Research Program.

U.S. Census Bureau. 2006. U.S. Department of the Interior, Fish and Wildlife Service, and U.S. Department of Commerce, *National Survey of Fishing, Hunting, and Wildlife Associated Recreation*. Washington, DC: U.S. Fish and Wildlife Service.

Williams, C.L., K. Blejwas, J.J. Johnston, and M.M. Jaeger. 2003. Temporal genetic variation in a coyote (*Canis latrans*) population experiencing high turnover. *Journal of Mammalogy* 84:177–184.

Williamson, R.D. 1983. Identification of urban habitat components which affect eastern gray squirrel abundance. *Urban Ecology* 7:345–356.

Whitaker, J.O. Jr. 1998. *National Audubon Society Field Guide to North American Mammals*. New York: Alfred. A. Knopf.

PART III

Urban Habitats and Hazards

CHAPTER 7

Urban Green Spaces

Environments can, by the judicious use of those tools employed in gardening or landscaping or farming, be built to order with assurance of attracting the desired bird.

—**Aldo Leopold,** *Game Management* **(1933)**

KEY CONCEPTS

1. There are three ecological categories of urban green spaces: remnant, successional, and managed habitat patches. One example of each is undeveloped natural areas, vacant lots, and cemeteries, respectively.
2. Urban green spaces should be classified based on ecological characteristics instead of by land use.
3. The ecological characteristics of urban green spaces select for different assemblages of wildlife species.
4. The different assemblages of wildlife species in urban green spaces present diverse wildlife management challenges that encourage or discourage their presence.

INTRODUCTION

This chapter focuses on the structural features of urbanization that are recognized by selected wildlife species as alternative habitats that provide food, water, shelter, and protection from the elements and predators. The structural features of urbanization (see Table 1 in McDonnell and Pickett 1990) can be placed into two categories. One category we have labeled as green spaces (e.g., plants are the dominant cover type) classified as either remnant, successional, or managed habitats. The second category we have labeled as gray spaces, which is the subject of Chapter 8. Examples of urban green spaces are discussed below in the context of the ecological interrelationships that result in the presence or absence of urban wildlife species.

GREEN SPACES

Developing a list of potential green spaces in the urban landscape is a challenging task. Even more challenging is constructing a biological or ecological classification of green spaces. Categorization of urban green spaces by land use (i.e., cemetery, park, golf course) is artificial and not reflective of ecological and biological features of the habitat. For example, housing subdivisions can exhibit different ecological characteristics (e.g., tree canopy height, species diversity, number of exotics, nutrient enrichment) depending on the socioeconomic status of the homeowners. It appears that exotic rodents such as house mice (*Mus musculus*) and Norway rats (*Rattus norvegicus*) are found more often in habitats close to poorer neighborhoods than in wealthier neighborhoods (Huckstep 1996, cited in Nilon and Paris 1997). Additionally, green spaces will reflect the features of the region in which they are found. For example, a golf course in west Texas will be structurally different from a golf course in western North Carolina.

Further evidence that urban green spaces should not be classified based on land use comes from urban wildlife studies. Hostetler and Knowles-Yanez (2003) found that land use was not an adequate predictor of bird distributions in the Phoenix metropolitan area. Rather, the authors hypothesized that the presence of birds in urban areas was most likely the result of their selection of a particular habitat structure such as the kinds of trees that were available. It is therefore warranted to classify urban green spaces based on ecological features rather than a human-derived construct such as land use. We have organized urban green spaces into three broad habitat types: remnant, successional, and managed habitats. This organization follows closely the urban habitat classification scheme described by Nilon and Paris (1997).

This chapter is organized around the concepts of habitat fragmentation. These concepts are reviewed in Sidebar 7.1.

REMNANT HABITAT PATCHES

Remnant habitat patches (RHPs) are sites that have not been cleared or heavily managed by humans and contain species that are typical to the geographic region. RHPs can be found in urban areas because their particular habitat features may impede development. For example, houses may be built outside of a riparian habitat because of frequent flooding. RHPs may also occur simply because development has not progressed enough to remove these habitats. Other times, humans will intentionally leave these habitats "undisturbed" (Figure 7.1). Sometimes, residential developers in the eastern United States will leave patches of forest interspersed within and between housing units.

As urbanization encroaches upon native habitats in different directions, RHPs become like islands embedded in the matrix of the urban/suburban landscape. While there is usually no direct intentional manipulation or management of the RHP, the type of land use surrounding the habitat (i.e., matrix) has real impacts on the numbers and types of wildlife found therein. The wildlife that typically inhabits an RHP is a function of the patch size, proximity and access to other natural vegetation or habitats, and the land use of the surrounding matrix (Hennings and Edge 2003; Melles, Glenn, and Martin 2003; Schiller and Horn 1997).

Scientists have long hypothesized that the species richness (number of species) found in a given area is positively correlated with the area size (Krebs 2001). Researchers have found this generalization to hold true for birds that live in habitat islands within the urban landscape (Fernandez-Juricic and Jokimaki 2001). For example, Soulé et al. (1988) found that the size of chaparral habitat patches in San Diego County, California, was positively correlated with the numbers of bird species. Crooks et al. (2001) later studied the same habitat patches and found that thirty populations of birds had gone extinct and only twelve colonizations had occurred since the original study. Extinctions were more likely to occur in small habitat patches. Additionally, colonization of the patches was positively correlated with size.

Theoretically, recolonization of RHPs will be dependent on the proximity of those patches to other natural habitats (MacArthur and Wilson 1967). These larger, natural habitats will serve as sources of wildlife that can potentially colonize the RHPs within the urban/suburban landscape (Marzluff and Ewing 2001). The farther the RHP is from the source habitat, the less likely the species will be able to disperse to the RHP. This is especially true for species that are poor dispersers, such as small mammals (Dickman and Doncaster 1989), amphibians, or soil invertebrates. Isolation

Figure 7.1 Remnant habitat patches in close proximity to housing development in Houston, Texas. (Courtesy Clark E. Adams)

of a habitat patch can occur in two fundamental ways. First, and most obviously, long distances between patches will produce an isolating effect. Therefore, scientists commonly recommend that corridors or "stepping stones" should be conserved in the urban landscape to facilitate movement between patches (Marzluff and Ewing 2001). Fernandez-Juricic (2000) found wooded streets that connect urban parks in Madrid, Spain, served as intermediate habitats for approximately one-half of the bird species usually found in the parks.

Second, the matrix surrounding the RHP can effectively isolate a patch if it is incompatible with dispersal of organisms (Jules and Shahani 2003). Imagine a gray squirrel dispersing one-half mile across a residential development between forest fragments. It likely will confront some obstacles to its dispersal, such as dogs and fifth graders with BB guns. However, another squirrel (*Sciurus* spp.) dispersing one-half mile between forest fragments across major highways and parking lots will certainly face a greater probability of death. It is important to realize that the matrix does not influence different species similarly. The matrix acts as a filter that allows the passage of some species, and prevents the dispersal of others. As a consequence, the composition of species in these isolated RHPs will be different than in patches that are connected to other natural habitats.

Crooks (2002) found that mesopredators (middle-sized predators such as domestic cats [*Felis catus*] and raccoons [*Procyon lotor*]) were unaffected by habitat fragmentation in an urban area. In contrast, other predators such as bobcats (*Lynx rufus*), badgers (*Taxidea taxus*), long-tailed weasels (*Mustela frenata*), coyotes (*Canis latrans*), and mountain lions (*Felis concolor*) were less likely to be found in small, isolated RHPs within the urban matrix. Crooks suggested that part of the reason predators such as raccoons and domestic cats are not affected by fragmentation is because they can move through the urban matrix easily. Other predators were sensitive to the urban matrix, and thus became stranded within the RHPs. These populations will probably disappear because recolonization is prevented by the urban matrix. Bolger et al. (1997) used the same habitat patches to study the effects of urban fragmentation on native rodent species. They demonstrated that small, isolated RHPs could not support viable populations of native rodents. Additionally, there was no evidence of recolonization in these patches after extinction of the local populations, leading the researchers to hypothesize that the urban matrix, even at short distances, was impenetrable by the rodents.

A study by Barko et al. (2003) was designed to assess the habitat features that determine the abundance of white-footed mice (*Peromyscus* sp.) in bottomland forest patches. Interestingly, they found the highest abundances in patches that were surrounded by urban land use in comparison to patches surrounded by upland forest habitat. While this finding could be interpreted to mean that these patches were the most suitable to white-footed mice populations, the researchers hypothesized that the urban matrix prevented emigration from the patches, thereby leading to an abnormally high population density.

Edge is defined as the area where two different habitat patches meet (R.L. Smith and Smith 2001). Volumes have been written on the effects (both negative and positive) of edge habitats on various species. It is not warranted to review here all the literature related to edge effects. However, it is important to note that edge can either reduce or enhance wildlife populations (Yahner 1988) within urban RHPs. For example, Bolger, Scott, and Rotenberry (1997) categorized breeding birds in coastal Southern California based on the species' sensitivities to the edge between natural shrub habitats and urban development. Edge/fragmentation-enhanced species including the house finch (*Carpodacus mexicana*), northern mockingbird (*Mimus polyglottos*), lesser goldfinch (*Spinus psaltria*), and Anna's hummingbird (*Calypte anna*) are readily adapted to the presence of humans and the consequential changes in habitat. The researchers found that populations of these species were actually higher in closer proximity to urban development than in the natural habitat patches. Edge/fragmentation-insensitive species including the California quail (*Callipepla californica*), California thrasher (*Toxostoma redivivum*), rufus-sided towhee (*Pepilo erythropthalmus*), wrentit (*Chamaea fasciata*), Bewick's wren (*Thyomanes bewickii*), California towhee (*Pipilo crissalis*), California gnat-catcher (*Polioptila californica*), scrub jay (*Aphelocoma coerulescens*), common bushtit (*Psaltriparus minimus*), and mourning dove (*Zenaida macroura*) showed no variation in their high abundance across the fragmented landscape. Edge/fragmentation-reduced species including the black-chinned sparrow (*Spizella atrogularies*), sage sparrow (*Amphispiza belli*), lark sparrow (*Chondestes grammacus*), rufus-crowned sparrow (*Aimophila ruficeps*), Costa's hummingbird (*Calypte costae*), and western meadowlark (*Sturnella neglecta*) showed marked variation in their abundance pattern across the landscape. Higher abundances were found in larger natural patches than in highly fragmented patches and near edges. The edge between a patch and the surrounding matrix is sometimes a combination of features of both habitats, gently changing from one into the other. Other times the edge is abrupt, as is the case in many urban habitats (Figure 7.2). Not only does the urban matrix present a dispersal barrier for many species, but also the edge area between remnant patches and the urban matrix can effectively reduce the size of the RHP for species that are sensitive to urban habitats.

In a study of neotropical migrant bird species, Friesen, Eagles, and Mackey (1995) found, as in other studies previously mentioned in this chapter, that size of the habitat fragment correlated with species richness and population abundances. However, they also found that diversity and abundance decreased with increasing development in the matrix regardless of the patch size. In fact, a small (4 hectare) woodlot with no surrounding development could support a richer bird community than a larger (25 hectare) woodlot surrounded by high housing density. Bock et al. (2002) found that native rodents in Boulder, Colorado, avoided the edges between grasslands and suburbs. Mechanisms that result in the sensitivity of some species to edge are poorly understood, although hypotheses suggest that edge habitats attract more avian nest predators such as blue jays (*Cyanocitta cristata*) and crows (*Corvus* sp.), and more mammalian mesopredators like raccoons, domestic cats, and striped skunks (*Mephitis mephitis*) which can reduce prey abundance and diversity. Additionally, nest parasites such as brown-headed cowbirds (*Molothrus ater*) *can* more easily access nests close to the edge (Paton 1994). Negative human disturbances are also more likely to occur at the edges of RHPs. Zipperer (2002) showed that remnant forest patches in urban areas were subject to several disturbances including dumping of trash and yard wastes, tree removal, and erosion.

Figure 7.2   Edge habitat in suburban areas can result from juxtaposition of remnant (woods), successional (tall and short grass prairie), and managed (lawn) habitat patches. (Courtesy Clark E. Adams)

SUCCESSIONAL HABITAT PATCHES

Successional habitat patches (SHPs) are sites that were previously cleared or managed by humans but have subsequently been abandoned. A vacant lot and abandoned shopping mall (Figure 3.7) are examples of successional habitats. We use the term successional because these habitats are no longer under direct management by humans, and are subject to the natural process of reverting back to habitat that existed before human disturbance. In fact, some define succession as the changes in the community structure over time following a disturbance (Ricklefs and Miller 1999). We defined succession in Chapter 3 as "the orderly and progressive replacement of one community type by another until a climax state is reached." By community structure, we mean both the biological and physical structure. The biological structure of a community is made up of the different types of plant and animals and the population sizes of those organisms.

An example of physical structure would be the vertical stratification of a forest (R.L. Smith and Smith 2001). If you have walked through different types of forests you have probably observed that some forests have several vertical layers, such as the herbaceous layer (wildflowers and grasses), the shrub layer (small shrubs and very young trees), the understory tree layer (small trees), and the canopy layer (the largest trees). Other forests, such as an urban wooded park that is maintained by mowing or a southern pine forest that burns periodically, however, will have only a canopy layer and the herbaceous layer.

The disturbance in an urban or suburban habitat that initiates succession is typically removal or clearing of all or part of the biological community. We represent a disturbance every week when we mow our lawn. Because of this weekly disturbance to our lawns by mowing, succession cannot proceed. However, if we stopped mowing our lawns, succession would bring about significant changes in the species of plants that live on our property. For example, the lawns of residents living in the eastern United States would eventually revert back to a forest habitat. A previously disturbed site near Washington, D.C. developed into an oak-beech forest in approximately seventy years (Ricklefs and Miller 1999).

A successional habitat patch can be thought of as a prime real estate spot. Different species are vying for access to the patch and competing with other species once they arrive. The first species to arrive at an SHP are those that have excellent dispersal abilities. For example, many grasses and other small herbaceous plants have light seeds that are carried great distances on wind currents. Have you ever blown the seeds of a dandelion on a warm summer day? Dandelions (*Taraxacum* spp.) are examples of fantastic dispersers. These same species are usually adapted to disturbed conditions and initially flourish following a disturbance event. (Mowing actually *encourages* the growth of dandelions.) Because these species have invested their limited energy into dispersal and rapid reproduction, they are usually poor competitors with other species. Consequently, over time, other species will replace them within the habitat patch. Additionally, some early colonizers can significantly change the abiotic or nonliving conditions of the habitat. This change can either make it easier or more difficult for other species to invade. For example, black locust trees (*Robinia pseudoacacia*) in the eastern United States enrich the soil with nitrogen. As a result of their presence, other tree species that otherwise would not have been able to cope with the nitrogen-poor soil are able to colonize and survive. Other species, however, have the opposite effect. Many exotic plants, such as kudzu (*Pueraria lobata*) or Japanese honeysuckle (*Lonicera japonica*), will change the physical structure (e.g., reduction of light) of a community in such a way that other species cannot invade or survive.

The rate at which succession occurs and the composition of the SHP will be a function of number and proximity of source species available for colonization (Dzwonko and Loster 1997). SHPs in urban areas in the eastern United States sometimes undergo succession at a relatively slow pace. One hypothesis suggests seed dispersal of woody species in these patches is limited because of low avian activity. Robinson and Handel (1993) tested this hypothesis by planting woody species to attract avian seed dispersers in an abandoned urban landfill. They predicted that by providing perching habitat for the birds, a positive feedback would result that would lead to more woody species recruitment. The plantings did attract avian seed dispersers that introduced twenty new plant species to the area. Another factor that has been found to influence forest habitat succession is the quantity of leaf litter on the forest floor. Kostel-Hughes, Young, and Carreiro (1998) demonstrated that forest floor leaf litter was higher in density, depth, and mass with increased distance from New York City. Small-seeded species were more likely than larger-seeded species to germinate in shallow leaf litter, leading the authors to conclude that physical changes to the forest floor from urbanization could change the trajectory of succession in urban forests. Additionally, small-scale disturbances by humans were identified as important factors in successional forest habitats in urban areas in Syracuse, New York (Zipperer 2002).

It should be obvious that, if the plant community of a given area changes over time, the animal community will change as well. Unfortunately little is known about how animal communities change over time in urban SHPs. From rural studies, we know that different wildlife species are adapted to the different stages of succession. For example, typical bird species of early successional habitats in the eastern deciduous forests include the grasshopper sparrow (*Ammodramus savannarum*). As habitats become shrubbier, species such as yellow-breasted chats (*Icteria virens;* C.F. Thompson and Nolan 1973), cardinals (*Cardinalis cardinalis*), and indigo buntings (*Passerina cyanea*) become more abundant. Typical birds of mid-successional forests include summer tanagers (*Piranga rubra*) and wood thrushes (*Hylocichla mustelina*). Black-throated green warblers (*Dendroica virens*) and ovenbirds (*Seiurus aurocapillus*) prefer more mature forests (Johnston and Odum 1956; Kricher 1973; Holmes and Sherry 2001).

MANAGED HABITAT PATCHES

Managed habitat patches (MHPs) are sites under direct and intense management by humans. Examples of urban or suburban MHPs are cemeteries (see Perspective Essay 7.1), parks, golf courses (Figure 7.3), nature centers, and residential yards. The physical and biological structure of MHPs is a result of several synergistic factors, including culture, history, economics, ecology, and

Figure 7.3 *(A color version of this figure follows page 158.)* Golf courses have the potential to provide habitat needs for several species of wildlife, including these deer. (Courtesy USDA, Ken Hammond)

regional flora and fauna. Additionally, as with the remnant and successional habitat patches, MHPs are also affected by their sizes, proximity to source habitats, and surrounding matrix. MHPs can mimic a broad range of "natural" habitat types such as deciduous forests, desert scrub, or grasslands. Additionally, many incorporate a diversity of habitats that may include both terrestrial and aquatic communities. A growing trend among homeowners is the construction of ponds or water gardens (see Chapter 5) in their yards, and every golf course will contain one or several small ponds or lakes.

Probably the most common habitat feature of many MHPs is trees, both deciduous and coniferous. Homeowners often plant trees for aesthetic or functional purposes—more often than not, without any consideration of whether the trees are native or exotics.

Cemeteries

In the Midwest, cemeteries sometimes contain fragments of remnant grassland communities (Stowe, Schmidt, and Green 2001). This results, in part, from the fact that many urban cemeteries were originally established on the outskirts of towns and were subsequently incorporated into the cities as they grew (Gilbert 1989). Most urban cemeteries possess "a structural framework of trees planted when they were first laid out" (Gilbert 1989), but the types of trees planted varies widely according to geographic region and the time period in which the cemetery was founded. Those established around the turn of the nineteenth century or a few decades earlier were often meant to conform to Romantic or Victorian aesthetic standards and often functioned secondarily as arboretums (Gilbert 1989). Consequently, many of these cemeteries will contain a mixture of exotic, ornamental, and popular native tree species. For example, Woodlawn Cemetery in Toledo, Ohio, founded in 1876, contains the exotics European larch (*Larix decidua*), European beech (*Fagus sylvatica*), and star magnolia (*Magnolia stellata*), as well as the natives red maple (*Acer rubrum*), tulip poplar (*Liriodendron tulipifera*), Ohio buckeye (*Aesculus glabra*), and the now-rare American elm (*Ulmus americana*). Various showy trees and shrubs are also very common, including magnolias (*Magnolia* sp.), Japanese maples (*Acer palmatum*), dogwoods (*Cornus* sp.), redbuds (*Cercis* sp.), crepe myrtles (*Lagerstroemia* spp.), hydrangeas (*Hydrangea* spp.), fruit trees, and willows (*Salix*

Figure 7.4   Many urban cemeteries are home to large, old trees, which can serve as important nesting and denning sites for wildlife. (Courtesy John M. Davis)

sp.) (Woodlawn Cemetery 2001–2003). Coniferous trees are also common features of MHPs. For homeowners, coniferous tree species provide shelter and aesthetics year round. Common coniferous species planted in urban MHPs include but are not limited to Douglas fir (*Pseudotsuga menziesii*), Norway spruce (*Picea abies*), blue spruce (*Picea pungens*), white pine (*Pinus strobus*), Scotch pine (*Pinus sylvestris*), and various cedars and junipers.

Trees planted in urban MHPs often reach very large sizes because they do not have to compete as much for resources as in a natural forest habitat. American Forests, a nonprofit citizen's conservation organization, maintains a registry of the largest trees in each state. Several state tree records occur in urban MHPs such as cemeteries (Figure 7.4) and residential yards (American Forests 2004). Large trees provide important habitat for several species of wildlife. For example, sharp-shinned hawks (*Accipiter striatus*) in and around Montreal used mature coniferous trees for nesting (Coleman, Bird, and Jacobs 2002). Another study (Mager and Nelson 2001) found red bats (*Lasiurus borealis*). These roosting sites are similar to sites selected by bats in natural habitats. The researchers noted that the urban habitats such as parks were important for bats because they provided large trees interspersed with open habitats for foraging. In fact, the researchers had the most success in trapping bats close to the edges of forests and fields and to streetlights. They hypothesized that the bats could find a high concentrated diversity of insect prey in these areas. Tree-trimming practices and natural senescence often result in the formation of cavities, which many types of birds and mammals will use for nesting, roosting, and shelter. Several species of birds use tree cavities for nesting, including bluebirds (*Sialia* spp.), woodpeckers (Picidae), owls (Strigiformes), ducks (Anseriformes), flycatchers (Tyrannidae), and chickadees (*Poecile* spp.).

Other studies show that the MHPs provide relatively easy access to resources needed for some species of wildlife. For example, studies (Boal and Mannan 1998; Mannan and Boal 2000) of Cooper's hawks (*Accipiter cooperii*) in Tucson, Arizona, showed that the hawks nested in groves of large exotic trees within residential or recreational areas. The authors hypothesized that the hawks selected urban areas for nesting because of the availability of water and prey (doves). As a result of the high density of doves in these areas, the hawks did not need to forage far from their nests, as they would normally do in rural habitats.

Studies have shown that urban cemeteries hold rich avian diversity (see Perspective Essay 7.1). Thomas and Dixon (1973) discovered ninety-five different species of birds frequenting Boston cemeteries. Thirty-four of the species used cemetery habitats for breeding and nesting. In addition to the blue jays (*Cyanocitta cristata*), robins (*Turdus migratorius*), and starlings (*Sturnus vulgaris*) commonly found in urban habitats, they also discovered a number of birds not typically found within densely populated cities, including sharp-shinned hawks, yellow-shafted flickers (*Colaptes auratus*), belted kingfishers (*Ceryle alcyon*), bobwhites (*Colinus virginianus*), ring-necked pheasants (*Phasianus colchicus*), and black-billed cuckoos (*Coccyzus erythropthalmus*). This last species seems especially surprising, given that it is assumed to be an inhabitant of dense, undisturbed forests (Alsop 2001). Lussenhop (1977) conducted a study in central Chicago and found that, instead of urban birds spilling over into the more diverse cemetery habitat, rural birds from the surrounding nonurbanized areas seemed to be making the most use of the cemetery. Species such as rock doves (*Columba livia*), house sparrows (*Passer domesticus*), and starlings were not found in higher densities than surrounding urban habitats. However, higher numbers of native, typically rural, birds such as red-headed woodpeckers (*Melanerpes erythrocephalus*), eastern wood-pewees (*Contopus virens*), red-winged blackbirds (*Agelaius phoeniceus*), and indigo buntings were found living and nesting within the cemetery than would have been expected in an urban area generally.

Harrison (1981) found that cemeteries in Macon, Georgia, were biologically diverse. Abundance of leaf litter invertebrates from the cemeteries was comparable to the samples taken from undisturbed forests in the region. Additionally the cemeteries were as diverse in avian species as a comparison study site, the Ocmulgee National Monument. However, it should be noted that the composition of the bird communities between the cemeteries and the Monument were different. Twenty species were unique to the National Monument and fourteen species were unique to the cemeteries.

Golf Courses

New golf courses are being built every year because of the public demand to participate in the game—a horrendously difficult, frustrating, and expensive recreational choice (C.E. Adams, personal observation). However, 24.5 million people spend 2.4 billion hours and $18 billion annually flaying away at a little ball that stands perfectly still until you hit it. At least 70 percent of the 4 million acres dedicated to golf courses in the United States is the "rough," or out-of-play areas that can be used to lose golf balls, but more importantly, to invite wildlife (Terman 1997). There is a growing body of evidence showing that the habitat diversity found in the rough does contain a rich faunal diversity not found in the surrounding urban matrix (Gange, Lindsay, and Shofield 2003). As stated previously some MHPs will contain remnant habitat patches and consequently have been identified as biological repositories within the urban landscape (Barrett and Barrett 2001). Golf courses in Kent, England, contain uncommon habitat types such as heath land, dune and slack communities, and rare populations of orchids, broomrape, and mosses (B.H. Green and Marshall 1987). The threatened Big Cypress fox squirrel (*Sciurus niger avicennia*) can be found on golf courses in Florida, and has provided researchers with valuable information regarding this normally secretive species (Jodice and Humphrey 1992).

The *Wildlife Society Bulletin* (2005, vol. 33, no. 2) contained a suite of nine research papers on the role of golf courses in bird conservation. A brief overview of each of these studies follows, but our readers are encouraged to give each study a more thorough examination than we provide here.

One study reviewed the use of created wetlands in golf course landscapes by waterbirds (White and Main 2005). They examined 183 ponds on twelve courses in southwest Florida for two years. During the course of the study they identified forty-two bird species among 10,474 individuals that visited the created wetlands. Another study by Jones et al. (2005) examined twenty-four courses in South Carolina during the summers of 2001 and 2002. They identified eighty-two bird species among 5362 bird sightings at the twenty-four courses. Bird species richness on the golf course was found to be positively influenced by a higher percentage of forested wetland, pine, and mixed-forest landscape patch types.

The study by Merola-Zwartjes and DeLong (2005) examined whether golf courses could serve as surrogate riparian habitats for southwestern birds. Their two-year study compared the avian communities on five golf courses and five natural areas in Arizona. Of the sixty-five species they identified on one or all five golf courses, forty-eight (74 percent) were often associated with riparian areas or wetlands. When compared to natural areas, golf courses supported a greater number of birds, greater avian species richness, and a higher diversity. They suggested that the value of golf courses in the desert regions would be improved by increasing landscape complexity and vertical structure, and the use of native plants.

Studies on the role of golf courses in bird conservation for specific species, for example, red-headed woodpecker (*Melanerpes erythrocephalus*), burrowing owl (*Athene cunicularia*), and two on the eastern bluebird (*Sialia sialis*) were conducted by Rodewald, Santiago, and Rodewald (2005), Smith, Conway, and Ellis (2005), Stanback and Seifert (2005), and LeClerc et al. (2005), respectively. All of these studies found golf courses could become alternative habitats for species extirpated from their natural habitats by providing nesting habitat, including artificial burrows, tree snags, and nest boxes. Golf courses provided as good as or better nesting and chick-rearing habitats as off-course natural habitats.

The role of golf courses in providing habitat for birds of conservation concern was the research objective of LeClerc and Cristol (2005). They identified sixty-nine species of conservation concern on eighty-seven golf courses in Virginia during the 2002 breeding season. They suggested a bird conservation management strategy that included high proportions of forest cover (the climax vegetation of the region) on or within 1.5 km of the course. The relationship between human-altered landscapes around six golf courses in Ohio and bird assemblages was studied by Porter, Bullock, and Blair (2005). They found that golf course size, provision of natural habitat or cover in and around the golf course (a buffer), and reduced human land-use buffers within 500 m and 1 km of the golf course were determinants of bird diversity on the golf course.

All of the above studies provided evidence of the value of golf courses in bird conservation. However, public perceptions of golf courses including many within the wildlife management profession are that they are bad for the environment. This is the predominant attitude of those who do not play golf. Nonplayers feel that golf course development destroys natural habitat, depletes water supplies, and contaminates soil and water with pesticides and fertilizers (Gange, Lindsay, and Shofield 2003). The golf industry has responded to these concerns through programs that promote green management practices that reduce or eliminate the use of exotic plants and pesticides. Two examples of green management programs are the United States Golf Associations Wildlife Links grants and Audubon International's Cooperative Sanctuary Program. Believe it or not, many golf courses have a rich diversity of vegetation types that invite an equally diverse fauna assemblage. There appears to be a movement toward naturalistic golf design wherein the management plan incorporates sustainable landscape plans developed in concert with the ecological structure of the region and enhances biodiversity. The Terra Verde golf course in Dallas, Texas, is an example of this approach, incorporating the restoration of natural vegetation into its management plan. Golf Links in Maryland turned adjacent old farmland into a complex of wetlands, offering habitat to many species of wildlife. The Phoenix Golf Links was the first course in Ohio to be developed on an abandoned landfill (Santiago and Rodewald 2004).

The above studies presented a fairly positive overview of the value of golf courses in wildlife conservation. Nevertheless, avian, conservation, and wildlife biologists want more long-term studies on:

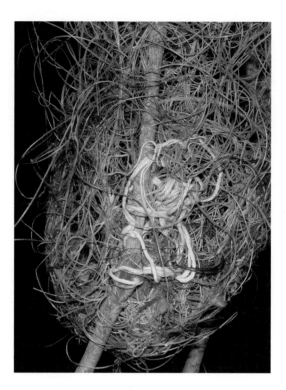

Figure 1.4 Urban birds use a variety of human artifacts to build their nests. Pictured here is an oriole nest made entirely out of nylon fishing line and some yarn. (Courtesy of the Texas Cooperative Wildlife Collection)

Figure 1.10 Baby raccoons (kits) in a tree den located right outside the bedroom window of an urban resident. (Courtesy Wes Orear)

Figure 5.2 Example of eutrophication in an inland freshwater pond. Note how the surface of the water is covered with a macrophyte, in this case a plant called giant salvina (*Salvinia molesta*). (Courtesy Michael Masser)

Figure 5.3 Material removed during the first step in preliminary wastewater treatment at the bar screen. (Courtesy Clark E. Adams)

Figure 5.7 Cattail Marsh, a reconstructed wetland built at the wastewater treatment facility in Beaumont, Texas. (Courtesy Clark E. Adams)

Figure 5.10 An example of stream channelization. (Courtesy John M. Davis)

Figure 5.12 An urban water garden. (Courtesy Michael Masser)

Figure 5.13 A storm-water catchment area consisting of about 10 acres of surface water located between a municipal golf course and urban neighborhood. Note the white signpost in the distance. (Courtesy Clark E. Adams)

Figure 7.3 Golf courses have the potential to provide habitat needs for several species of wildlife, including these deer. (Courtesy USDA, Ken Hammond)

Figure 8.1 Some experts estimate up to 100 million birds die each year in collisions caused by the artificial lights of tall buildings, communication towers, and airports. This photo shows a sample of birds collected beneath high-rise buildings in Toronto during one migration season. (Courtesy Mark Jackson and Fatal Light Awareness Program)

Figure 8.6   Cliff swallows (*N* = 102) nesting under a highway bridge. (Courtesy Clark E. Adams)

Figure 9.3   A Texas rattlesnake roundup holding pen for captured rattlesnakes. (Courtesy Clark E. Adams)

Figure 9.6 Roosting birds—and their droppings—can become an annoying problem. A one night accumulation of bird guano leads to a difficult, daily, and costly clean-up process. (Courtesy of Clark E. Adams)

Figure 9.7 Coyote. (Courtesy Sara Ash)

Figure 11.2 Some of the items confiscated under the Lacey Act. (Courtesy Steve Hillebrand/USFWS)

Figure 12.9 Damage by feral hogs to homeowner's lawn. (Courtesy Linda Tschirhardt)

1. The effects of pesticides and other forms of chemical management on survival, condition, and reproduction on golf course wildlife;
2. Reproductive rates and survivorship on and off golf courses to reveal whether golf courses are population sources or sinks;
3. Differences in the number of predators and nest predation rates on and off golf courses;
4. Wildlife uses of golf course habitats during nonbreeding seasons (are they real homes or temporary resting and nesting sites?);
5. Whether golf courses can serve as habitat-surrogates providing the specific ecological features required by some declining wildlife specialist species;
6. The value of golf courses in terms of contributions to community economies, land reclamation, and as wildlife habitat;
7. Methods to evaluate management outcomes (Cristol and Rodewald 2005; Burdge and Cristol 2008).

This wish list of research needs is a logical next step in the determination of the value of golf courses in wildlife conservation. But try to imagine the costs involved in equipment, time, and personnel in an era of tight research budgets, bottom-line agendas, and an ever-growing population of golfers in the United States. Even right-minded wildlife biologists need a reality check regarding the cultural, political, and economic ramifications of research agendas. Furthermore, any future research endeavors will need to be sensitive to the size of the landscape that can provide results applicable to at least a biotic region level of analysis.

Nature Centers

Nature centers were a concept that gained momentum in the early 1960s, referred to as the "Golden Age of Nature Centers" in the Fort Worth Nature Center's Master Plan. In 1961, the National Audubon Society merged with a group named Nature Centers for Young America, Inc. They then formed the Nature Center Division (NCD) of the Audubon Society. Their purpose was to encourage American communities to set aside natural land and use it to teach conservation and natural history while allowing people to develop an understanding and appreciation of nature. In post-World War II America, families were leaving rural settings and moving into more urban and suburban areas. Within just one generation, people were losing touch with nature. The philosophy of the NCD was that people needed to know and understand something in order to appreciate and value it. Their most simple definition of a nature center was a parcel of natural land where people, particularly young people, and nature can meet (Shomon 1962). The NCD acted as an educational service offering advice and guidance to communities that wanted to develop nature centers. The official definition of a nature center was:

A nature center can be defined as an area of undeveloped land near or within a city or town and having on it the facilities and services designed to conduct community outdoor programs in natural sciences, nature study and appreciation and conservation. It is, in essence, an outdoor focal point where the citizens of a community, both young and old, can enjoy a segment of the natural world and learn something about the interrelationship of living and non-living things, including man's place in the ecological community. (Shomon 1962)

The goal of a nature center, as defined by the NCD, was to provide educational, scientific, cultural, and recreational experiences for the community. A nature center had three basic elements, including:

1. Land—undeveloped with as much local plant and animal life as possible.
2. Buildings—educational buildings where people can assemble and exhibits can be used to teach about the area.
3. People—staff and visitors.

Shomon (1962) considered it almost a patriotic goal for cities to create nature centers: "A wise investment in America's future. It is one of the most worthy and noble and unselfish projects that any group can undertake and pursue in and around an expanding city." The NCD suggested that the setting of the nature center should be a representative sample of the natural landscape of a community, and also provide "breathing" space for a city (Shomon, 1962). Byron L. Ashbaugh (1963), associate director of the Nature Centers Divisions, in the Audubon Society's second education bulletin, stated that "The basic purpose of a nature center is to provide a green island for every community where there is open space still available." The NCD recognized early on that one of the growing problems associated with rapid urbanization was going to be loss of natural green spaces that would enable citizens to stay connected to nature.

The NCD of the National Audubon Society was discontinued sometime in the 1970s.

Nature centers are an example of an MHP and RHP in the sense that most are connected to habitats under intense conservation programs (managed) of native flora and fauna (remnant). A close examination of the prevalence of nature centers in the United States revealed an astonishing array of examples (Table 7.1, which is a summary of information provided at the website http://www.nationmaster.com/encyclopedia/List-of-nature-centers-in-the-United-States.)

The authors' knowledge of the total numbers and locations of nature centers in Nebraska and Missouri gave credibility to this website as an accurate source of information in a national overview

Table 7.1   Census of Nature Centers in the United States in 2008

State	Nature Centers	State	Nature Centers
Alabama	5	Montana	3
Alaska	2	Nebraska	8
Arizona	6	Nevada	3
Arkansas	5	New Hampshire	10
California	56	New Jersey	27
Colorado	17	New Mexico	4
Connecticut	28	New York	84
Delaware	6	North Carolina	28
District of Columbia	1	North Dakota	1
Florida	54	Ohio	72
Georgia	18	Oklahoma	17
Hawaii	2	Oregon	10
Idaho	1	Pennsylvania	30
Illinois	51	Rhode Island	5
Indiana	33	South Carolina	7
Iowa	49	South Dakota	8
Kansas	11	Tennessee	13
Kentucky	12	Texas	46
Louisiana	4	Utah	7
Maine	12	Vermont	10
Maryland	27	Virginia	23
Massachusetts	25	Washington	6
Michigan	39	West Virginia	3
Minnesota	27	Wisconsin	51
Mississippi	4	Wyoming	1
Missouri	18	TOTAL	991

Source: http://www.nationmaster.com/encyclopedia/List-of-nature-centers-in-the-United-States.

of the numbers and locations of nature centers in the fifty states. There are at least 991 nature centers in the fifty states. Of note are the differences in the numbers of nature centers in each of the states. The differences beg the question of why there are eighty-four and seventy-two nature centers in New York and Ohio, respectively, and so few in other states, for example, one each in Idaho, North Dakota, and Wyoming. One explanation was that some states have made environmental education a significant component in their state mandated science curriculum, which required a proliferation of nature centers and staff (Rob Denkhaus, Fort Worth Nature Center, personal communication). Another explanation of the difference is that those states with few nature centers already have vast expanses of open space, thus precluding the need to set aside natural areas for public use and education (Sara Ash, personal communication).

Perhaps an equally important consideration is the value of nature centers in urban wildlife management and the conservation of urban and exurban green spaces. From the perspective of urban wildlife management, other information provided on the above website made it possible to determine the total land area associated with the 991 nature centers, dominant habitat types, governance, and involvement in public education and/or conservation programs. This extensive analysis revealed that these 991 nature centers were associated (owned, leased, or shared) with 12.6 million acres of land area represented by every type of native habitat characteristic of the geographical location of the nature center. Nearly all, 94 percent, provided education programming, formal and informal, indoor and outdoor, and for youth and adults. Conservation of habitat and wildlife rehabilitation were activities included on 60 percent of the nature center websites. Governance of the 991 nature centers included city, county, state, and federal agencies; private nonprofits (e.g., Audubon, Nature Conservancy); park districts; colleges and universities; Lower Colorado River Authority; and branches of the U.S. military. Natural resource disciplines represented by nature center personnel include the Society for Ecological Restoration, The Wildlife Society (TWS), Natural Area Associations, Society for Range Management, Native Prairies Association, and all of the taxonomic organizations (e.g., mammalogists, ornithologists, herpetologists, and ichthyologists).

RURAL VERSUS URBAN WILDLIFE POPULATIONS

Despite growing interest in urban MHPs as habitats for wildlife, little is known about the differences in life history characteristics between rural and urban populations. D.C. Thompson (1977, 1978) made several observations about a gray squirrel (*Sciurus carolinensis*) population living in an urban cemetery. First, urban squirrels reached sexual maturity at a significantly older age than did rural squirrels (Longley 1963). The mechanism resulting in this disparity is unclear but may be a result of sampling differences. Breeding season and number of young born per litter were equal between urban and rural squirrels; however, survival of the young was higher among urban squirrels. Dykstra et al. (2000) found equal reproductive rates between suburban and rural red-shouldered hawks (*Buteo lineatus*). However, Boal and Mannan (1999) found that although urban Cooper's hawks (*Accipiter cooperii*) exhibited larger clutch sizes, the nestling mortality rates were higher (50 percent) compared to their rural counterparts (5 percent). The primary cause of nestling death in urban areas was the parasitic disease called trichomoniasis.

Other studies have demonstrated that behavioral differences exist between rural and urban populations of species. For example, great tits (*Parus major*) sing at higher frequencies in noisy urban areas than do individuals in quieter areas (Slabbekoorn and Peet 2003). Estes and Mannan (2003) found that male Cooper's hawks in urban areas delivered significantly more prey biomass to the nest than did rural males. Additionally, the urban males were more likely to deliver prey directly to the nest, and urban females rejected deliveries more often than did rural females. Finally, both rural males and females vocalized more often than did urban individuals. The authors attributed these contrasting nesting behaviors to the differences in prey abundance between rural and urban locations. Moreover, tolerance levels of some birds, as indicated by their flushing distances, increases in areas with high human visitation (Fernandez-Juricic and Jokimaki 2001).

While it appears that urban MHPs offer a wealth of resources for several species of wildlife, we believe that a word of caution is warranted. Urban habitats, whether remnant fragments of natural communities or highly managed patches, cannot support the same level of biodiversity found in native biological communities. Additionally, they will support a vastly different *composition* of species, with rarer, more specialized species being absent. Formation of urban habitats selects for those species that are adapted to the specific biological and physical structures of urban environments and to a constant stream of anthropogenic disturbances. The high rate of visitors to urban parks in Madrid, Spain, reduced the avian richness of those sites (Fernandez-Juricic and Jokimaki 2001).

The physical and biological structures of MHPs that have trees as the dominant cover are radically different from natural forests. First, vertical layers are removed, which results in the absence of escape cover for small vertebrates or nesting habitats for shrub- and ground-nesting birds (Livingston, Shaw, and Harris 2003; Blair 1996; Marzluff and Ewing 2001; Zalweski 1994). In fact, the existence of open lawns in MHPs dramatically reduces the species diversity of those habitats. Additionally, maintenance of lawns is ecological homicide because it requires the use of pesticides, fertilizers, and the burning of fossil fuels, all contributing factors to losses in the diversity and function of ecosystems.

Exotic species (including domestic pets) can have dramatic effects on the species composition of MHPs. Many MHPs are dominated by exotic plantings. While these exotic plants provide additional physical structure, they are usually not preferred by native wildlife species. Many researchers (Germaine et al. 1998; D.M. Green and Baker 2003) have emphasized the use of native plants in restoring avian communities within urban areas. Domestic pets, primarily domestic cats, are significant factors determining the distribution and abundance of some wildlife species within the urban landscape (Lepczyk, Mertig, and Liu 2003; Soulé et al. 1988). Baker et al. (2003) found that wood mice abundance in residential gardens in Bristol, England, was negatively correlated with the presence of cats, and Churcher and Lawton (1987) found that cats played a major role in the population dynamics of urban sparrows.

SIDEBAR 7.1   A Primer on Habitat Fragmentation

One of the most noticeable effects of urbanization is the degree to which natural habitats are sliced, diced, and divided, a phenomenon called "habitat fragmentation" (Figure 7.5). The results of habitat fragmentation are habitat patches of various shapes and sizes that have different origins as defined and discussed in this chapter, that is, remnant, successional, and managed. Habitat fragmentation needs to be considered in terms of number and size of resultant habitat patches, the edge effect, corridors that connect habitat patches, and types of matrix that surround habitat patches. All of these considerations present a different set of ecological conditions and structure that predetermine species survival or the lack thereof.

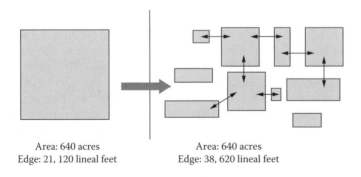

Area: 640 acres
Edge: 21, 120 lineal feet

Area: 640 acres
Edge: 38, 620 lineal feet

Figure 7.5   A diagrammatic representation of habitat fragmentation illustrating variations of patch sizes, connecting corridors, increase in edge, and surrounding matrix. (Diagram by Clark E. Adams)

Patch size makes a difference! There seems to be a direct relationship between patch size and species diversity within the patch, that is, as patch size decreases so does species diversity and vice versa. Patch size is also a determinant of overall species survivability in terms of providing all of the resources necessary for life and reproduction. For example, small patches compromise the availability of adequate food, water, shelter, and spatial requirements required for survival.

Another significant result from habitat fragmentation is the multiplication of lineal feet of edge, which introduces an "edge effect." By fragmenting the square mile of land into the habitat patches as illustrated in Figure 7.5, the 21,120 lineal feet of edge increases to 38,620 lineal feet. The ecosystem structural change of increasing edge is a natural by-product of cutting the habitat into smaller pieces, and challenges urban wildlife with another form of natural selection. Several examples of species sensitivity (positive and negative) to edge are discussed in this chapter.

The arrows in Figure 7.5 represent corridors between habitat patches, which could consist of rights-of-way, streets, tree-lined streets, power lines, stream beds, residential lawns, fence lines, or any other natural or man-made structure that allows passage through, within, and over the patches illustrated here. Note the two small patches that have no corridors. The species occupying these two patches are pretty much doomed, particularly if they are poor dispersers in the first place. Again, species survival is a direct result of the ability to negotiate these corridors to conduct the activities associated with living.

The space between the habitat patches is another important consideration for species survival in terms of distance between patches and what is actually in the space—also called the "surrounding matrix." The matrix could consist of any type of built (see Chapter 8) or natural habitats, for example, urban neighborhoods or park lands. This chapter discusses matrix structure in terms of its influence on species recolonization within habitat patches.

PERSPECTIVE ESSAY 7.1   Birds in Texas Cemeteries

I was interested in learning more about the types of birds that frequented cemeteries in this state, so I solicited the TexBirds mailing list for information on bird sightings in Texas cemeteries, as well as for information on the cemeteries themselves. From the birders who responded to this request for information I received a large amount of information concerning the bird species within Texas that frequently utilize urban cemeteries either year round or as a stop on their migration route during the spring and fall.

Much of this data was collected in two cemeteries in Dallas and Houston, Texas. The 33-acre Greenwood Cemetery in Dallas (Figure 7.4) was founded in 1850 and contains a large number of mature trees, mostly white pine, with a scattering of juniper and small amount of crepe myrtle serving as the understory. This cemetery was birded frequently from 1977 to 1979 by Homer Klonis of Dallas, primarily during the spring and fall migratory periods. Klonis observed sixty-seven different bird species (Table 7.2).

Most sightings were common urban birds (e.g., species 1 and 2, Table 7.1). A number of typically rural birds were also observed, including little blue herons (*Egretta caerulea*), great egret (*Ardea alba*), sharp-shinned hawk, common nighthawks (*Chordeiles minor*), summer tanager (*Piranga rubra*), yellow-billed cuckoo (*Coccyzus americanus*), and loggerhead shrike (*Lanius ludovicianus*). Many neotropical migrants were also spotted, most notably a wide variety of different warblers and sparrows that were passing through.

The second cemetery for which a significant amount of data was collected was Forest Park Cemetery in west Houston, a 115-acre area surrounded by a mixture of apartments and homes on the west side and office buildings on the north. The vegetation within this cemetery consists of sparsely planted trees with a few flowering shrubs. Although live oak (*Quercus virginiana*) is the dominant tree species, water oak (*Quercus nigra*), hackberry (*Celtis occidentalis*), redbud (*Cercis canadensis*), eastern red cedar (*Juniperus virginiana*), chestnut oak (*Quercus prinus*), Chinese tallow (*Sapium sebiferum*), willows, palm trees, and various conifers are also represented. There were also a few open areas consisting entirely of unmown grass; these sections are most likely slated for future development. Data was collected by Harry Elliott of Houston, who observed twenty-nine different bird species on two days in March (Table 7.3). Again, urbanized species (e.g., 1 to 9, Table 7.3) were present in sizable numbers, but migrants and species less commonly found in typical urban habitats were also observed, including loggerhead shrikes, purple martins, various warblers, and a large hawk, possibly a Cooper's hawk.

Several birders contributed information on other cemeteries throughout the state and some of the more interesting avian species that could be found in them (Table 7.4). The city cemetery

Table 7.2 Total Bird Observations by Sight in Greenwood Cemetery, Dallas, Texas, from April 1977 to September 1979

Species	Total Number Observed
Starling	>706
House sparrow	142
Chipping sparrow	131
Blue jay	>94
Northern mockingbird	>92
Northern cardinal	>65
Mourning dove	64
American robin	64
Nashville warbler	62
Great-tailed grackle	>50
Common grackle	>45
Chimney swift	>36
American crow	>34
Ruby-crowned kinglet	>28
Brown-headed cowbird	21
Brown thrasher	>19
Cedar waxwing	17
Lincoln's sparrow	16
Empidonax spp.	14
Black-and-white warbler	12
Field sparrow	>11
Great crested flycatcher	10
Northern flicker	9
Brown creeper	9
Little blue heron	8
Black-throated green warbler	8
American redstart	7
Junco	7
Eastern Phoebe	6
Red-breasted nuthatch	6
Solitary vireo	6
Bay-breasted warbler	6
Swainson's thrush	5
Yellow-rumped warbler	5
Baltimore oriole	5
Red-winged blackbird	5
Red-headed woodpecker	4
Carolina chickadee	>4
Tennessee warbler	4
Common nighthawk	3
Red-bellied woodpecker	3
Hairy woodpecker	3
Great horned owl	2
Downy woodpecker	2
Yellow warbler	2

Table 7.2   Total Bird Observations by Sight in
Greenwood Cemetery, Dallas, Texas,
from April 1977 to September 1979
(*Continued*)

Species	Total Number Observed
Blackburnian warbler	2
Wilson's warbler	2
Spotted towhee	2
Song sparrow	2
Great egret	1
Sharp-shinned hawk	1
Yellow-billed cuckoo	1
Yellow-bellied sapsucker	1
Eastern kingbird	1
Scissor-tailed flycatcher	1
Tufted titmouse	1
Loggerhead shrike	1
Carolina wren	1
Bewick's wren	1
Golden-crowned kinglet	1
Blue-gray gnatcatcher	1
Orange-crowned warbler	1
Magnolia warbler	1
Canada warbler	1
White-crowned sparrow	1
White-throated sparrow	1
Summer tanager	1

Source: Data courtesy of Homer Klonis of Dallas, Texas.
Note: Most observations occurred during the spring and
fall migratory periods.

in Weslaco, Texas, was especially well known for its sightings of rare or unusual subtropical birds (including the infrequently observed blue mockingbird). In comparison, one could find large numbers of shorebirds such as curlews and godwits in cemeteries in Galveston, Texas. Cemeteries in West Texas and the panhandle were cited as being good roosting and nesting places for owls due to the high density of large conifers in otherwise relatively treeless regions.

Ardath Lawson

**Table 7.3   Total Bird Observations by Sight and Song
in Forest Park Cemetery, Houston, Texas, in
March 2003**

Species	Total Number Observed
Yellow-rumped warbler	175
American robin	150
Red-winged blackbird	150
Mourning dove	45
Northern mockingbird	27
European starling	23
Rock dove	20
Great-tailed grackle	>15
House sparrow	15
Blue-gray gnatcatcher	13
White-winged dove	12
Common grackle	>10
Northern cardinal	>9
Killdeer	>8
Blue jay	8
Ruby-crowned kinglet	8
American crow	6
Purple martin	>5
Loggerhead shrike	5
Red-bellied woodpecker	4
Accipiters (probably Cooper's hawk)	2
White-eyed vireo	>2
Carolina wren	>2
Eurasian collared dove	>1
Inca dove	>1
Yellow-bellied sapsucker	>1
Canada warbler	>1
Orange-crowned warbler	1
Eastern meadowlark	1

Source: Data courtesy of Harry Elliott, Houston, Texas.
Note: Numbers of some birds are estimates.

Table 7.4   Other Bird Species Observed in Texas Cemeteries

Cemetery	Species Sighted
Weslaco City Cemetery, Welsaco, Hidalgo Co.	Green parakeet
	Red-crowned parrot
	Black-bellied whistling duck
	White-tailed kite
	Great kiskadee
	Green jay
	Blue mockingbird
Oakwood Cemetery, Comanche, Comanche Co.	Mississippi kite
Unknown, Galveston Co.	Plain chachalaca
	Long-billed curlew
	Veery
	American goldfinch
Various West Texas Cemeteries	Marbled godwit
	Barn owl
	Eastern screech owl
	Great horned owl
	Long-eared owl
State Cemetery, Austin, Travis Co.	Golden-fronted woodpecker
Old Fairview Cemetery, Bastrop, Bastrop Co.	Black-throated gray warbler

SPECIES PROFILE: RED FOX (*VULPES VULPES*)

Suburban foxes are not simply tame towards men. They are also damn supercilious. One pads among the azaleas in our garden at night, staring through the lounge windows to watch the *News at Ten*.

—Richard Gordon

The embodiment of cunning, red foxes in most of the United States are a combined strain derived from the interbreeding of native lines with foxes imported in the mid-eighteenth century from England and released in Delaware, Maryland, New York, New Jersey, and Virginia to support hunting on horseback. Landowners were dissatisfied with the sport provided by the native gray fox (*Urocyon cinereoargenteus*), which treed rather than running, or ran in a much smaller area. Conflicting records and the fact that, historically, red foxes were divided into two species (*Vulpes vulpes* in the Old World, *Vulpes fulva* in the New World) caused confusion regarding early accounts of the distribution and introduction of the species (now simply *Vulpes vulpes*) in North America (New Jersey DEP website). Unregulated trapping for the fur trade, combined with bounties, took a toll on fox populations. They have since recovered and are now found throughout much of North America, with the exception of most of the American West.

The red fox is another iconic species that requires only a brief description here. The coat is primarily rusty red above with white chin, throat, and belly. Other color phases exist, including black, silver, cross (reddish brown with a dark cross on the shoulders), and intermediate phases; all have a white-tipped tail. At 7.5 to 15 lbs (3.4 to 6.8 kg) red foxes are similar in size to a Boston terrier—if the dog had an extraordinarily long bushy tail. Add the pointy ears, slender muzzle, and slanted eyes and you have an animal that is almost instantly recognizable (Whitaker 1998; New Jersey DEP Website).

Mating occurs from January to early March and one litter of one to ten kits is born within fifty-one to fifty-three days. The maternity den is often an enlarged woodchuck or badger den. The family group stays together until autumn, when the kits disperse. Sexual maturity is ten months for both males and females. Red foxes are primarily nocturnal and/or crepuscular. Their diet is highly

Figure 7.6   Red fox. (Courtesy Ronald Laubenstein/USFWS)

adaptable, which is always handy if you're trying to make a living in an urban habitat. Specific food items include fruit, berries, grasshoppers, caterpillars, crayfish, small reptiles, amphibians, birds, and mammals. Foxes are cat-like when stalking prey. They will cache extra food in snow, leaves, or soil (Whitaker 1998).

Much of the research that's been done on urban red foxes comes out of Europe, primarily Great Britain (e.g., Harris and Rayner 1986; Doncaster, Dickman, and McDonald 1990; Woollard and Harris 1990; Doncaster and MacDonald 1997). Human-fox conflicts tend to center on the animals' scavenging behavior, damage to gardens, and loss of companion animals (dogs, cats, and birds).

As with most other urban species, the most effective management approach is conflict prevention. This can be accomplished through exclusion (e.g., fencing), frightening or harassment (e.g., strobe lights, noise-makers), and educating human residents to protect pets and other domesticated animals, and to avoid inadvertently attracting foxes. Lethal management methods exist, but as has been explained in other chapters, the results tend to be short-lived (yes, we recognize the pun). Additionally, foxes are charismatic enough to garner sympathy from nearly everyone but the most directly impacted individuals, and wildlife managers may pay a heavy price in negative public sentiment for choosing lethal rather than preventative options.

Urban habitats often are, for many species, islands, with various barriers to dispersal between habitats—roads are a prime example. In one interesting study, Wandeler et al. (2003) investigated two hypotheses: (1) that the fox population of Zurich, Switzerland, was isolated from adjacent rural fox populations, and (2) that urban habitat acted as a constant sink for rural dispersers. Their results suggested the two urban areas studied were independently founded by a small number of individuals from adjacent rural areas resulting in genetic drift and genetic differentiation between rural and urban fox populations. Based on the observed levels of migration between urban and rural populations, the authors predicted genetic differentiation over time.

ACKNOWLEDGMENT

Thanks to our colleague Sara Ash for her contributions to this chapter.

CHAPTER ACTIVITIES

1. Survey a small segment of your city for the three types of urban green spaces. Rank them in order from most to least common. How much of the total land area does each of the habitat types cover? Would you expect the ratios of habitat types to change depending on which part of the city you sample (e.g., suburb vs. downtown vs. business district)?

2. Interview someone in charge of a managed habitat patch (e.g., groundskeeper of a cemetery, golf course manager, or city park manager). Ask them questions about their goals and objectives for the land. Do their objectives include wildlife issues? If so, are these issues related to the promotion and/ or prevention of wildlife populations?

3. Identify two successional habitat patches of different sizes in your city. Predict how the size of the patch will influence how many species and what types of species you find there. Identify additional factors other than size that could determine differences in species richness and composition between successional habitat patches.

4. Contact your local birdwatchers club/organization. Ask them where they bird watch within the city. Visit these sites and hypothesize about why birds occur there. Of the sites identified by the birdwatchers, what is the ratio of remnant, successional, and managed habitat patches?

LITERATURE CITED

Alsop, F.J. 2001. *Birds of North America: Eastern Region*. New York: DK Publishing.

American Forests. 2004. *National Register of Big Trees*. Accessed at http://www.americanforests.org/resources/bigtrees/.

Ashbaugh, B.L. 1963. *Planning a Nature Center* (Information – Education Bulletin No. 2). Nature Centers Division, National Audubon Society. New York, NY.

Baker, P.J., R.J. Ansell, P.A.A. Dodds, C.E. Webber, and S. Harris. 2003. Factors affecting the distribution of small mammals in an urban area. *Mammal Review* 33(1):95–100.

Barko, V.A., G.A. Feldhamer, M.C. Nicholson, and D.K. Davie. 2003. Urban habitat: A determinant of white-footed mouse (*Peromyscus leucopus*) abundance in southern Illinois. *Southeastern Naturalist* 2(3):369–376.

Barrett, G.W. and T.L. Barrett. 2001. Cemeteries as repositories of natural and cultural diversity. *Conservation Biology* 15(6):1820–1824.

Blair, R.B. 1996. Land use and avian species diversity along an urban gradient. *Ecological Applications* 6(2):506–519.

Boal, C.W. and R.W. Mannan. 1998. Nest-site selection by Cooper's hawks in an urban environment. *Journal of Wildlife Management* 62(3):864–871.

Boal, C.W. and R.W. Mannan. 1999. Comparative breeding ecology of Cooper's hawks in urban and exurban areas of southern Arizona. *Journal of Wildlife Management* 63(1):77–84.

Bock, C.E., K.T. Vierling, S.L. Haire, J.D. Boone, and W.W. Merkle. 2002. Patterns of rodent abundance on open-space grasslands in relation to suburban edges. *Conservation Biology* 16(6):1653–1658.

Bolger, D.T., A.C. Alberts, R.M. Sauvajot, P. Potenza, C. McCalvin, D. Tran, S. Mazzoni, and M.E. Soule. 1997. Response of rodents to habitat fragmentation in coastal southern California. *Ecological Applications* 7(2):552–563.

Bolger, D.T., T.A. Scott, and J.T. Rotenberry. 1997. Breeding bird abundance in an urbanizing landscape in coastal southern California. *Conservation Biology* 11(2):406–421.

Burdge, R. and D. Cristol. 2008. Fore! Fairways for wildlife. *Wildlife Professional* 2:32–36.

Churcher, P.B. and J.H. Lawton. 1987. Predation by domestic cats in an English village. *Journal of Zoology* 212:439–455.

Coleman, J.L., D.M. Bird, and E.A. Jacobs. 2002. Habitat use and productivity of sharp-shinned hawks nesting in an urban area. *Wilson Bulletin* 114(4):467–473.

Crooks, K.R. 2002. Relative sensitivities of mammalian carnivores to habitat fragmentation. *Conservation Biology* 16(2):488–502.

Crooks, K.R., A.V. Suarez, D.T. Bolger, and M.E. Soule. 2001. Extinction and colonization of birds on habitat islands. *Conservation Biology* 15(1):159–172.

Cristol, D.A. and A.D. Rodewald. 2005. Introduction: Can golf courses play a role in bird conservation? *Wildlife Society Bulletin* 33:407–410.

Dickman, C.R. and Doncaster, C.P. 1989. The ecology of small mammals in urban habitats. II. Demography and dispersal. *Journal of Animal Ecology* 58:119–129.

Doncaster, C.P., C.R. Dickman, and D.W. MacDonald. 1990. Feeding ecology of red foxes (*Vulpes vulpes*) in the City of Oxford, England. *Journal of Mammalogy* 7(2):188–194.

Doncaster, C.P. and D.W. MacDonald. 1997. Activity patterns and interactions of red foxes (*Vulpes vulpes*) in Oxford city. *Journal of Zoology* 241(1):73–87.

Dykstra, C.R., J.L. Hays, F.B. Daniel, and M.M. Simon. 2000. Nest site selection and productivity of suburban red-shouldered hawks in southern Ohio. *Condor* 102:401–408.

Dzwonko, Z. and S. Loster. 1997. Effects of dominant trees and anthropogenic disturbances on species richness and floristic composition of secondary communities in southern Poland. *Journal of Applied Ecology* 34:861–870.

Estes, W.A. and R.W. Mannan. 2003. Feeding behavior of Cooper's hawks at urban and rural nests in southeastern Arizona. *Condor* 105:107–116.

Fernandez-Juricic, E. 2000. Avifaunal use of wooded streets in an urban landscape. *Conservation Biology* 14(2):513–521.

Fernandez-Juricic, E. and J. Jokimaki. 2001. A habitat island approach to conserving birds in urban landscapes: Case studies from southern and northern Europe. *Biodiversity and Conservation* 10:2023–2043.

Friesen, L.E., P.F.J. Eagles, and R.J. Mackey. 1995. Effects of residential development on forest-dwelling neotropical migrant songbirds. *Conservation Biology* 9(6):1408–1414.

Gange, A.C., D.E. Lindsay, and J.M. Shofield. 2003. The ecology of golf courses. *Biologist* 50:63–68.

Germaine, S.S., S.S. Rosenstock, R.E. Schweinsburg, and W.S. Richardson. 1998. Relationships among breeding birds, habitat, and residential development in greater Tucson, Arizona. *Ecological Applications* 8(3):680–691.

Gilbert, O.L. 1989. *The Ecology of Urban Habitats*. New York: Chapman and Hall.

Green, B.H. and I.C. Marshall. 1987. An assessment of the role of golf courses in Kent, England, in protecting wildlife and landscapes. *Landscape and Urban Planning* 14:143–154.

Green, D.M. and M.G. Baker. 2003. Urbanization impacts on habitat and bird communities in a Sonoran desert ecosystem. *Landscape and Urban Planning* 63:225–239.

Harris, S. and J.M.V. Rayner. 1986. Urban fox (*Vulpes vulpes*) population estimates and habitat requirements in several British cities. *Journal of Animal Ecology* 55(2):575–591.

Harrison, J.O. 1981. Older urban cemeteries as potential wildlife sanctuaries. *Georgia Journal of Science* 39:117–126.

Hennings, L.A. and W.D. Edge. 2003. Riparian bird community structure in Portland, Oregon: Habitat, urbanization, and spatial scale patterns. *Condor* 105:288–302.

Holmes, R.T. and T.W. Sherry. 2001. Thirty-year bird population trends in an unfragmented temperate deciduous forest: Importance of habitat change. *Auk* 118(3):589–609.

Hostetler, M. and K. Knowles-Yanez. 2003. Land use, scale, and bird distributions in the Phoenix metropolitan area. *Landscape and Urban Planning* 62:55–68.

Jodice, P.G.R. and S.R. Humphrey. 1992. Activity and diet of an urban population of Big Cypress fox squirrels. *Journal of Wildlife Management* 56(4):685–692.

Johnston, D.W. and E.P. Odum. 1956. Breeding bird populations in relation to plant succession on the Piedmont of Georgia. *Ecology* 37:50–62.

Jones, S.G., D.H. Gordon, G.M. Phillips, and B.R.D. Richardson. 2005. Avian community response to a golf-course landscape unit gradient. *Wildlife Society Bulletin* 33:422–434.

Jules, E.S. and P. Shahani. 2003. A broader ecological context to habitat fragmentation: Why matrix habitat is more important than we thought. *Journal of Vegetation Science* 14:459–464.

Kostel-Hughes, F., T.P. Young, and M.M. Carreiro. 1998. Forest leaf litter quantity and seedling occurrence along an urban-rural gradient. *Urban Ecosystems* 2:263–278.

Krebs, C.J. 2001. *Ecology: The Experimental Analysis of Distribution and Abundance*. Benjamin Cummings, San Francisco, CA.

Kricher, J.C. 1973. Summer bird species diversity in relation to secondary succession on the New Jersey Piedmont. *American Midland Naturalist* 89(1):121–137.

LeClerc, J.E., J.P.K. Che, J.P. Swaddle, and D.A. Cristol. 2005. Reproductive success and developmental stabil-ity of eastern bluebirds on golf courses: Evidence that golf courses can be productive. *Wildlife Society Bulletin* 33:483–493.

LeClerc, J.E. and D.A. Cristol. 2005. Are golf courses providing habitat for birds of conservation concern in Virginia? *Wildlife Society Bulletin* 33:463–470.

Leopold, A. 1993. *Game Management.*

Lepczyk, C.A., A.G. Mertig, and J. Liu. 2003. Landowners and cat predation across rural-to-urban landscapes. *Biological Conservation* 115:191–201.

Livingston, M., W.W. Shaw, and L.K. Harris. 2003. A model for assessing wildlife habitats in urban landscapes of eastern Pima County, Arizona (USA). *Landscape and Urban Planning* 64:131–144.

Longley, W.H. 1963. Minnesota gray and fox squirrels. *American Midland Naturalist* 69(1):82–98.

Lussenhop, J. 1977. Urban cemeteries as bird refuges. *Condor* 709:456–461.

MacArthur, R. and E.O. Wilson. 1967. *The Theory of Island Biogeography.* Princeton, NJ: Princeton University Press.

Mager, K.J. and T.A. Nelson. 2001. Roost-site selection by eastern red bats (*Lasiurus borealis*). *American Midland Naturalist* 145:120–126.

Mannan, R.W. and C.W. Boal. 2000. Home range characteristics of male Cooper's hawks in an urban environ-ment. *Wilson Bulletin* 112(1):21–27.

Marzluff, J.M. and K. Ewing. 2001. Restoration of fragmented landscapes for the conservation of birds: A general framework and specific recommendations for urbanizing landscapes. *Restoration Ecology* 9(3):280–292.

McDonnell, M.J. and S.T.A. Pickett. 1990. Ecosystem structure and function along urban-rural gradients: An unexploited opportunity for ecology. *Ecology* 71(4):1232–1237.

Melles, S., S. Glenn, and K. Martin. 2003. Urban bird diversity and landscape complexity: Species-environment associations along a multiscale habitat gradient. *Conservation Ecology* 7(1):5; Accessed at http://www.consecol.org./vol7/iss1/art5.

Merola-Zwartjes, M. and J.P. DeLong. 2005. Avian species assemblages on New Mexico golf courses: Surrogate riparian habitat for birds? *Wildlife Society Bulletin* 33:435–447.

Nilon, C.H. and R.C. Paris. 1997. Terrestrial vertebrates in urban ecosystems: Developing hypotheses for the Gwynns Falls watershed in Baltimore, Maryland. *Urban Ecosystems* 1:247–257.

Paton, P.W.C. 1994. The effect of edge on avian nest success: How strong is the evidence? *Conservation Biology* 8:17–26.

Porter, E.E., J. Bulluck, and R.B. Blair. 2005. Multiple spatial-scale assessment of the conservation value of golf courses for breeding birds in southwestern Ohio. *Wildlife Society Bulletin* 33:494–506.

Ricklefs, R.E. and G.L. Miller. 1999. *Ecology.* New York: W.H. Freeman.

Rodewald, P.G., M.J. Santiago, and A.D. Rodewald. 2005. Habitat use of breeding red-headed woodpeckers on golf courses in Ohio. *Wildlife Society Bulletin* 33:448–453.

Robinson, G.R. and S.N. Handel. 1993. Forest restoration on a closed landfill: Rapid addition of new species by bird dispersal. *Conservation Biology* 7(2):271–278.

Santiago, M.J. and A.D. Rodewald. 2004. Considering wildlife in golf course management. Extension Fact Sheet W-15-04. Ohio State University Extension, Columbus, OH.

Schiller, A. and S.P. Horn. 1997. Wildlife conservation in urban greenways of the mid-southeastern United States. *Urban Ecosystems* 1:103–116.

Shomon, J.J. 1962. *A Nature Center for Your Community* (Information-Education Bulletin No. 1). Nature Centers Division, National Audubon Society, New York, NY.

Slabbekoorn, H. and M. Peet. 2003. Birds sing at a higher pitch in urban noise. *Nature* 424:267.

Smith, M.D., C.J. Conway, and L.A. Ellis. 2005. Burrowing owl nesting productivity: A comparison between artificial and natural burrows on and off golf courses. *Wildlife Society Bulletin* 33:454–462.

Smith, R.L. and T.M. Smith. 2001. *Ecology and Field Biology.* Benjamin Cummings, San Francisco, CA.

Soulé, M.E., D.T. Bolger, A.C. Alberts, J. Wright, M. Sorice, and S. Hill. 1988. Reconstructed dynamics of rapid extinctions of chaparral-requiring birds in urban habitat islands. *Conservation Biology* 2(1):75–92.

Stanback, M.T. and M.L. Seifert. 2005. A comparison of eastern bluebird reproductive parameters in golf and rural habitats. *Wildlife Society Bulletin* 33:471–482.

Stowe, J.P., Jr., E.V. Schmidt, and D. Green. 2001. Toxic burials: The final insult. *Conservation Biology* 15(6):1817–1819.

Terman, M.R. 1997. Natural links: naturalistic golf courses as wildlife habitat. *Landscape and Urban Planning* 38:183–197.

Thomas, J.W. and R.A. Dixon. 1973. Cemetery ecology. *Natural History* 82(3):61–67.

Thompson, C.F. and V. Nolan, Jr. 1973. Population biology of the yellow-breasted chat (*Icteria virens*) in southern Indiana. *Ecological Monographs* 43(2):145–171.

Thompson, D.C. 1977. The social system of the grey squirrel. *Behaviour* 64:305–328.

Thompson, D.C. 1978. Regulation of a northern grey squirrel (*Sciurus carolinensis*) population. *Ecology* 59(4):708–715.

Wandeler, P., S.M. Funk, C.R. Largiader, S. Gloor, and U. Breitenmoser. 2003. The city-fox phenomenon: Genetic consequences of a recent colonization of urban habitat. *Molecular Ecology* 12(3):647–656.

Whitaker, J.O., Jr. 1998. *National Audubon Society Field Guide to North American Mammals*. New York: Alfred. A. Knopf.

White, C. L. and M.B. Main. 2005. Waterbird use of created wetlands in golf-course landscapes. *Wildlife Society Bulletin* 33:411–421.

Woodlawn Cemetery. 2001–2003. *Historic Woodlawn Cemetery*. Accessed at: http://www.historic-woodlawn.com/index.html.

Woolard, T. and S. Harris. 1990. A behavioural comparison of dispersing and non-dispersing foxes (*Vulpes vulpes*) and an evaluation of some dispersal hypotheses. *Journal of Animal Ecology* 59:709–722.

Yahner, R.H. 1988. Changes in wildlife communities near edges. *Conservation Biology* 2(4):333–339.

Zalweski, A. 1994. A comparative study of breeding bird populations and associated landscape character, Torun, Poland. *Landscape and Urban Planning* 29:31–41.

Zipperer, W.C. 2002. Species composition and structure of regenerated and remnant forest patches within an urban landscape. *Urban Ecosystems* 6:271–290.

CHAPTER **8**

Urban Gray Spaces

The thing scared me to death.

**—New Yorker who encountered a wild turkey on the balcony
of his 28th floor apartment on West 70th Street**

KEY CONCEPTS

1. Gray spaces are the human constructed features unique to urbanization.
2. Wildlife encounters with buildings, windows, and towers can have both positive and negative consequences.
3. The disruption of natural habitats by highways has measurable impacts on the population dynamics, survivability, and natural history of many wildlife species.
4. Birds and bats have adopted bridges as alternative sites for rearing young, resting, and as safe havens from predation and adverse weather conditions.
5. Landfills, as repositories for municipal solid wastes, provide a food source for many wildlife species, resulting in human-wildlife conflicts at landfills.
6. The primary function of an airport is to ensure public safety, but airport designs and siting often attract certain species of wildlife, which causes human-wildlife conflicts at airports.
7. Peregrine falcons have been brought back from the brink of extinction using urban gray spaces.

INTRODUCTION

Over the past decade, several ecologists have recognized the need to apply the science of ecology to urban environments. "Urbanization can be characterized as an increase in human habitation, coupled with increased per capita energy and resource consumption and extensive modification of the landscape, creating a system that does not depend principally on local natural resources to persist" (McDonnell and Pickett 1990:1231). Their list of structural features, unique to urbanization, included dwellings, factories, office buildings, warehouses, roads, pipelines, power lines, railroads, channelized stream beds, reservoirs, sewage disposal facilities, landfills, and airports.

Lost natural habitats are replaced by four types of altered habitat that become progressively more common toward the urban core. The four types of habitat are presented below in terms of increasing habitability to most native species and decreasing proportion of coverage toward the urban core.

1. Built habitat: buildings and sealed surfaces, such as roads and parking lots, cover over 80 percent of central urban areas.
2. Managed vegetation: residential, commercial, and other regularly maintained green spaces.
3. Ruderal vegetation: empty lots, abandoned farmland, and other green spaces that are cleared but not managed.
4. Natural remnant vegetation: remaining islands of original vegetation (usually subject to substantial nonnative plant invasion) (McKinney 2002).

Habitat types 2, 3, and 4 were discussed extensively in Chapter 7. This chapter examines how various wildlife species utilize the "built" habitats within urban ecosystems. Our synthesis of the literature demonstrated that wildlife attempt to occupy nearly every "nook and cranny" in urban structures. This should not be surprising given the remarkable flexibility of urban adapters (Chapter 6) in taking advantage of the new structures provided in urban settings.

BUILDINGS, WINDOWS, AND TOWERS

Wildlife is most noticed by urban residents when the animal appears in or is in close proximity to the areas where we live, work, or recreate. The famous cliché "build it and they will come" is correctly applied to the response of wild animals to the urban structural features of homes, high-rise office buildings, warehouses, and athletic stadiums, among others. Some wild animals are also selective in terms of what part of the building they prefer (e.g., roof, attic, walls, basement, inside or outside), others are generalists and will occupy any available spaces to which they have access (see Figure 1.12).

Buildings

Wild animals occupy urban buildings because they have been invited by the structural design, are opportunistic in seeking shelter, or need an area to rest (e.g., hibernate) or raise young. More often than not, humans who occupy the buildings adopted by wild animals do not even know they are present until the tell-tale signs of fecal droppings, structural damage from gnawing, smells, or sounds tip them off concerning the animal's presence. The assemblage of vertebrate animals that have accepted urban structures and features associated with these structures (e.g., lawns and gardens) as alternative substrates to conduct their life cycle activities include 204 bird, 50 mammal, and 41 amphibian or reptile species (http://www.enature.com/).

On the other hand, birds and buildings are a lethal combination. In Chicago, experts estimate that any single, tall building could be killing 2000 birds a year during peak migration (Figure 8.1). From 1968 to 1998, more than 26,000 migrating birds died crashing into a single building along the Chicago lakefront. Birds migrate at night using the stars as a navigational tool and often following a corridor along a body of water such as Lake Michigan. Lighted buildings can disorient them, attracting them to their deaths. Bird biologists speculate that, from a vantage point over the lake, the low, dark mass of McCormick Place may appear to be a cluster of trees. It might look, to birds, like a haven of food and shelter (De Vore 1998).

Windows

Human-built structures often pose serious hazards to birds (Figure 8.2). Windows on residential homes account for at least one hundred million bird fatalities (migrants and resident birds) each year in the United States. Plate glass collisions account for 34 percent of all avian mortality second only to hunting (42 percent). Collisions occur during all seasons, all times of day, and with windows facing any direction (Klem 1991). Approximately 25 percent (225/917) of the avian species in the

Figure 8.1   **(A color version of this figure follows page 158.)** Some experts estimate up to 100 million birds die each year in collisions caused by the artificial lights of tall buildings, communication towers, and airports. This photo shows a sample of birds collected beneath high-rise buildings in Toronto during one migration season. (Courtesy Mark Jackson and Fatal Light Awareness Program)

Figure 8.2   This immature Northern goshawk (*Accipiter gentillis*) is just one of millions of avian fatalities caused each year when birds mistake reflective windows for open sky. (Courtesy Fatal Light Awareness Program)

Table 8.1   Species Most Frequently Reported Striking Windows in the United States and Canada

American robin (*Turdus migratorius*)	White-throated sparrow (*Zonotrichia albicollis*)
Dark-eyed junco (*Junco hyemalis*)	Ruby-throated hummingbird (*Archilochus colubris*)
Cedar waxwing (*Bombycilla cedrorum*)	Tennessee warbler (*Vermivora peregrine*)
Ovenbird (*Seiurus aurocapillus*)	Yellow-bellied sapsucker (*Sphvrapicus varius*)
Swainson's thrush (*Catharus ustulalus*)	Purple finch (*Carpodacus purpureus*)
Northern flicker (*Colaptes auratus*)	Common yellowthroat (*Geothlvpis trichas*)
Hermit thrush (*Catharus guttatus*)	Rose-breasted grosbeak (*Pheucticus ludovicianus*)
Yellow-rumped warbler (*Dendroica coronata*)	Gray catbird (*Dumetella carolinensis*)
Northern cardinal (*Cardinalis cardinalis*)	Wood thrush (*Hylocichla mustalina*)
Evening grosbeak (*Coccothraustes vespertinus*)	Indigo bunting (*Passerina cyanea*)

Source: D. Klem, Jr. 1989. *Wilson Bulletin* 101:606–620. With permission.

United States and Canada have been documented striking windows. The twenty most reported species are listed in Table 8.1.

There seems to be a greater vulnerability for those species whose activities occur on or near the ground, such as several species of thrushes (Turdidae), wood warblers (*Phylloscopus sibilatrix*), and finches (Fringillidae). The oven bird was the neotropical migrant reported most often as a window kill. Birds fail to see windows as barriers in their flight patterns and are vulnerable to window strikes wherever they coexist. Any factor (e.g., bird feeders) that increases the density of birds near windows will account for strike frequency (Klem 1989). The effects of window strikes for birds range from no visible damage to being knocked unconscious to fractured bones, and superficial and internal bleeding (Klem 1990).

Suggested methods to prevent bird window strikes involve decreasing bird density near the window and/or increasing window visibility for birds. For example, decreasing bird density can be accomplished by moving feeders and birdbaths twenty to thirty feet from windows. Window visibility can be enhanced using window screens, using interior vertical blinds or leaded glass decorations, placing cutout decals or vertical strips on the windows, and covering the window with soap or planting shade trees outside windows to reduce reflection (Klem 1991).

Communication Towers

Communication towers, wind turbines, smoke stacks, and high-rise buildings cause bird and bat mortalities. Banks (1979) estimated that 1.2 million birds per year were killed by communications towers across the United States (Figure 8.3a). The Federal Aviation Administration (FAA) tracks the number of towers across the continent to monitor aviation hazards. Generally, once a tower reaches 200 feet or higher, the FAA considers it a potential aviation hazard. As of November 2, 1998, the FAA's Digital Obstacle File listed 39,530 towers in height classes ranging from 200 to over 1000 feet high distributed across the lower forty-eight states. The real total is actually higher because towers that are close together often get lumped as one aviation obstruction. Since the 1990s, the birth of the cell phone and personal communication service industry has accelerated tower construction to over 5000 new towers per year. Considering the greater proliferation of new towers, the annual estimated bird mortality at communication towers could be over 5 million birds per year (http://www.towerkill.com/). In addition to those already mentioned, that is, FAA and Towerkill, several other organizations (listed below) have ongoing programs that monitor bird kills at communication towers, and/or advance the status of information on this topic.

1. American Bird Conservancy (ABC): http://www.abcbirds.org/conservationissues/threats/towers.html
2. Fatal Light Awareness Program (FLAP): http://www.flap.org/

(a)

(b)

Figure 8.3 There are many lethal bird and bat encounters with communication towers (a) and wind turbines (b) that extend anywhere from 200 to over 1000 feet high. (Courtesy John M. Davis and Clark E. Adams)

Table 8.2   Bird Fatalities Collected at the WMSV Television Tower: 1960–1997

Rank	Species	Number	Rank	Species	Number
1	Ovenbird	4362	11	Blackburnian warbler	337
2	Tennessee warbler	3579	12	Gray catbird	328
3	Magnolia warbler	1992	13	Yellow-breasted chat	227
4	Red-eyed vireo	1618	14	Philadelphia vireo	205
5	Black-and-white warbler	1177	15	Northern waterthrush	203
6	Chestnut-sided warbler	953	16	Palm warbler	192
7	Bay-breasted warbler	855	17	Indigo bunting	164
8	American redstart	555	18	Kentucky warbler	160
9	Black-throated green warbler	367	19	Rose-breasted grosbeak	146
10	Common yellowthroat	357	20	Yellow-rumped warbler	131

Source: J.D. Nehring. 1998. Assessment of avian population change using migration casualty data from a television tower in Nashville, TN. M.Sc. Thesis, Middle Tennessee State University, Murfreesboro, TN. With permission.

3. Federal Communications Commission (FCC): http://www.fcc.gov/
4. Fish & Wildlife Service (FWS): http://www.fws.gov/habitatconservation/communicationtowers.htm

It is difficult to obtain an accurate count of bird mortality at towers because of scavengers that consume the dead birds before they can be identified, and the lack of long-term longitudinal studies at specific sites. One study summarized the data collected over a 37-year period at a television tower in Nashville, Tennessee. The study identified the overall numbers and species of birds that were killed (Table 8.2). Shire, Brown, and Winegrad (2000) provided a summary of forty-seven studies of bird kills at communication towers, which included 184,797 birds of 230 different species (approximately one quarter of the number of species in the United States). There was an 80 percent correlation between the top twenty species listed by Nehring (1998) and Shire, Brown, and Winegrad (2000). The majority (92 percent) of birds killed at towers are migratory, and they are killed predominantly or frequently at night. Of note was the prevalence of oven birds (*Seiurus aurocapillus*) as tower victims as well as window victims.

Birds die by direct impact (called "blind collision" mechanism) with the tower or its guy wires or with other birds flying in circles around a lighted tower. Most fatalities occur when cloud cover prevents moonlight, and birds' ability to use their night-time navigation systems (called "phototactic" mechanisms). Birds attempt to use the tower lights as a reference point during flight, and fly in circles around the tower until they run into something or fall to the ground in exhaustion.

The number of tower kills of birds appears to be highly correlated with: (1) tower height, (2) light (steady rather than flashing, and wavelength), (3) guy wires, topographic position, (4) season (during spring and fall migrations), (5) time of day (night time), and (6) overcast weather conditions (Avery, Springer, and Cassel 1976; Gauthreaux and Belser 2006; Evans et al. 2007; Longcore, Rich and Gauthreaux 2008). At this time, there are no management strategies available to mitigate bird mortality at communication towers. Perhaps the day will come when the towers are no longer needed to facilitate the mass communication needs of the public. Even then, it is doubtful that the towers would be dismantled very quickly.

Wind Towers

Wind power development is a relatively new technology to produce electrical energy that could replace less environmentally friendly technologies such as coal-fired generating plants. The wind power industry is growing rapidly in response to alternative energy technology demands in the United States and other countries. There are many wildlife management issues related to wind power technology—so important that The Wildlife Society (TWS) commissioned a technical review on the subject (Arnett et al. 2007). For example, there is a growing body of evidence associating the deaths of birds and bats with collisions into 340 foot high wind turbines (Figure 8.3b). A total of 1157 birds representing fifty different species were killed by wind turbines at the Altamont Pass Wind Resources Area from May 1998 to May 2003 (Smallwood and Thelander 2008: Table 1). Of the 33,000 bird fatalities reported in the references used by Erickson et al. (2001), 34 percent were diurnal raptors, 32 percent protected passerines, 14 percent non-protected birds, 9 percent owls, and 4 percent water birds and/or waterfowl. Rugge (2001) recorded 316 bird fatalities of which 55 percent (174) were raptors (more than half were red-tailed hawks, *Buteo jamaicensis*). Recent studies of raptor mortalities by Smallwood and Thelander (2004, as cited by Arnett et al. 2007), extrapolated to the entire wind resource area, estimated from 881 to 1300 fatalities each year. Fatality estimates included from 75 to 116 golden eagles (*Aquila chrysaetos*), 300 red-tailed hawks, 73 to 333 American kestrels (*Falco sparverius*), and 99 to 380 burrowing owls (*Athene cunicularia*), which is a species of special concern in California. Factors that influenced raptor fatalities at wind turbines included:

1. Geographic and habitat locations, that is, topographic relief
2. Season and types of available prey species
3. Wind conditions that promoted flight characteristics, for example, soaring or kiting
4. Flight behavior while foraging
5. Raptor densities, particularly immigrants into the area
6. Turbine type (older) and site conditions

Hoover and Morrison (2005) and Arnett et al. (2007) provided some proactive management strategies to reduce raptor fatalities at wind turbines. They recommended decision making based on improved scientific data gathering, more rigorous and consistent permitting requirements, the development of effective animal damage mitigation strategies, and better public education, information exchange, and participation. New wind turbine site assessments should also take into consideration topographical and/or weather conditions, habitat fragmentation effects, and *a priori* wildlife assessments.

Bat mortality at wind turbines has received considerable attention due to a large bat kill at a West Virginia wind farm in 2003 (Williams 2003, as cited by Johnson et al. 2004). Although speculative, bat fatalities at wind turbines may be associated with migration. It is largely unknown why bats collide with turbines, but there is some speculation concerning whether the bats turn off their echolocation abilities during migration. No research has been done on the sound wave effects of wind turbines traveling at 100 miles per hour, and the possible effect on the echolocation capabilities of migratory bats. "It is important to develop and verify models that allow prediction of impacts to individuals and populations of both birds and bats" (Arnett et al. 2007:36).

An ongoing study by Johnson et al. (2003, 2004) and Johnson and Strickland (2003) identified six different species of bats that collided with turbines (Table 8.3). Most of the fatalities were comprised of three species of tree bats that migrate long distances and do not hibernate, including the hoary bat, eastern red bat, and silver-haired bat. As with bird fatalities at communication towers, one would expect that geographic and habitat locations and season may be factors that influenced species compositions of bat fatalities at wind turbines. However, the species composition was consistent in Tennessee, Wisconsin, Washington, Oregon, Colorado, Wyoming, Pennsylvania, California, Minnesota, and West Virginia (Table 8.3). Additionally, the highest numbers of bat fatalities occurred at three wind plants, including Buffalo Ridge, Minnesota ($N = 420$), Backbone Mountain, West Virginia ($N = 242$), and Foote Creek Rim, Wyoming ($N = 135$). There may be some relationship between number of bats found and the level of investigation by researchers at the various wind plants. Some bats could not be identified because of heavy insect scavenging. There is some irony in the phenomenon of insect-eating bats being eaten by insects. No bat fatalities

Table 8.3   Number and Total Proportion of Bats Collected at the Wind Towers in Ten Different States from 1998 to 2002

Species	No. of Carcasses	% Identified Fatalities
Hoary bat (*Lasiurus cinereus*)	601	41
Red bat (*Lasiurus borealis*)	441	30
Eastern pipistrelle (*Pipistrellus subflavus*)	142	10
Silver-haired bat (*Lasionycteris noctivagans*)	113	8
Little brown bat (*Myotis lucifugus*)	109	7
Unidentified	25	2
Big brown bat (*Eptesicus fuscus*)	22	1
Northern long-eared bat (*Myotis septentrionalis*)	7	<0.5
Mexican free-tailed bat (*Tadarida brasiliensis*)	1	—
Long-eared myotis bat (*Myotis evotis*)	1	—
Total	1,462	100

Source: Johnson et al. (2003, 2004); Johnson and Strickland (2003).

classified as threatened or endangered species have been documented at wind plants (Johnson and Strickland 2003).

ROADS AND HIGHWAYS

Urban sprawl could not have occurred without the Highway Trust Fund (HTF) created by the Highway Revenue Act of 1956 (Pub. L. 84-627), primarily to ensure a dependable source of financing for the National System of Interstate and Defense Highways and also as the source of funding for the remainder of the Federal-aid Highway Program. The HTF provided each state with the resources to build a massive interstate highway system throughout the United States.

One of the most widespread forms of modifications of the natural landscape during the past century has been 3.9 million miles of road construction. Trombulak and Frissell (2000) estimated an average loss of 11.8 million acres of land and water bodies that formerly supported plants, animals, and other organisms. With the continuing growth in the size of highways (e.g., number of lanes) and higher traffic volumes there is a growing threat to wildlife that affects a wider range of wildlife species, and presents an almost impassable barrier for many species of reptiles, amphibians, and small mammals (Jackson 2000). At first, Forman (2000) estimated a "road-effect zone" (area affected ecologically by roads and associated vehicular traffic) that encompassed 19 percent of the total area of the continental United States. In another publication, Forman et al. (2002) estimated that the area covered by roads, roadsides, and medians was equivalent to 1 percent of the land base of the United States, a surface area equivalent to the state of South Carolina. The long-term consequences of highway construction include:

1. Animal mortality, that is, road kills and secondary effect on carrion feeders.
2. Loss and change of habitat and biological communities.
3. Habitat fragmentation and secondary effect on dispersal and vagility and/or isolation of selected species.
4. Degradation of habitat quality.
5. Increase in human exploitation through poaching and hunting.
6. Disruption of social structure.
7. Reduced access to vital habitats.
8. Population fragmentation and isolation.
9. Disruption of dispersal processes that maintain gene flow within species populations.

10. The need to develop structures that mitigate the impacts of highways on wildlife populations (Jackson 2000; Spellerberg 1998).

Some of these consequences are discussed in more detail below.

Animal Mortality

All wildlife species have the basic need to find adequate food and water, shelter and mates. The movement necessary to fulfill these basic needs is usually what compels an animal to cross the extremely inhospitable landscape of a highway (Jacobson 2002). The highways of America are littered with road-killed animals (Figure 8.4). It would not be an understatement to characterize the frequency and volume of wildlife losses due to road-kill as absolute carnage—perhaps as many as one million individuals per day (Lalo 1987). The census of road-killed animals by C.E. Adams (1983; Table 8.4) showed a preponderance of large ungulates (mostly white-tailed deer, *Odocoileus virginianus*). Road-kill victims include a variety of different species and numbers of each species, particularly amphibians (Glista, DeVault, and DeWoody 2008; Table 8.5).

Deer road-kills increased in twenty-six states from 1993 to 2007. The national deer road-kill for 1991 conservatively totaled at least 726,000 deer, causing $1.1 billion in property damage (an average of $1,577 per accident), 29,000 human injuries, and 211 human fatalities (Conover et al. 1995; Romin and Bissonette 1996).

However, deer are not the only victims of highway traffic. A road survey of small vertebrate road-killed animals in Canada by Clevenger et al. (2003) identified 677 animals (56 different species), including 313 mammals (18 species), 316 birds (36 species), and 48 amphibians (two species). Vehicular collisions were the cause of at least 50 percent of bobcat (*Lynx rufus*) and coyote (*Canis latrans*) mortalities (Tigas, Van Vuren, and Sauvajot 2002). Nearly half of the reported deaths of the endangered Florida panther (*Puma concolor coryi*) are from collisions with vehicles (Foster and Humphrey 1995). A survey of road-killed snakes in Arizona by Rosen and Lowe (1994) identified 368 snakes, which included two species of conservation interest. They estimated that from tens to hundreds of millions of snakes have been killed by automobiles in the United States. In eastern

Figure 8.4 Road-killed wildlife. (Courtesy John M. Davis)

Table 8.4   Vertebrate Animals and Numbers Listed as Traffic Victims in Nineteen Journal Articles

Vertebrate Animals	Numbers	Vertebrate Animals	Numbers	Vertebrate Animals	Numbers
Frogs/toads	1,129	Cottontail	14,401	Porcupine	1
Salamanders	100	Coyote	933	B.T. prairie dog	1
Lizards	95	Deer	90,150	Raccoon	4,037
Snakes	957	Domestic cat	429	Rat—cotton	56
Turtles	223	Cow	2	Rat—Norway	17
Doves/pigeons	97	Dog	98	Rat—kangaroo	316
Domestic fowl	266	Pig	10	Skunk	3,353
Grouse/quail	94	Franklins ground squirrel	28	Tree squirrel	1,683
Pheasant	11,351	13-lined ground squirrel	320	Weasel	2
Raptors	214	Jackrabbit	1,782	Woodchuck	8
Songbirds	2,403	Mice	11		
Waterfowl	15	Mink	1		
Armadillo	50	Mole	5		
Badger	617	Muskrat	1,256		
Chipmunk	8	Opossum	2,517		

Source: C.E. Adams. 1983. *American Biology Teacher* 45:256–261. With permission.

Texas road mortality may have caused the loss of the timber rattlesnake (*Crotalus horridus*) populations and large snake populations (Jackson 2000). Mumme et al. (2000) examined the impact of road mortality on the demographic make up of the Florida scrub jay (*Aphelocoma coerulescens*), a threatened species.

Traffic mortality had a significant negative effect on the local densities of frogs and toads (Fahrig et al. 1995). The census of road-killed animals by Glista, DeVault, and DeWoody 2008 (see Table 8.5) is particularly disturbing because of the species diversity and numbers ($N = 9950$) of amphibians and reptiles (herptofauna) killed over relatively short (7.4 miles total) stretches of road in Indiana. They combined their tally of road-killed herptofauna with two other studies, which resulted in a total of 42,502 dead amphibians and reptiles. These data, when interpreted in light of the global decline of herptofauna, suggested that road-kills may be having as much of an affect on the loss of herptofauna populations as habitat destruction, climate change, infectious diseases, and UV radiation.

Roads and automobiles are relatively recent environmental variables in the evolutionary history of all wildlife animals. As such, there is no genetic program that enables wildlife to deal with a potentially lethal encounter with automobile traffic. For example, some raptors (e.g., owls) swoop low to the ground to capture prey. Swooping is not an adaptive trait for survival when done over a busy interstate highway. Most animals are not accustomed to the light intensity of automobile headlights, which causes them to freeze in place rather than move out of the path of an approaching vehicle at night. Some animals cannot get across the road fast enough (e.g., turtles, amphibians, and small mammals). Armadillos (*Dasypus novemcinctus*) have the strange habit of jumping straight up when a car passes over them, which is why road-killed armadillos usually display massive trauma to the back. The classic "playing possum" response to danger is one reason why opossums (*Didelphis virginiana*) are well represented among road-kill victims (Jacobson 2002).

The presence of a road may modify an animal's behavior through home range shifts, as well as altered movement patterns, reproductive success, escape responses, and physiological states (Trombulak and Frissell 1999). On the other hand, both turkey vultures (*Cathartes aura*) and black vultures (*Coragyps atratus*) preferentially establish home ranges in areas with greater road densities (Coleman and Fraser 1989), probably because of the increase in carrion from road kills.

November 25, 2009

Dear Customer:

Thank you for your purchase of *Urban Wildlife Management, Second Edition,* by Clark E. Adams and Kieran J. Lindsey.

Part of Table 8.5 was omitted during the production process. The missing portion appears below.

Scientific Name	Common Name	Total	Scientific Name	Common Name	Total
C. Amphibia			**D. Reptilia**		
Ambystoma tigrinum	Eastern Tiger Salamander	142	*Chelydra serpentina*	Snapping Turtle	23
Bufo americanus	American Toad	111	*Chrysemys picta*	Midland Painted Turtle	28
Hyla spp.	Tree Frog	1	*Elaphe obsoleta*	Black Rat Snake	5
Pseudacris crucifer	Spring Peeper	8	*Elaphe vulpina*	Fox Snake	9
Rana catesbeiana	Bullfrog	1,671	*Graptemys geographica*	Northern Map Turtle	1
Rana clamitans	Green Frog	172	*Nerodia sipedon*	Northern Water Snake	1
Rana palustris	Pickerel Frog	18	*Storeria dekayi wrightorum*	Midland Brown Snake	19
Rana pipiens *	Northern Leopard Frog	74	*Terrapene carolina*	Eastern Box Turtle	1
Rana spp.	unknown ranid	7,602	*Thamnophis sirtalis*	Common Garter Snake	35
?	unknown frog	10	*Trachemys scripta*	Red-eared Slider	13
Total		**9,809**	?	unknown snake	4
			?	unknown turtle	2
			Total		**141**

We sincerely regret any inconvenience this may have caused you. Please let us know if we can be of any assistance regarding this title or any other titles that Taylor & Francis publishes.

Best regards,

Taylor & Francis

K10251
1-4398-0460-5
978-1-4398-0460-5

URBAN GRAY SPACES 183

Table 8.5 Vertebrate Species Recorded along Four Tippecanoe County, Indiana, Survey Routes*

Scientific Name	Common Name	Total	Scientific Name	Common Name	Total
A. Mammalia			**B. Aves**		
Blarina brevicauda	Northern short-tailed shrew	14	Agelaius phoeniceus	Red-winged blackbird	8
Canis familiaris	Domestic dog	1	Branta canadensis	Canada goose	2
Canis latrans	Coyote	1	Butorides virescens	Green heron	1
Didelphis virginiana	Opossum	79	Cardeulis tristis	American goldfinch	1
Felis catus	Domestic cat	5	Cardinalis cardinalis	Northern cardinal	9
Lasiurus borealis*	Eastern red bat	1	Chaetura pelagica	Chimney swift	36
Marmota monax	Woodchuck	1	Colaptes auratus	Northern flicker	1
Mephitis mephitis	Striped skunk	16	Dumetella carolinensis	Gray catbird	1
Microtus ochrogaster	Prairie vole	1	Eremophila alpestris	Homed lark	1
Microtus pennsylvanicus	Meadow vole	15	Hirundo rustica	Barn swallow	5
Mus musculus	House mouse	2	Melanerpes erythrocephalus	Red-headed woodpecker	2
Mustela vison	Mink	6	Melospiza melodia	Song sparrow	9
Odocoileus virginianus	White-tailed deer	4	Molothrus ater	Brown-headed cowbird	2
Ondatra zibethicus	Muskrat	10	Otus asio	Eastern screech owl	6
Peromyscus spp.	Deer/white-footed mouse	39	Passer domesticus	House sparrow	15
Procyon lotor	Raccoon	43	Passerina cyanea	Indigo bunting	3
Scalopus aquaticus	Eastern mole	4	Phasianus colchicus	Ring-necked pheasant	2
Sciurus carolinensis	Eastern gray squirrel	23	Porzana carolina	Sora	1
Sciurus niger	Eastern fox squirrel	27	Quiscalus quiscula	Common grackle	6
Sorex cinereus	Masked shrew	1	Spizella passerina	Chipping sparrow	1
Spermophilus tridecemlineatus	13-lined ground squirrel	6	Sturnella magna	Eastern meadowlark	2
Sylvilagus floridanus	Eastern cottontail	37	Sturnus vulgaris	European starling	11
Tamiasciurus hudsonicus	Red squirrel	6	Tachycineta bicolor	Tree swallow	1
Tamias striatus	Eastern chipmunk	7	Troglodytes aedon	House wren	1
Vulpes vulpes	Red fox	1	Turdus migratorius	American robin	18
?	Unknown bat	2	Zenaida macroura	Mourning dove	4
?	Unknown mammal	8	?	Unknown bird	56
Total		360	Total		205

Source: D.J. Glista et al. 2008. *Hepetological Conservation and Biology* 3:77–87. With permission.
* Survey conducted March 8, 2005, to July 31, 2006. Overall total = 10,515 road-kills.
a Indicates species of special conservation concern in Indiana.

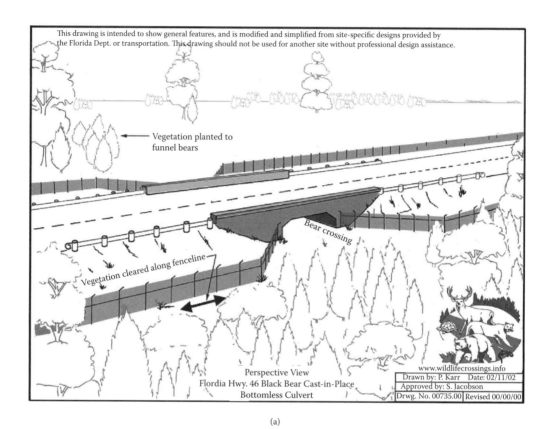

This drawing is intended to show general features, and is modified and simplified from site-specific designs provided by the Florida Dept. or transportation. This drawing should not be used for another site without professional design assistance.

Vegetation planted to funnel bears

Bear crossing

Vegetation cleared along fenceline

Perspective View
Flordia Hwy. 46 Black Bear Cast-in-Place
Bottomless Culvert

www.wildlifecrossings.info

Drawn by: P. Karr	Date: 02/11/02
Approved by: S. Jacobson	
Drwg. No. 00735.00	Revised 00/00/00

(a)

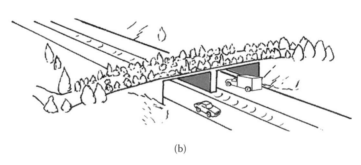

(b)

Figure 8.5 Three examples of wildlife crossing designs. (Provided by USDA Forest Service in their Wildlife Crossings Toolkit.)

Overpasses, Underpasses, and Escape Routes

Highway construction affects wildlife through the direct loss and fragmentation of habitat, and by disrupting animal movement and dispersal (e.g., migratory routes and home range activities). From a landscape ecology perspective, highways have the potential to undermine ecological processes through the fragmentation of wildlife populations, restriction of wildlife movements, and the disruption of gene flow and metapopulation dynamics (Jackson and Griffin 2000). In order to mitigate highway impacts on wildlife, alternative structural designs have been used to provide safe passage for large and small animals over or under the highway or to provide escape routes (e.g., fencing) that direct wildlife away from the highway or toward under- or overpasses (Transportation Research

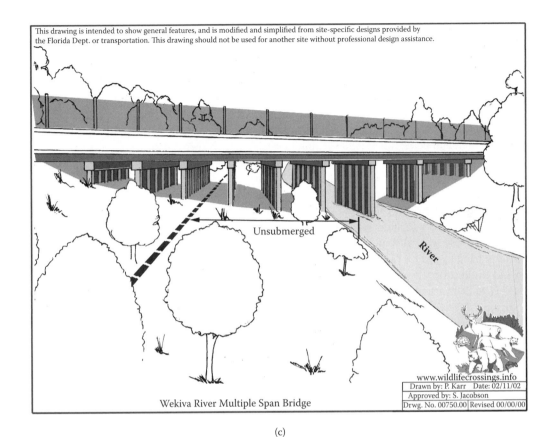

This drawing is intended to show general features, and is modified and simplified from site-specific designs provided by the Florida Dept. or transportation. This drawing should not be used for another site without professional design assistance.

Unsubmerged

River

www.wildlifecrossings.info

Drawn by: P. Karr	Date: 02/11/02
Approved by: S. Jacobson	
Drwg. No. 00750.00	Revised 00/00/00

Wekiva River Multiple Span Bridge

(c)

Figure 8.5 *Continued.*

Board 2002; Figures 8.5a, b, and c). Florida has been particularly active in developing mitigation measures to reduce highway mortality of herptofauna, Florida panthers, and other wildlife species (Foster and Humphrey 1995; Dodd, Barichivich, and Smith 2004; Aresco 2005).

The Wildlife Crossings Toolkit (www.wildlifecrossings.info), an online information source developed by the U.S. Department of Agriculture's Forest Service, was designed for professional wildlife biologists and engineers faced with integrating the highway infrastructure and wildlife resources. The Toolkit is a searchable database of case histories of mitigation measures and articles on decreasing wildlife mortality and increasing animals' ability to cross highways. Professional wildlife biologists and engineers can use the Toolkit to creatively solve challenges associated with highways.

There is an International Conference on Ecology and Transportation (ICOET). Conducted every two years, ICOET is designed to address the broad range of ecological issues related to surface transportation development, providing the most current research information and best practices in the areas of wildlife, fisheries, wetlands, water quality, overall ecosystems management, and related policy issues. ICOET is a multidisciplinary, interagency supported event, administered by the Center for Transportation and the Environment (http://www.icoet.net). They have held five conferences since 2001; the last one in 2009. The website contains additional details concerning these conferences. A proceedings is produced after each conference that contains reports on research conducted to determine the impacts of the highway infrastructure on wildlife and their habitats.

Structural Design Considerations

Bridge overpasses, underpasses, and fencing are designed to provide safe passage for a large assemblage of vertebrates, including amphibians, reptiles, and mammals, or whole faunal communities within a region. In addition, many of the bridge designs provide other benefits to some animals, including nesting, resting, brood-rearing, and hibernation sites. The bridge structures also provide alternative dispersal sites for large and small animals, particularly when young mammals are leaving parental home ranges and seeking to establish their own territories.

It is important to maintain connectivity between the island populations of some animals created by the habitat fragmentation caused by highway construction. Connectivity is important in terms of reversing: (1) loss through dispersal, (2) chance extinctions, and (3) the effects of inbreeding such as genetic drift and the loss of genetic variability. However, the potential negative consequences of increased connectivity include: (1) disease or parasite transmission into a population that may have been disease-free (e.g., chronic wasting disease in deer), (2) allowing exotic or other competitors to enter a habitat, and (3) outbreeding depression in some species that may have adapted to isolation and could suffer the negative effects from the introduction of distantly related genetic material (McKelvey, Schwartz, and Ruggiero 2002).

In some cases bridge design changes were serendipitous for some animals (see Chapter 9). For example, a design change in the reconstruction of the multiple-span bridge over the Colorado River in Austin, Texas, created a wider crevice size between the bridge's expansion joints. New spaces between the expansion joints were three-quarters to four inches wide and about sixteen inches deep. Mexican free-tailed bats found the new spaces irresistible and moved in soon after bridge construction was finished (Murphy 1990). Crevice width and depth are important bridge design considerations for bats. Textured surfaces on concrete and wood bridges are important features for day and night roosting by bats and nest building by swallows. The rough textured surfaces and temperature range inside box culvert bridges make ideal hibernacula for some bat species (Walker et al. 1996). Listed below are some general considerations for wildlife crossing structures designed to mitigate highway impacts on wildlife.

1. Place structures in areas of known migratory and dispersal routes of selected species and/or areas of high highway mortality.
2. Size of the underpass should be relative to the width of the highway—bigger is better.
3. Consider the following:
   a. Lighting—some species are hesitant to enter underpasses that lack sufficient ambient light, but avoid artificially lit areas.
   b. Moisture conditions—wet substrates are important for some amphibian species.
   c. Sustained temperature regimens and air flow.
   d. Traffic noise, as it is a problem for some mammals sensitive to human disturbance.
   e. Substrates that provide tactile security.
   f. Placement away from a high level of human disturbances.
   g. Potential interaction among species (e.g., predator and prey) at crossing sites.
4. Design approaches (covered or clear) that provide visual security for the design species.
5. Include fencing to help guide animals to the passage system and prevent them from circumventing the system (Transportation Research Board 2002; Jackson and Griffin 2000).

The Highway Bridge Replacement and Rehabilitation Program (HBRRP) is authorized by the federal Transportation Equity Act for the 21st Century (TEA21). The purpose of the program is to replace or rehabilitate public highway bridges over waterways, other topographical barriers, other highways, or railroads when the state and the Federal Highway Administration determine that a bridge is significantly important and is unsafe because of structural deficiencies, physical deterioration, or functional obsolescence (http://www.fhwa.dot.gov/bridge/hbrrp.htm). This program offers a unique and timely opportunity to rebuild, reconstruct, or refurbish bridges to accommodate wildlife using designs such as those illustrated in the Wildlife Crossings Toolkit (Figures 8.5a, b, and c). These accommodations could be the types of structures that now allow safe passages for wildlife. If these accommodations are not part of the HBRRP, the opportunity to provide them will be missed until the next bridge replacement cycle fifty to seventy years in the future (personal communication, S.L. Jacobson, Wildlife Biologist, USDA Forest Service).

Bridges, Birds, and Bats

Some animals have taken advantage of the alternative nesting, roosting, and dispersal opportunities provided by various highway structures (e.g., bridges, underpasses, overpasses, and culverts). For example, in the past 100 to 150 years, cliff swallows (*Hirundo pyrrhonota*), have expanded their range across the Great Plains and into eastern North America, a range expansion coincident with the wide-spread construction of highway culverts, bridges, and buildings that provide abundant alternative nesting sites (Brown and Brown 1995). Originally nesting primarily in caves, the barn swallow (*Hirundo rustica*) has almost completely converted to breeding under the eaves of or inside artificial structures such as buildings and bridges (Brown and Brown 1999). Since the mid-1980s, the cave swallow (*Hirundo fulva*) has undergone a dramatic range expansion in Texas and also colonized south Florida. In each of these cases, invasion of new territory has been facilitated by the adoption of bridges and culverts for nesting, with new colonies often springing up along highways (West 1995; Figure 8.6).

Human structures may often be superior to natural nest sites in important ways: predators may be less likely to gain access to nests, nest substrates may be superior for the long-term attachment of nests, and the thermal environments may be more favorable. By adopting human structures for nest sites, barn swallows are able to take advantage of localized food sources not otherwise profitably exploited (Speich, Jones, and Benedict 1986). Other birds that have occupied bridges (personal communication, R.J. Reynolds, Virginia Department of Game and Inland Fisheries) include the eastern phoebe (*Sayornis phoebe*), northern rough-winged swallow (*Stelgidopteryx serripennis*), pigeon (*Columba livia*), and osprey (*Pandion haliaetus*). Some birds (e.g., phoebes), will occupy the nests built by other birds under bridges rather than build their own nest.

Figure 8.6 *(A color version of this figure follows page 158.)* Cliff swallows (*N* = 102) nesting under a highway bridge. (Courtesy Clark E. Adams)

**Table 8.6   Bat Species Known to Occupy Highway Bridges
in the United States**

Common Names	Scientific Names
Big brown bat	*Eptesicus fuscus*
Big free-tailed bat	*Tadarida molossa*
California leaf-nosed bat	*Macrotus californicus*
California myotis	*Myotis californicus*
Cave myotis	*Myotis velifer*
Eastern long-eared myotis	*Myotis evotis*
Eastern pipistrelle	*Pipistrellus subflavus*
Evening bat	*Nycticeius humeralis*
Fringed myotis	*Myotis thysanodes*
Gray myotis	*Myotis grisescens*
Indiana bats	*Myotis sodalis*
Little brown myotis	*Myotis lucifugus*
Long-legged myotis	*Myotis volans*
Mexican free-tailed bats	*Tadarida brasiliensis*
Mexican long-tongued bat	*Leptonycteris nivalis*
Northern long-eared myotis	*Myotis septentrionalis*
Pallid bats	*Antrozous pallidus*
Rafinesque's big-eared bat	*Plecotus rafinesquii*
Silver-haired bat	*Lasionycteris noctivagans*
Southeastern myotis	*Myotis austroriparius*
Western or Townsend's big-eared bat	*Plecotus townsendi*
Western pipistrelle	*Pipistrellus hesperus*
Western small-footed myotis	*Myotis subulatus*
Yuma myotis	*Myotis yumanensis*

Source: B.W. Keeley and M.D. Tuttle. 1999. *Bats in American Bridges.* Austin, TX: Bat Conservation International Inc. With permission.

Many different species of bats have adapted to the various structural designs of highway bridges for day and night roosting, migratory rest stops, brood rearing, and as hibernacula during the winter (Adam and Hayes 2000; Keeley 1995; Kiser et al. 2001; Lewis 1994; Walker et al. 1996). "Bridges used as night roosts may serve several functions, including conservation of energy (thermoregulation), protection from predators, and locations for information transfer, social interaction, and consumption and digestion of prey" (Adam and Hayes 2000:402). The assemblage of bat species (Table 8.6) that use bridges includes twenty-four of the forty-five U.S. species of bats, and another thirteen species are likely to do so.

The diversity of bat species that occupy bridges is impressive, as well as the numbers of some species. For example, the Congress Avenue Bridge in Austin, Texas, is the summer home for 1.5 million Mexican free-tailed bats. The spectacular evening emergence of the bats has become a wildlife-watching opportunity for Austin residents and tourists alike (see Figure 9.2). Crowds of several hundred are usual. A public education campaign by Bat Conservation International (BCI) changed public opinion regarding the ecological importance (they consume 10,000 to 30,000 pounds of insects nightly) of the Congress Avenue Bridge bat colony, and corrected misinformation concerning bat natural history (Murphy 1990). The economic impact to the community is discussed in Chapter 9. There is now a collaborative effort between BCI and the Texas Department of Transportation to develop bat friendly bridge designs.

LANDFILLS, DUMPSTERS, AND GARBAGE CANS

Organic Waste Accumulations: A Concept Unique to Urban Ecosystems

Examples of urban organic wastes would include garbage or municipal solid waste (MSW), yard clippings and leaves, and human, pet, and urban wildlife excrements. Estimates of the total volume of various organic wastes within urban communities include the following.

As mentioned in Chapter 3, a city with one million inhabitants is estimated to consume 25,000 tons (t) of water and 2000 t of food per day, and produces 50,000 t of effluent water and 2000 t of waste material daily (Deelstra 1989). Approximately 231.9 million tons of MSW were generated in the United States in 2000. Between 1960 and 2006 the amount of waste each person creates has almost doubled from 2.7 to 4.4 pounds per day (U.S. Environmental Protection Agency 2002). The rich diversity of MSW waiting to be picked up on city streets is an invitation for a free meal for any species of wildlife, vertebrate or invertebrate.

The volume of pet wastes can only be estimated given the large varieties of pet types and sizes in urban communities. Nearly 1.4 billion tons of animal manure are produced annually in the United States. The New Jersey Department of Health estimated that there are over 500,000 dogs in the state. Adding in cats and other smaller pets with dogs results in a significant volume of waste daily. The EPA estimated that for watersheds of up to twenty square miles draining to small coastal bays, two or three days of droppings from a population of about 100 dogs would contribute enough bacteria and nutrients to temporarily close a bay to swimming and shell fishing. In the Four Mile Run watershed in Northern Virginia, a dog population of 11,400 is estimated to contribute about 5000 pounds of solid waste every day, and has been identified as a major contributor of bacteria to the stream. A single gram of dog feces can contain 23 million fecal coliform bacteria. A general lack of public recognition about the water quality and health consequences of dog waste calls for improved watershed education efforts (U.S. Environmental Protection Agency 1993).

There are thousands of Canada geese in some communities, for example, Minneapolis and St. Paul, Minnesota. Canada geese eat and excrete at a high frequency daily. For example, the number of droppings per goose may range from 28 to 92 per day, weighing from 1.17 to 1.9 grams. Geese are not very concerned where they defecate, whether on lawns, golf courses, athletic fields, park lawns, swimming pools, or city lakes. Goose droppings are organic and require bacterial decomposition, which in lakes increases biological oxygen demand, thereby reducing oxygen levels. The nitrogen and phosphorus in goose manure represents another form of fertilizer. The amount of nitrogen per goose per year ranges from 1.15 to 3.11 pounds. The amount of phosphorus per goose per year ranges from 0.36 to 1.41 pounds. These fertilizer loads will cause eutrophication in city waterways. The resident Canada goose problem is covered in more detail in Chapter 14.

Overall, the average family throws away 1.28 pounds of food per day, for an annual total of 470 pounds per household, or 14 percent of all food brought into the house (unpublished data, Tim Jones, Bureau of Applied Research in Anthropology, University of Arizona). There is no estimate of the food discarded by restaurants and grocery stores.

The process of removing organic wastes from urban communities is a primary consideration in city planning because there is so much of it. There is so much waste because urban residents, their pets, and urban wildlife are concentrated in a relatively small area (e.g., 5000 humans per square mile), and do not know how to or do not want to recycle the various forms of matter they encounter on a day-to-day basis. Most of it can be recycled.

There are many advanced forms of technologies to recycle urban organic wastes, but cost and convention prevent these technologies from being implemented in most urban communities. Conventional technologies generally involve a process of "filing by piling," that is, nothing is actually done with the wastes, they are just allowed to accumulate in different piles of matter, straining the ability of local ecosystems to assimilate them.

Garbage accumulation in urban environments has had a profound effect on the legendary predator-prey relationships between cats and rats (Childs 1991). He found that inner-city rats grow faster, reproduce earlier, and have many more offspring than parkland rats. He observed cat predation only on rats seven ounces or less, too young to contribute to the rat recruitment rate in the inner city. By weight, at least 30 percent of inner-city garbage contains edible material for rats and cats. Some inner-city rats can grow as large as one and a half pounds. Inner-city cats were more likely to be observed feeding side by side with the inner-city rat on the same nutrient-rich garbage resource rather than preying on the rat as a food source. An even more bizarre change in cat versus rat behavior was the observation of rats feeding on young cats (Sullivan 2004).

The habitats adjacent to landfills, dumpsters, and garbage cans will predict, in part, the type of wildlife that will exploit the organic resources for food. The list of wildlife attracted to residential and/or landfill garbage is impressive. The cast of garbage consumers includes gulls (*Laurus* spp.), vultures (*Cathartes aura* and *Coragyps atratus*), bears (*Ursus* spp.), rats (*Rattus norvegicus*), free-ranging cats (*Felis catus*) and dogs (*Canis familiaris*), coyotes (*Canis latrans*), deer (*Odocoileus* spp.), raccoons (*Procyon lotor*), opossums (*Didelphis marsupialis*), bobcats (*Felis rufus*), and ravens (*Corvus corax*). Predatory hawks and owls appear at landfills to hunt rodents, birds, and young cats (Belant et al. 1995; Belant, Ickes, and Seamans 1998; Childs 1991; Eberhard 1954; Horton, Brough, and Rochard 1983; Hutchings 2003; Manley and Williams 1998; Restani, Marzluff, and Yates 2001; Slate et al. 2000; Tigas, Van Vuren, and Sauvajot 2002; Williams 2002).

An understanding of the full assemblage of wildlife species attracted to concentrations of human food waste is limited by several factors. First, research on wildlife attracted to garbage sources does not have a lot of prestige in the scientific community so it is difficult to find investigators and the financial resources to conduct the research. Second, merely identifying what species utilize garbage is not as important a research objective as determining the impact of this type of feeding behavior on other aspects of the animals "normal" existence. For example, Childs (1991) observed three large rats and four cats feeding side by side from an overturned trash can. Feral cats can convert to scavenging rather than predation to obtain food. However, a well-fed cat will still kill other wildlife species if the opportunity presents itself (Hutchings 2003). Williams (2002) reported on animal addiction to human food leading to loss of interest in natural foods and digestive disorders. Bears can become aggressive panhandlers for human food (Manley and Williams 1998). At landfill sites, gull and raven numbers increase beyond the carrying capacity of their natural habitats (Belant et al. 1995; Restani, Marzluff, and Yates 2001). Estimates of gulls using landfills ranged from several hundred to 50,000, but the actual number of gulls may be underestimated (Slate et al. 2000; Belant 1997). Finally, research on animal utilization of garbage as a food source would fall under an urban wildlife management paradigm, which introduces a complex set of problems resulting from frequent contacts with people during the research process (VanDruff, Bolen, and San Julian 1996).

Factors That Promote the Presence of Wildlife at Landfills

As mentioned in the Introduction, the classic image of urban wildlife is raccoons emptying garbage cans in an urban area (Figure 8.7). One of the most critical problems to solve in urban areas is household garbage disposal. There are three basic steps in the garbage disposal process beginning with the residential collection in garbage cans, followed by municipal collection, and finally deposition at a landfill site (i.e., city dump). The first and last steps provide food accumulations for wildlife. As mentioned earlier, the average family throws away an enormous amount of food daily and annually. Excluding the categories of liquids, slop, and other, by weight, the food group frequencies were grain (20 percent), meats (16 percent), fruits (30 percent), vegetables (32 percent), and fats (2 percent; unpublished data, Tim Jones, Bureau of Applied Research in Anthropology,

Figure 8.7   Raccoon emptying a garbage can. (Courtesy USDA)

University of Arizona). This is just the amount and types of food that end up in the garbage can. It does not include food fed to pets, that goes down garbage disposals, is composted, or fed directly to wildlife in the neighborhood. Nor does it include food discarded by restaurants or grocery stores. Nevertheless, urban residents provide a buffet of food resources that attract many species of wildlife to landfills.

Standards Used in Landfill Siting

Few city and county politicians want to be in office when the job of siting a new landfill becomes necessary. The task can become embroiled in conflicts related to land acquisition negotiations, consideration of environmental impacts, public dissent if the new landfill is too close to their property, property values, quality of life, and changes in land-use patterns. The landfill siting process gives priority consideration to proximity to floodplains or areas that have critical habitats, historical/archeological features, and wetlands. The siting is also evaluated in terms of the required setback distances to navigable waters, state and federal highways, public parks, airports, and water supply wells. Other important considerations in the siting process are proposed landfill life and disposal capacity, municipalities and industries to be served, anticipated waste types, characteristics and amount of waste to be handled, and regional geotechnical characteristics of proposed location (http://www.dnr.state.wi.us/org/aw/wm/solid/landfill/siting.htm).

The landfill siting process involves many environmental concerns not found in the airport siting process discussed later in this chapter. As pointed out above, several environmental impacts need to be investigated before landfill construction can take place. Of special note is the impact of the landfill on the hydrology of the area because of the high potential of leachate generation and groundwater contamination. However, both neglect to include impacts on surrounding habitats and endemic plants and animals. Like airports, landfills are normally built on the outer fringe of urban or suburban communities, which results in the most direct and influential impact on the natural ecosystem and wildlife of the area.

Types of Habitats Found in and around Landfills

As with airports, the types of ecosystems in and around landfills are a function of the dominant ecosystem in the region of construction. Landfills (or open dumps) can be found wherever there is a concentration of humans, and this includes nearly all terrestrial ecosystems. Depending on the eco-system where the landfill is located, different wildlife species will be attracted to the food sources that are available at landfills. For example, polar bears (*Ursus maritimus*) in Alaska, black bears (*Ursus americanus*) in the northeastern United States, gulls (*Laurus* sp.) along coastal states, and rats (*Rattus rattus*) throughout the United States represent ecological equivalents for landfill feeding in their geographical regions.

Human-Wildlife Conflicts at Landfills

Anytime wild animals associate human presence with a food supply and become food-conditioned to the resource, there will be human-wildlife conflicts. These conflicts can range from the aggra-vation of having to pick up loose garbage spread across your lawn by a raccoon to being seriously mauled by a bear. The most publicized human-wildlife conflicts are airplane strikes (covered later in this chapter) and bear attacks, particularly if the bear is a female with cubs. Slate et al. (2000) listed several direct human-wildlife conflicts with gulls, which included safety concerns for equip-ment operators because of reduced visibility as gulls move in large flocks at the working face, inability to control gulls as potential disease vectors, corrosion of equipment, and increased bacte-rial counts from gull fecal material (Figure 8.8).

Belant (1997) listed the types of human-wildlife conflicts that occurred when gulls leave the landfill and move into the urban community. For example, roof-nesting gulls harassed maintenance personnel, deposited fecal wastes on roofs and nearby vehicles, plugged up roof drainage systems with debris, were noisy, and caused structural damage to buildings. Gull transmission of bacteria that cause enteric disease in humans is a growing concern with large colonies of roof nesters. In a survey of municipalities in the United States regarding vertebrate pests, gulls were ranked ninth by Fitzwater (1988) and the fourth most frequently occurring nuisance species in the Middle Atlantic States (see Table 2.1).

Figure 8.8 Gulls at a landfill in Charlotte, North Carolina. (Courtesy Jeffery S. Pippen)

Three events have led to an ever-increasing number of human-bear conflicts. The first event was the urban/suburban sprawl into traditional bear habitat forcing a daily coexistence. The second event was the bears' attraction to unnatural foods (e.g., residential or landfill garbage) to which bears became food-conditioned. Bears obtain food from humans by scavenging garbage at dumps; raiding garbage containers, camps, or buildings; stealing food directly from people; or being given handouts by people. Bears that feed on garbage are larger and tend to have higher reproductive rates (Stringham 1989). The third event, in concert with the second, was bear habituation to the presence of humans. Human and marauding bear conflicts are of particular importance because of the potential of physical or mortal danger to both (Rogers 1989; Peine 2001).

Wildlife Management at Landfills

Wildlife management at landfills can begin by addressing the root cause of the problem rather than the symptoms. Addressing the root cause requires creative siting away from transportation centers (e.g., airports), wildlife refuges, wetlands, major flyways of migratory birds, and sensitive habitats. The next requirement is the utilization of modern landfill technologies that reduce the exposure of MSW to wildlife in the surrounding habitats. For example, most modern landfills compact the MSW and deposit it in refuse cells that are covered with six to eight inches of fill dirt.

Other technologies that reduce or eliminate altogether the exposure of MSW to wildlife are waste-to-energy (WTE) facilities that separate the MSW into recoverable and combustible categories. Overall, about 80 percent of the MSW is burned for energy, 12 percent is recovered, and 8 percent is put in landfills (Wright 2004:498).

Many communities are turning to nontraditional waste management facilities such as yard waste compost facilities, construction and demolition landfills, and trash transfer stations (Gabrey 1997). He found that nontraditional waste-management facilities do not appear to attract birds or small mammals at higher than background levels, and would not pose a significant nuisance problem to the community or be a hazard to aircraft if located near airports. A key to wildlife management at landfills is waste reduction, particularly food wastes. Food waste is primarily (70 percent) liquid and can be converted into a dry pulpy material using a dehydrating machine (http://www.wastereductionsystems.net/Food_Waste_Reduction.html). Composting has been an extremely successful food waste reduction option in many communities (http://www.ilsr.org /recycling/wrrs/food/food.html). Composting of food wastes can be done at home using a garbage can as the composting apparatus (see Perspective Essay 4.2). Finally, and probably most importantly, wildlife management at landfills begins at home with the individual in terms of how he or she purchases, prepares, and consumes food. In general, the majority of affluent urbanites could purchase, prepare, and consume less food.

The management alternatives to control gull populations at landfills included nonlethal pyrotechnics, high grass management on gull loafing areas, overhead wires on active and capped portions of landfills, gull distress calls, bird of prey kites and predator-eye or other balloons, repellants, and shooting gulls (Slate et al. 2000). None of these alternatives can be expected to have any long-term effects given the lack of control over gull reproductive success, limitations imposed by the Migratory Bird Treaty Act, the continued presence of food waste at landfills, and the ineffectiveness of natural controls, for example, using birds of prey in large gull populations. In fact, unless the primary factor that attracts any animal to garbage dumps or landfills is removed, control strategies will only provide short-term reductions in animal numbers.

The management of bears at landfills presents a unique set of problems given its human value as a charismatic species. The valuation forces influencing management policies for nuisance bears are extremely varied and usually invoke strong emotions (Peine 2001). Consider how the bear has captured the human emotion for the species in terms of children's books (e.g., Winnie the Pooh, Smokey the Bear, and teddy bears). Some people value the bear for its parts, a picture, a glimpse, as spiritual deities, to attract customers, or as a symbol of identity of place at communities or schools. Nevertheless, nuisance bears have to be dealt with when they have crossed the line and have a negative impact on human health or economics (Schmidt 1997). Bear management strategies included garbage control, public education on the link between human waste-handling and bear problems, physical and chemical aversive conditioning, trap and translocate, and killing the bear (Peine 2001; Clark, Van Manen, and Pelton 2002; Ternent and Garshelis 1999). Unlike gulls, bears are less likely to rebound from or even survive management programs that do not provide alternative habitats where humans do not exist.

AIRPORTS

Recall the discussion of food chains in Chapter 3 and how the abiotic conditions of an area predict the dominant plant communities, which in turn predict the dominant animal communities (Figure 3.3). Different wildlife species would be present in and around an airport given its proximity to wetland, grassland, tropical forests, or desert ecosystems. On a smaller scale, grasslands attract rodents, which, in turn, attract birds of prey. Other factors that promote the presence of wildlife in and around airports are the presence of wildlife attractants, including wetlands, agriculture, wastewater treatment facilities, wildlife refuges, landfills, and aquaculture activities.

For example, agriculture and landfills are two attractants that provide abundant alternative food sources for selected species that will probably multiply beyond the carrying capacity of their natural environment. Gallaher (2003) reported a correlation between wildlife attractants and aircraft strikes with wildlife. Airports often are situated in outlying areas surrounded by woods, agricultural fields, and early successional habitats. In addition, landing fields are planted with grasses and forbs. These habitat conditions provide prime grazing locations for deer (Wright, Dolbeer, and Montoney 1998).

Standards for Airport Siting and Zoning

The development, movement, or expansion of a public or private airport is not a trivial matter. Decisions have to be made regarding location (siting), zoning rules have to be reviewed or written, and forums for public input need to be established. An examination of several siting and zoning ordinances across the country revealed certain standards that were common to all. The common siting standards included airport construction considerations, proximity to and required clearance heights over other transportation networks and waterways, and scheduled public hearings. Common zoning considerations dealt with identifying the existence of natural and man-made obstructions to air navigation.

The specific hazards to landing and taking off include trees and towers. Particulate (e.g., smoke) air pollutants have to be identified. Height of the landing field above sea level was another common consideration. Compatibility with existing land uses, and future county and/or regional planning commission projects was a common ingredient in zoning ordinances. In general, airport zoning ordinances focused on anything that reduced the size of the area for landing and taking off without mention of how various wildlife species might have an impact on this consideration (e.g., deer on the runway).

Public hearings on airport siting or zoning usually focused on impacts on property values, quality of life changes, and the "not in my backyard" (NIMBY) syndrome, that is, public distrust of the political agenda. There were no ordinances that addressed some important ecological considerations, including endangered or threatened species, flyways of migratory waterfowl, and potential impacts on critical habitats. Environmental impact statements (EIS) submitted to the U.S. Fish and

Wildlife Service become a critical part of the airport siting and zoning planning process when ecological issues need to be addressed.

Due to the real and increasing danger of animal-airplane strikes at airports and liability concerns, the Federal Aviation Administration (1997) published an advisory circular (AC No: 150/5200-33) titled "Hazardous Wildlife Attractants on or Near Airports." The circular specifies the types of land uses, adjacent to or on airport property, that attract birds and other types of wildlife that are potentially hazardous to aircraft. Types of land uses specifically mentioned were landfills, wastewater treatment facilities, water retention ponds, wetlands, agricultural practices, aquaculture activities, and surface mining operations, among others. The siting criteria of the circular specify the minimum distance of the airport operations from the above types of land uses.

Types of Habitats Found in and around Airports

The types of ecosystems in and around airports are a function of the dominant ecosystem in the region of construction. Airports have been built in every ecosystem type, including hot and cold deserts, grasslands, temperate deciduous forests, subtropical and tropical forests, and coastal wetlands. Airports are normally built on the outer fringe of urban or suburban communities, which results in the most direct and influential impact on the natural ecosystem and wildlife of the area. For example, in some ecosystem types, compliance with the standard zoning and siting ordinances requires dramatic habitat alterations, including clear-cutting, draining wetlands, landscape alterations, removal of endemic plants and animals, introduction of exotic plants and animals, and the structural changes associated with building the airport.

In order for the human activities associated with air travel to take place, natural habitats need to be simplified to accommodate their conversion into urban airports. For this to take place the natural ecosystem succession processes are altered or delayed on all or parts of the airport property. There are alternative airport development designs that focus on sustainable resource use in and around airports (Dalzell 2004).

Wildlife Species Attracted to Airport Habitats

Wetlands can be found in the vicinity of numerous airports because wetlands are left undeveloped and may provide for airport approaches involving less risk to the public than approaches over developed areas. The dominant avifauna in wetland habitats includes migratory waterfowl, wading birds, and shorebirds (Kennamer 1999). Birds can also be drawn to airports by open short grass and warm pavement where they find security from predators and humans and a place to rest, loaf, or feed (MacKinnon 1997).

The three dominant habitat types at an airport are short grass, agricultural, and old field succession (e.g., abandoned farmland; Baker and Brooks 1981). The three habitat types harbor an assemblage of rodents, rabbits, and birds, which are the prey species for resident raptor (e.g., hawks and owls) populations. A high ratio of edge-to-airport area including a scrub community, sugarcane field, and golf course and resort supported over sixteen different avian species (Linnell, Conover, and Ohashi 1996).

Mowed and unmowed airport grasslands provided habitat for twenty-two and twenty-nine different vertebrate species, respectively (Barras et al. 2000). Small mammal abundance was greater in unmowed than mowed areas. Twice as many raptors were observed in unmowed areas than mowed areas. The authors concluded that maintaining mowed vegetation on the entire airport will decrease the rodent population and, in turn, the raptor population. However, some grassland habitats in or around airports represent the only remaining habitat for threatened or endangered bird species (Rossi-Linderme and Hoppy 2000). Airport grasslands are recognized as substitute habitats, lost to urban development and agriculture, for six species of grassland birds (Kershner and Bollinger 1996).

Human-Wildlife Interactions at Airports

From 1990 to 2000, 33,000 bird and about 500 deer strikes to civil aircraft were reported in the United States (Figure 8.9). These estimates are based on a 20 percent air strike reporting rate. "Aggressive programs by natural resource and environmental agencies during the past 50 years (e.g., expansion of wildlife refuge system), coupled with land-use changes, have resulted in dramatic increases in populations of many wildlife species that are a threat to aviation" (Dolbeer, Wright, and Cleary 2000:39). They demonstrated population increases of Canada geese (*Branta*

(a)

(b)

Figure 8.9   Birds surround aircraft as they take off and land. (Courtesy USDA APHIS Wildlife Services)

canadensis), ring-billed gulls (*Larus delawarensis*), red-tailed hawks (*Buteo jamaicensis*), turkey vultures (*Cathartes aura*), and white-tailed deer (*Odocoileus virginianus*).

There is a detailed summary of the degree to which these same species are involved in strikes with aircraft (Cleary, Dolbeer, and Wright 2003). For example, waterfowl, raptors, gulls, doves, and deer caused the highest number of strikes resulting in substantial aircraft damage. Migratory waterfowl and deer caused the highest number of strikes resulting in human injury or death. Over a thirteen-year period, reported losses from bird and mammal strikes totaled 364,626 hours of aircraft downtime, and $170.9 million in monetary losses. White-tailed deer, vultures, and geese were ranked one, two, and three, respectively, as the most hazardous species groups to aviation (Dolbeer, Wright, and Cleary 2000).

A 2.5-pound herring gull, struck by an aircraft traveling 290 miles per hour, has an impact of 34 tons.

—Rossi-Linderme and Hoppy (2000)

Wildlife Management Priorities at Airports

Airport wildlife management plans center on minimizing aircraft strikes, and in some cases preserving habitat for threatened or endangered species. The body of literature on animal control strategies at airports has been amply reported by Conover (2002). In general, the strategies for controlling the numbers of each species hazardous to aircraft flight depend on (1) the offending species, (2) legal ramifications concerning the species in question and some control options (see Chapter 12), (3) the desired level of population control, (4) impact on other species, (5) cost effectiveness, and (6) public reactions.

Management strategies can be site specific or conducted at the landscape level. The standard selection of management techniques is nonlethal, which includes habitat management, removal of food sources, frightening devices, exclusion, repellents, and capture and translocation. Lethal control techniques include nest or egg disturbance, toxicants, shooting, and predators (Belant 1997).

Sometimes wildlife management at airports may present a paradox of priorities. There are cases when the wildlife management plan to mitigate air strikes are in direct conflict with the management plan to sustain populations of threatened and endangered species (Rossi-Linderme and Hoppy 2000). The Bird Aircraft Strike Hazard (BASH) Team concluded that the best way to reduce the number of birds roosting in and around the base was to reduce the attractiveness of the airfield, which would include mowing the surrounding grasslands. However, these grasslands were considered the premier breeding habitat for the upland sandpiper (*Bartramina longicauda*) and grasshopper sparrow (*Ammodramus savannarum*). Partnering with other federal agencies and working with local and regional groups resulted in a tentative airfield management compromise.

PERSPECTIVE ESSAY 8.1   The Peregrine Story

On a positive note, high-rise buildings and power plants, bridges and overpasses, power plant smoke stacks, and other structures have provided alternative nesting and brood rearing sites for the peregrine falcon (*Falco peregrine*; Figure 8.10). The primary reason for species extinctions is a sudden curtailment of reproductive ability often caused by habitat loss or change (pollution). The falcon was one of many predatory birds that were adversely affected by the indiscriminant use of DDT and PCBs from 1940 to 1970. Predatory birds are at the top of the food chain, and therefore will receive the highest concentrations of each chemical due to ever increasing concentrations as the chemical moves up the food chain, a process called biomagnification. The chemicals accumulated in the body fat of female falcons, which resulted in a decreased ability to metabolize calcium carbonate. The shell gland, which is part of the reproductive tract of the female, could not deposit enough calcium in the egg shell. This resulted in cracked eggs

Figure 8.10   Peregrine falcon with chicks. (USFWS photo)

when the female attempted to incubate them. The loss of falcon chicks on a regular basis soon caused a precipitous decline in falcon numbers throughout the United States. By the 1970s none remained in the eastern United States. In 1970, the U.S. Fish and Wildlife Service listed the peregrine falcon as endangered under the Endangered Species Conservation Act of 1969.

Once the use of DDT and PCBs was banned in 1972, a program to reintroduce and propagate the peregrine falcon began in New York City. The structural and biotic features of large cities were ideal for falcon reintroduction. For example, high-rise buildings contain nesting ledges similar to those found on rocky cliffs in the falcon's natural environment. Furthermore, there is an abundant prey base for falcons in cities. Cade et al. (1996) reported 104 different bird species as falcon food, and Bell, Gregoire, and Walton (1996) reported 34 species. Common peregrine falcon food species were pigeons (rock doves), northern flicker, blue jay, mourning dove, European starlings, and American robins.

The technique scientists used to help bring back the peregrines is called "hacking." Hacking is a delicate method of releasing birds back into the wild. Young falcons, still unable to fly, are placed in a specially designed box at the site from which they are to be released. This location is known as the hack site.

Site attendants care for the birds twenty-four hours a day. Peepholes are built into the box so the attendants can watch the peregrines without disturbing them. Food is provided through a chute, again eliminating any view of humans. Keeping actual interactions with the falcons to a minimum ensures the birds are kept wild and not imprinted on people. The front of the box, which is covered with bars, allows the young birds to become accustomed to their surroundings; while at the same time providing them with protection.

The box is opened after a week. When the young falcons are six weeks old, they are ready to take their first flight. Peregrines instinctively know how to fly and hunt without the aid of an adult! They spend about three weeks learning how to fly and the next three weeks learning how to hunt. However, during this time they continue to return to the hack site for food. As the immature peregrines become better at hunting they will wander further from the site (http://www.chias.org/biology/cprr_rec.html).

The reintroduction of peregrine falcons has occurred in many cities across North America (Cade et al. 1996; Figure 8.11). The reintroductions have been so successful in increasing the

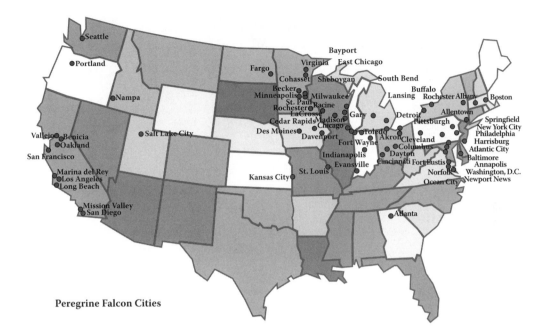

Peregrine Falcon Cities

Figure 8.11   Cities where peregrine falcons live. (Concept by Clark E. Adams, map by Linda Causey)

number of breeding pairs that the bird was removed from the federal list of threatened and endangered species in 1999. The USFWS has now initiated a monitoring plan for peregrine falcons in forty states beginning in 2003 and ending in 2015.

SPECIES PROFILE: MEXICAN FREE-TAILED BAT (*TADARIDA BRASILIENSIS*)

Of all small wildlife species, bats may be the species with the best chance to challenge coyotes as the most reviled by humans. Bats have been persecuted by people for generations and only recently have begun to garner respect among nonscientists for their unique and vital role in our ecosystems. Many people do not realize that bats call urban and suburban habitats home. However, several North American species have been found to both forage and reproduce in developed areas (Barbour and Davis 1974). Usually conflict arises at a localized scale, with an occasional small group of bats or a single bat taking up residence in the attic of a suburban home or warehouse building. However, because some species will roost in very large groups, the potential for major conflict exists.

Bats are one of the most diverse groups of mammals, with over 900 species (Wilson and Ruff 1999). North American bats, and specifically bats found in urban and suburban locations, are usually insectivores. Some of these bat species will migrate south when their food supply runs out. However, other species of bats in North America hibernate during the fall and winter months. This behavior allows them to reduce their metabolic needs until food becomes available again.

Some bat species form large communal roosts and/or hibernacula (roost sites for hibernation). For example, the Mexican free-tailed bat (Figure 8.12) is well known for forming large colonies, some of which contain millions of bats (Davis et al. 1962). Scientists do not definitively know why bats form such large colonies, but their hypotheses are similar to the explanations for the formation of large bird roosts (see Chapter 9, Species Profile). Because of their small size and specialized

Figure 8.12   Mexican free-tailed bat. (Courtesy National Park Service)

thermoregulatory requirements, bats are limited in where they can roost and/or hibernate. Some species (e.g., *Myotis grisescens*) need to roost close to foraging areas to reduce the energy costs associated with flying (Tuttle 1979). As a result, when suitable sites are found, several bats will take up residence. By roosting in close proximity to other individuals, bats also reduce their energy costs both in and out of hibernation (Herreid 1963). Transfer of information about foraging sites and protection from predators may also be benefits of communal roosting (Altringham 1996).

Many bat species, including the Mexican free-tailed bat, form what are referred to as maternity colonies. These roosts are largely comprised of females and their young (McCracken 1999). By roosting together in warm sites, the young bats are able to direct most of their energy into rapid growth (Altringham 1996).

The Congress Avenue Bridge in Austin, Texas, is home to a famous maternity colony of approximately 1.5 million female Mexican free-tailed bats and their offspring during the spring and summer months. This colony has received extensive attention from scientists and the media (e.g., *National Geographic*) extolling its ecological importance. However, when the bats first colonized the bridge in the early 1980s, public reaction was not positive. Bats have been historically feared by people and they've had a hard time shaking this negative image. Media reports in the *Austin American-Statesman* played into these fears by focusing on issues such as rabies transmission.

Visit Austin today and you're likely to hear only positive comments about the bats. The bridge has become a tourist hotspot, with approximately 100,000 people visiting during summer months to picnic and watch the bats emerge from their roost. The city named its professional hockey team the Ice Bats. What brought about this change? How has the city of Austin capitalized on an animal that usually provokes fear and disgust in so many people?

The human-bat conflict in Austin was resolved largely as a result of an expertly organized educational campaign championed by the private organization Bat Conservation International (BCI). BCI's founder, Merlin Tuttle, moved the organization's headquarters from Minneapolis to Austin in 1986 (Murphy 1990) and immediately began to educate the residents of Austin about the ecological virtues of the bats. For example, the group estimated that on a typical summer night the bats eat between 10,000 and 20,000 pounds of insects, many of which are agricultural pests (BCI website). Mexican free-tailed bats have been reported to fly 50 km to their foraging sites (McCracken 1999), so both urbanites and farmers in the area also benefit from the existence of this urban roost. BCI also worked to dispel myths about bats, especially concerning rabies. Rabies is certainly a potential risk; however, less than 1 percent of all bats are infected with the disease. Handling a sick or grounded bat is the most common way that humans are exposed to rabies through bats (Greenhall and Frantz 1994).

CHAPTER ACTIVITIES

1. Conduct a census of the road-killed animals in your community, noting:
   a. Species;
   b. Age and sex;
   c. Where road-kills are most likely to occur and why;
   d. Time of year when road-kills are most frequent; and
   e. Whether there were human costs involved.
2. Investigate how buildings, windows, and towers affect wildlife in your community. This will be particularly evident during the spring and fall migration of neotropical migrants.
3. Examine and record how the bridges in your community are used by various species of wildlife.
4. Investigate the wildlife issues that are prevalent at your community landfill and airport.

LITERATURE CITED

Adam, M.D. and J.P. Hayes. 2000. Use of bridges as night roost by bats in the Oregon coast range. *J. Mammalogy* 81:402–407.

Adams, C.E. 1983. Road-killed animals as resources for ecological studies. *American Biology Teacher* 45:256–261.

Altringham, J.D. 1996. *Bats: Biology and Behavior.* Oxford, UK: Oxford University Press.

Aresco, M.J. 2005. Mitigation measures to reduce highway mortality of turtles and other herpetofauna at a north Florida lake. *Journal of Wildlife Management* 69:549–560.

Arnett, E.B., D.B. Inkley, D.H. Johnson, R.P. Larkin, S. Manes, A.M. Manville, J.R. Mason, M.L. Morrison, M.D. Strickland, and R. Thresher. 2007. Impacts of Wind Energy Facilities on Wildlife and Wildlife Habitat. Wildlife Society Technical Review 07-2. Bethesda, MD: The Wildlife Society.

Avery, M., P.F. Springer, and J.F. Cassel. 1976. The effects of a tall tower on nocturnal bird migration: A portable ceilometer study. *Auk* 93:281–291.

Baker, J.A. and R.J. Brooks. 1981. Raptor and vole populations at an airport. *Journal of Wildlife Management* 45:390–396.

Banks, R.C. 1979. Human related mortality of birds in the United States. Spec. Sci. Rep. 215. Washington, DC: U.S. Fish and Wildlife Services.

Barbour, R.W. and W.H. Davis. 1974. *Mammals of Kentucky.* Lexington: University of Kentucky Press.

Barras, S.C., R. Dolbeer, R.B. Chipman, G.E. Bernhardt, and M.S. Carrara. 2000. Bird and small mammal use of mowed and unmowed vegetation at John F. Kennedy International airport, 1998 to 1999. In *Proc. 19th Vertebrate Conference*, ed. T.P. Salmon and A.C. Crabb, 31–36. University of California, Davis.

Belant, J.L. 1997. Gulls in urban environments: Landscape-level management to reduce conflict. *Landscape and Urban Planning* 38:245–258.

Belant, J.L., S.K. Ickes, and T.W. Seamans. 1998. Importance of landfills to urban-nesting herring and ring-billed gulls. *Landscape and Urban Planning* 43:11–19.

Belant, J.L., T.W. Seamans, S.W. Gabrey, and R.A. Dolbeer. 1995. Abundance of gulls and other birds at landfills in northern Ohio. *American Midland Naturalist* 134:30–40.

Bell, D.A., D.P. Gregoire, and B.J. Walton. 1996. Bridge use by peregrine falcons in the San Francisco Bay area. In *Raptors in Human Landscapes*, ed. D.M. Bird, D.E. Varland, and J.J. Negro, 14–23. London: Academic Press.

Blair, R.B., and A.E. Launer. 1997. Butterfly diversity and human land use: Species assemblages along an urban gradient. *Biological Conservation* 80:113–125.

Brown, C.R. and M.B. Brown. 1995. Cliff swallows (*Hirundo pyrrhonota*). In *The Birds of North America*, No. 149, ed. A. Poole and F. Gill. Philadelphia, PA: The Birds of North America Inc.

Brown, C.R. and M.B. Brown. 1999. Barn swallows (*Hirundo rustica*). In *The Birds of North America*, No. 452, ed. A. Poole and F. Gill. Philadelphia, PA: The Birds of North America Inc.

Cade, T.J., M. Martell, P. Redig, G. Septon, and H. Tordoff. 1996. Peregrine falcons in urban North America. In *Raptors in Human Landscapes,* ed. D.M. Bird, D.E. Varland, and J.J. Negro, 3–13. London: Academic Press.

Childs, J.E. 1991. And the cat shall lie down with the rat. *Natural History,* June, 16–19.

Clark, J.E., F.T. van Manen, and M.R. Pelton. 2002. Correlates of success for on-site releases of nuisance black bears in Great Smoky Mountains National Park. *Wildlife Society Bulletin* 30:104–111.

Cleary, E.C., R.A. Dolbeer, and S.E. Wright. 2003. *Wildlife Strikes to Civil Aircraft in the United States 1990–2002.* Serial Report Number 9. Washington, DC: U.S. Department of Transportation, Federal Aviation Administration. http://wildlife.pr. erau.edu/Bash90-02b.pdf.

Clevenger, A.P., B. Chruszcz, and K.E. Gunson. 2003. Spatial patterns and factors influencing small vertebrate fauna road-kill aggregations. *Biological Conservation* 109:15–26.

Coleman, J.S. and J.D. Fraser. 1989. Habitat use and home ranges of black and turkey vultures. *Journal of Wildlife Management* 53:782–792.

Conover, M. 2002. *Resolving Human-Wildlife Conflicts: The Science of Wildlife Damage Management.* Boca Raton, FL: Lewis Publishers.

Conover, M.R., W.C. Pitt, K.K. Kessler, T.J. DuBow, and W.A. Sanborn. 1995. Review of human injuries, illnesses, and economic losses caused by wildlife in the United States. *Wildlife Society Bulletin* 23:407–414.

Dalzell, S. 2004. Airport sustainable design. Massport. http://www.aci-na.org/docs/1.

Davis, R.B., Herreid, C.F. II, and Short, H.L. 1962. Mexican free-tailed bats in Texas. *Ecological Monographs* 32:311–346.

Deelstra, T. 1989. Can cities survive: Solid waste management in urban environments. *AT Source* 18(2): 21–27.

De Vore, S. 1998. Birds and buildings: A lethal combo. *Chicago Wilderness Magazine*, Spring.

Dodd, C.K., Jr., W.J. Barichivich, and L.L. Smith. 2004. Effectiveness of a barrier wall and culverts in reducing wildlife mortality on a heavily traveled highway in Florida. *Biological Conservation* 118:619–631.

Dolbeer, R.A., S.E. Wright, and E.C. Cleary. 2000. Ranking the hazard level of wildlife species to aviation. *Wildlife Society Bulletin* 28:372–378.

Eberhard, T. 1954. Food habits of Pennsylvania house cats. *Journal of Wildlife Management* 18:284–286.

Erickson, W.P., G.D. Johnson, M.D. Strickland, D.P. Young, Jr., K.J. Sernka, and R.E. Good. 2001. *Avian Collisions with Wind Turbines: A Summary of Existing Studies and Comparisons to Other Sources of Avian Collision Mortality in the United States.* Cheyenne, WY: Western EcoSystem Technology Inc.

Evans, W.R., Y. Akashi, N.S. Altman, and A.M. Manville II. 2007. Response of night-migrating songbirds in cloud to colored and flashing light. *North American Birds* 60:476–488.

Fahrig, L., J.H. Pedlar, S.E. Pope, P.D. Taylor, and J.F. Wegner. 1995. Effect of road traffic on amphibian density. *Biological Conservation* 73:177–182.

Fitzwater, W.D. 1988. Solutions to urban bird problems. *Proceedings Vertebrate Pest Conference* 13:254–259.

Forman, R.T.T. 2000. Estimate of the area affected ecologically by the road system in the United States. *Conservation Biology* 14:31–35.

Forman, R.T.T., D. Sperling, J. Bissonette, A.P. Clevenger, C.D. Cutshall, V.H. Dale, L. Fahrig, R. France, C.R. Goldman, K. Heanue, J.A. Jones, F.J. Swanson, T. Turrentine, and T.C. Winter. 2002. *Road Ecology: Science and Solutions.* Washington, DC: Island Press.

Foster, M.L. and S.R. Humphrey. 1995. Use of highway under passes by Florida panthers and other wildlife. *Wildlife Society Bulletin* 23:95–100.

Gabrey, W.W. 1977. Bird and small mammal abundance at four types of waste-management facilities in northeast Ohio. *Landscape and Urban Planning* 37:223–233.

Gallaher, P.A. 2003. The airport environment and its effect on wildlife damage to aircraft. M.S. Thesis. Embry-Riddle Aeronautical University, Daytona Beach, FL.

Gauthreaux, S.A., Jr. and C.G. Belser. 2006. Effects of artificial night lighting on migrating birds. In *Ecological Consequences of Artificial Night Lighting*, ed. C. Rich and T. Longcore, 67–93. Covelo, CA: Island Press.

Glista, D.J., T.L. DeVault, and J.A. DeWoody. 2008. Vertebrate road mortality predominantly impacts amphibians. *Hepetological Conservation and Biology* 3:77–87.

Greenhall, A.M. and S.C. Frantz. 1994. Bats. In *Prevention and Control of Wildlife Damage,* ed. S.E. Hygnstom, R.M. Timm, and G.E. Larson, D-5–D-24. University of Nebraska Cooperative Extension, U.S. Department of Agriculture-Animal Plant Health Inspection Service-Animal Damage Control. Lincoln, NE.

Herreid, C.F. II. 1963. Temperature regulation of Mexican free-tailed bats in cave habitats. *Journal of Mammalogy* 44:560–573.

Hoover, S.L. and M.L. Morrison. 2005. Behavior of red-tailed hawks in wind turbine development. *Journal of Wildlife Management* 69:150–159.

Horton, N., T. Brough, and J.B.A. Rochard. 1983. The importance of refuse tips to gulls wintering in an inland area of south-east England. *Journal of Applied Ecology* 20:751–765.

Hutchings, S. 2003. The diet of feral house cats (*Felis catus*) at a regional rubbish tip, Victoria. *Wildlife Research* 30:103–110.

Jackson, S.D. 2000. Overview of transportation impacts on wildlife movements and populations. In *Wildlife and Highways: Seeking Solutions to an Ecological and Socio-Economic Dilemma,* ed. T.A. Messmer and B. West, 7–20. The Wildlife Society. Bethesda, MD.

Jackson, S.D. and C.R. Griffin. 2000. A strategy for mitigating highway impacts on wildlife. In *Wildlife and Highways: Seeking Solutions to an Ecological and Socio-Economic Dilemma,* ed. T.A. Messmer and B. West, 143–159. The Wildlife Society. Bethesda, MD.

Jacobson, S. 2002. Using wildlife behavioral traits to design effective crossing structures. Wildlife Crossings Toolkit. .http://www.wildlifecrossings.info/.

Johnson, G.D., W.P. Erickson, M.D. Strickland, M.F. Shepherd, and D.A. Shepherd. 2003. Mortality of bats at a large-scale wind power development at Buffalo Ridge, Minnesota. *American Midland Naturalist* 150:332–342.

Johnson, G.D., M.K. Perlik, W.P. Erickson, and M.D. Strickland. 2004. Bat activity, composition, and collision mortality at a large wind plant in Minnesota. *Wildlife Society Bulletin* 32:1278–1288.

Johnson, G.D. and M.D. Strickland. 2003. *Biological Assessment for the Federally Endangered Indiana Bat* (Myotis sodalist) *and Virginia Big-Eared Bat* (Corynorhinus townsendii virginianus). Cheyenne, WY: Western Ecosystems Technology, Inc.

Keeley, B.W. 1995. Why do bats use this bridge but not that one? *Bats* 13:18.

Keeley, B.W. and M.D. Tuttle. 1999. *Bats in American Bridges.* Austin, TX: Bat Conservation International, Inc.

Kennamer, R. 1999. Wetlands, birds, and airports. http://www.uga.edu/srel/Graphics /Airports_birds.pdf.

Kershner, E.L. and E.K. Bollinger. 1996. Reproductive success of grassland birds at east-central Illinois airports. *American Midland Naturalist* 136:358–366.

Kiser, J.D., J.R. MacGregor, H.D. Bryan, and A. Howard. 2001. The use of concrete bridges as night roosts by Indiana bats in south central Indiana. *Bat Research News* 42:33.

Klem, D., Jr. 1989. Bird-window collisions. *Wilson Bulletin* 101:606–620.

Klem, D., Jr. 1990. Bird injuries, cause of death, and recuperation from collisions with windows. *Journal of Field Ornithology* 6:115–119.

Klem, D., Jr. 1991. Glass and bird kills: An overview and suggested planning and design methods of preventing a fatal hazard. In *Wildlife Conservation in Metropolitan Environments*, ed. L.W. Adams and D.L. Leedy, 99–104. Natl. Inst. Urban Wildl. Symp. Ser. 2.

Lewis, S.E. 1994. Night roosting ecology of pallid bats (*Antrozous pallidus*) in Oregon. *American Midland Naturalist* 132:219-226.

Lalo, J. 1987. The problem of road-kill. *American Forests* 50:50–52.

Linnell, M.A., M.R. Conover, and T.J. Ohashi. 1996. Analysis of bird strikes at a tropical airport. *Journal of Wildlife Management* 60:935–945.

Longcore, T., C. Rich, and S.A. Gauthreaux, Jr. 2008. Height, guy wires, and steady-burning lights increase hazard of communication towers to nocturnal migrants: A review and meta-analysis. *Auk* 125(2):485–492.

MacKinnon, B. 1997. Airport wildlife management: Beyond airport boundaries. *Airport Wildlife Management Bulletin*, no. 20.

Manley, T. and J. Williams. 1998. Bearproofing solid waste containers for grizzly and black bears in lake and city counties. *Intermountain Journal of Science* 4:101.

McCracken, G.F. 1999. Brazilian free-tailed bat. In *The Smithsonian Book of North American Mammals,* ed. D.E. Wilson and S. Ruff, 127–129. Washington, DC: Smithsonian Institution Press.

McDonnell, M.J. and S.T.A. Pickett. 1990. Ecosystem structure and function along urban-rural gradients: An unexploited opportunity for ecology. *Ecology* 71:1232–1237.

McKelvey, K.S., M.K. Schwartz, and L.F. Ruggiero. 2002. Why is connectivity important for wildlife conservation. http://www.wildlifecrossing.info/.

McKinney, M.L. 2002. Urbanization, biodiversity, and conservation. *Bioscience* 52:883–890.

Mumme, R.L., S.J. Shoech, G.E. Woolfenden, and J.W. Fitzpatrick. 2000. Life and death in the fast lane: demographic consequences of road morality in the Florida scrub-jay. *Conservation Biology* 14:501–512.

Murphy, M. 1990. The bats at the bridge. *Bats* 8:5–7.

Nehring, J.D. 1998. Assessment of avian population change using migration casualty data from a television tower in Nashville, TN. M.Sc. Thesis, Middle Tennessee State University, Murfreesboro, TN.

Peine, J.D. 2001. Nuisance bears in communities. *Human Dimensions* 6:223–237.

Restani, M., J.M. Marzluff, and R.E. Yates. 2001. Effects of anthropogenic food sources on movement, survivorship, and sociality of common ravens in the Arctic. *Condor* 103:399–404.

Rogers, L.L. 1989. Black bears, people, and garbage dumps in Minnesota. In *Bear-People Conflicts: Proceedings of a Symposium on Management Strategies*, ed. M. Bromley, 43–46. Northwest Territories Department of Renewable Resources, 6–10 April 1987, Yellowknife, Canada

Romin, L.A. and J.A. Bissonette. 1996. Deer-vehicle collisions: Status of state monitoring activities and mitigation efforts. *Journal of Wildlife Management* 24:276–283.

Rosen, P.C., and C.H. Lowe. 1994. Highway mortality of snakes in the Sonoran desert. *Biological Conservation* 68:143–148.

Rossi-Linderme, G. and B.K. Hoppy. 2000. Natural resource management and the bird aircraft strike hazard at Westover air reserve base, Massachusetts. *Proceedings 19th Vertebrate Conference*, ed. T.P. Salmon and A.C. Crabb, 63–67. University of California, Davis.

Rugge, L. 2001. An avian risk behavior and mortality assessment at the Altamont Wind Resource Area in Livermore, California. Thesis. California State University, Sacramento.

Schmidt, R. 1997. Drawing the line for wildlife. *Wildlife Control Technology*, December, 6–7.

Shire, G.G., K. Brown, and G. Winegrad. 2000. *Communication Towers: A Deadly Hazard to Birds*. Washington, DC: American Bird Conservancy.

Slate, D., J. McConnel, M. Barden, R. Chipman, J. Janicke, and C. Bently. 2000. Controlling gulls at landfills. Proceedings 19th Vertebrate Pest Conference, ed. T.P. Salmon and A.C. Crabb. University of California, Davis.

Smallwood, K.S. and C.G. Thelander. 2004. Developing methods to reduce bird mortality in the Altamont Pass Wind Resources Area. Final report to the California Energy Commission, PIER-EA contract No. 500-01-019, Sacramento, California

Smallwood, K.S. and C.G. Thelander. 2008. Bird mortality in the Altamont Pass Wind Resource Area, California. *Journal of Wildlife Management* 72:215–223.

Speich, S.M., H.L. Jones, and E.M. Benedict. 1986. Review of the natural nesting of the barn swallow in North America. *American Midland Naturalist* 115:248–254.

Spellerberg, I.F. 1998. Ecological effects of roads and traffic: A literature review. *Global Ecology and Biogeography Letters* 7:317–333.

Stringham, S.F. 1989. Demographic consequences of bears eating garbage at dumps: An overview. In *Bear-People Conflicts: Proceedings of a Symposium on Management Strategies*, ed. M. Bromley, 35–42. Northwest Territories Department of Renewable Resources, 6–10 April, 1987, Yellowknife, Canada

Sullivan, R. 2004. *Rats: Observations on the History and Habitat of the City's Most Unwanted Inhabitants.* New York: Bloomsbury.

Ternent, M.A. and D.L. Garshelis. 1999. Taste-aversion conditioning to reduce nuisance activity by black bears in a Minnesota military reservation. *Wildlife Society Bulletin* 27:720–728.

Tigas, L.A., D.H. Van Vuren, and R.M. Sauvajot. 2002. Behavioral responses of bobcats and coyotes to habitat fragmentation and corridors in an urban environment. *Biological Conservation* 108:299–306.

Transportation Research Board. 2002. *Interactions between Roadways and Wildlife Ecology: A Synthesis of Highway Practice.* Washington, DC: National Cooperative Highway Research Program.

Trombulak, S.C. and C.A. Frissell. 2000. Review of ecological effects of roads on terrestrial and aquatic communities. *Conservation Biology* 14:18–30.

Tuttle, M.D. 1979. Status, causes of decline, and management of endangered gray bats. *Journal of Wildlife Management* 43(1):1–17.

U.S. Environmental Protection Agency. 1993. Guidance specifying management measures for sources of nonpoint pollution in coastal waters. Office of Water. Washington, DC.

U.S. Environmental Protection Agency. 2002. Municipal solid wastes in the United States: 2000 facts and fig-
    ures. Office of Solid Waste and Emergency Response. Washington, DC.

VanDruff, L.W., E.G. Bolen, and G.J. San Julian. 1996. Management of urban wildlife. In *Research and
    Management Techniques for Wildlife and Habitats*, 5th ed. rev., ed. T. A. Bookhout, 507–530. Bethesda,
    MD: The Wildlife Society.

Walker, C.W., J.K. Sandel, R.L. Honeycutt, and C.E. Adams. 1996. Winter utilization of box culverts by ves-
    pertilionid bats in Southeast Texas. *Texas Journal of Science* 48:166–168.

West, S. 1995. Cave swallows (*Hirundo fulva*). In *The Birds of North America*, No. 141, ed. A. Poole and F.
    Gill. Philadelphia, PA: The Birds of North America Inc.

Williams, D. 2002. Conspicuous consumption. *National Parks* 76:40–45.

Wilson, D.E. and S. Ruff. 1999. *The Smithsonian Book of North American Mammals.* Washington, DC:
    Smithsonian Institution Press.

Wright, R. T. 2004. *Environmental Science: Toward a Sustainable Future.* 9th ed. Pearson Education, Upper
    Saddle River, NJ.

Wright, S.E., R. A. Dolbeer, and A.J. Montoney. 1998. Deer on airports: An accident waiting to happen. In
    Proc. 18th Vertebrate Pest Conference, ed. R.O. Baker and A.C. Crabb, 90–95. University of California,
    Davis.

PART IV

Sociopolitical Issues

CHAPTER 9

Human Dimensions in Urban Wildlife Management

You see, one of the most important things to managing wildlife has nothing to do with wild-life at all. It's managing people!

—Neil A. Waer, Wildlife Biologist

KEY CONCEPTS

1. There are many definitions for "human dimensions" (HD) of wildlife management.
2. HD research topics investigate people's activities, attitudes, expectations, and knowledge concerning wildlife.
3. There are at least three journals that publish papers on HD research.
4. The design of an HD study involves at least seven sequential steps.
5. People view wildlife in terms of at least nine different wildlife value typologies.
6. The consequences of providing wildlife with supplemental feeding (intentional and/or unintentional) can be positive, negative, or neutral.
7. The challenges of managing human-wildlife conflicts in human-made environments consist of complex interrelationships.
8. There are many factors that limit urban wildlife biologists' abilities to address urban wildlife conflict issues.
9. Conflict resolution using public involvement consists of at least five different approaches.

INTRODUCTION

As Americans have become more urbanized, their relationship with wildlife has changed considerably. Traditional consumptive-use activities, such as hunting and fishing, continue to be popular forms of recreation, albeit for a decreasing percentage of the population. Wildlife watching, on the other hand, has become increasingly popular. Activities such as observing, feeding, and photographing wild animals are now a big business, accounting for expenditures of over $45.7 billion in 2006 alone (U.S. Department of the Interior 2006). Expenditures grew 19 percent in the five years since the previous survey. During the same period fishing and hunting accounted for $42.0 billion and $22.9 billion, respectively.

Not all interactions between people and wildlife, however, are enjoyable. As human development sprawls into formerly wild habitat, and as wildlife species attempt to adapt to the presence of humans, conflicts between humans and wildlife have increased. Some of the most common forms of conflict include damage to homes and gardens, crops, and livestock, and threats to the health and safety of people and their companion animals. As a result, conflicts may arise between the general public and wildlife managers as well. Several regional studies have indicated that urban residents often have conflicting goals, such as the desire to reduce human-wildlife conflict with particular species while at the same time enhancing wildlife viewing opportunities of the same species (Conover 1997). In addition, the American public has widely diverse attitudes, expectations, and tolerance levels concerning wildlife. Developing management solutions and communication strategies to meet the needs of all stakeholders has proven to be a challenge for both wildlife management agencies and the wildlife professionals they employ.

THE "PEOPLE FACTOR"

Wildlife professionals have long joked that their job is really about managing people. The importance of the "people factor" was recognized by no less notable a figure than Aldo Leopold in his groundbreaking text on *Game Management,* published in 1933. But it wasn't until the 1970s that the wildlife profession as a whole began to give serious consideration to the effect of human activities and perceptions on wildlife management.

Human dimensions (HD) is the study of the human considerations in wildlife management. Stout (1998) found several definitions of human dimensions research, including the following.

1. The application of social-science principles about human behavior for biological and ecological aspects of natural resource management. (Gigliotti and Decker 1992)
2. An approach consisting of an explanation of human actions based on social science concepts and methods and the application of these findings in wildlife decision-making, with emphasis on theory-building. (Manfredo, Vaske, and Decker 1995)
3. A science which integrates multiple disciplines into natural resource management; specifically, HD research is "the scientific investigation of the physical, biological, sociological, psychological, cultural, and economic aspects of natural resource utilization at the individual and community levels." (Ewert 1996:5)

It is evident that the term covers a broad set of research topics, including communication, behavior of both groups and individuals, program assessments, public participation in management planning and implementation, as well as economic, social, aesthetic, and other types of wildlife values.

The key element in HD research is a diversified approach that integrates sociology, psychology, anthropology, economics, ecology, and education in natural resource management. Stout (1998) provided a summary of the types of study themes and management applications addressed in HD research (Table 9.1). The first book dedicated to an aspect of human dimensions appeared in 1964, Douglas Gilbert's *Public Relations in Natural Resources Management.* This seminal text was followed by a small but steady stream of publications from the 1980s through today (Table 9.2).

Presently, there are many journals publishing human dimensions research, including *Fisheries* (American Fisheries Society), *Human Dimensions of Wildlife* (Taylor & Francis Group), *Journal of Leisure Research* (National Recreation and Park Association), *Leisure Sciences Journal* (Taylor & Francis Group), *Proceedings of the Southeastern Association of Fish and Wildlife Agencies, Journal of Wildlife Management* (The Wildlife Society), *Conservation Biology* (Wiley-Blackwell), *Biodiversity and Conservation* (Springer), and *Transactions of the North American Wildlife and Natural Resources Conference.* Clearly, interest in human dimensions of wildlife management has grown substantially since Leopold first suggested the concept.

Table 9.1 Examples of Research Themes and Management Applications of Human Dimensions Studies

Research Themes	Management Applications
Attitudes, beliefs, motivations, satisfactions, preferences, and ethics of user groups	Agency and program needs assessments, design specifications, and evaluation
Modeling of recreational participation, specialization, and involvement	Natural resources education, interpretation, and communication
Quantification of socioeconomic and noneconomic values (cost and benefit)	Public involvement, conflict management, and dispute resolution strategies
Risk perception and communication	Human impacts on resources, situational analyses, and demand projections
Contributions of wildlife to human quality of life	Human aspects of species management
Identification of socialization and social worlds	Market segmentation, stakeholder identification and characterization
	Integrated resource planning and policy analysis
	Appraisals of natural resource professionals

Source: Bryan (1996); Decker, Brown, and Knuth (1996); Ditton (1996); Stout (1998).

Table 9.2 Books on the Human Dimensions of Wildlife Management, 1964–2008

Year	Title	Authors or Editors
1964	*Public Relations in Natural Resources Management*	D. Gilbert
1987	*Valuing Wildlife: Economic and Social Perspectives*	D.J. Decker and G.R. Goff
1992	*American Fish and Wildlife Policy: The Human Dimension*	W.R. Mangun
1993	*Wildlife and People: The Human Dimensions of Wildlife Ecology*	G.G. Gray
1996	*Natural Resource Management: The Human Dimension*	A. W. Ewert
2001	*Human Dimensions of Wildlife Management in North America*	T.L. Brown, W.F. Siemer, and D.J. Decker (Editors)
2008	*Wildlife and Society: The Science of Human Dimensions*	M.J. Manfredo, J.J. Vaske, P.J. Brown, D.J. Decker, and E.A. Duke (Editors)
2008	*Who Cares About Wildlife?*	M.J. Manfredo

CONDUCTING HUMAN DIMENSIONS RESEARCH

The complexity of HD studies makes them difficult to conduct, and in many cases this type of research is held to a more rigorous standard than traditional wildlife research. For example, human dimensions researchers can easily find reliable data on the size of the human population and their subset of interest (e.g., birdwatchers, hunters, anglers, hikers). Strict attention must be paid to these numbers in order to draw a sample that will accurately represent the entire population; failure to adhere to the rules of sampling protocol will cause the study to be labeled as "biased" by peer reviewers. Conversely, researchers involved in the study of wildlife species rely on population estimates as a matter of course. The fact that methods used for estimating wildlife numbers can have huge margins of error often is ignored.

Other standards that must be met when undertaking a human dimensions study include: (1) a study sample that is a large enough subset of the target population to reduce sampling error to ±3 percent, (2) factors that contribute to bias must be anticipated and controlled, and (3) the study must include measurable objectives and research hypotheses. Measurable objectives and hypotheses focus on "who, what, where, and when" questions. For example, a human dimensions study can reasonably expect to measure whether or not gender plays a role in acceptance of a particular management technique, but it cannot reasonably expect to measure why men and women react differently to a proposed method.

Researchers attempting to design a human dimensions study should follow seven basic steps (Salant and Dillman 1994):

1. *Determine the target audience and the best method for reaching them.* Considerations may include
   - Whether the target audience is broad (e.g., general public) or narrow (e.g., birders);
   - How potential respondents will be identified;
   - How contact will be made (e.g., by telephone, by mail, or in-person interviews);
   - Available resources (e.g., money, people, and time).
2. *Identify control groups for comparison of responses with target group.* For example, if a study involves resident attitudes concerning urban deer management, both neighborhoods with and neighborhoods without deer conflicts should be included in the research design.
3. *Develop the survey instrument.* This step takes both time and skill to do well and should use the study objectives and research hypotheses as the blueprint for development.
4. *Determine appropriate statistical measures to use in data analysis.* First establish whether the study is attempting to establish a qualitative or quantitative difference. Keep in mind there are situations where statistical tests are irrelevant. For example, if a survey of all state natural resource agencies is conducted, and responses are received from all fifty agencies, then the study has captured all of the possible respondents. Any difference in response is statistically significant.
5. *Administer the survey instrument.* Response rates need to be high enough (e.g., >50 percent) to allow for a statistically significant representation of the target audience. The smaller the target audience, the larger the sample needed to provide significant representation.
6. *Follow up with nonrespondents.* This step is necessary to determine if there is some inherent bias in the nonrespondent group. For example, there may be a common factor (e.g., age, gender) within the nonrespondent group that prevents them from participating in the study.
7. *Communicate results.* The final step of any research project is to share the results in peer-reviewed journals, popular literature, the news media, and at professional and public meetings.

It is possible to short-circuit the rigor in HD research by cutting corners or adopting a study design that isn't grounded in measurable objectives or testable hypotheses. A common misconception by individuals not schooled in reputable HD research methods is that survey instrument development involves little more than coming up with some questions to ask constituents—in other words, an attitude of "anyone can create an effective questionnaire." Not so! Such simplistic attempts will provide quick answers, in particular, the answers desired, but with little consideration given to sampling protocols, control of bias, nonresponse follow-ups, hypothesis testing, and communicating the data to anyone beyond those who initiate the study. The data reported are usually in the form of frequencies (percent of this or that) without an examination of connections to other factors that might cause a response. Information gathered in this way—often at considerable expense—usually cannot stand the rigor of HD peer review. Those who do not have a background in questionnaire development and/or statistical analysis should take advantage of the personnel resources available through universities and even the private sector rather than use the "do-it-yourself" approach.

SURVEYING WILDLIFE RECREATIONISTS

The National Survey of Fishing, Hunting, and Wildlife-Associated Recreation (FHWAR Survey), conducted by the U.S. Department of the Interior every five years since 1955, is one of the oldest and most comprehensive continuing human dimensions surveys available. Wildlife-associated recreation, formerly referred to as "nonconsumptive wildlife-related recreation," includes observing, photographing, and feeding fish and wildlife. The FHWAR Survey's three interest categories (fishing, hunting, and wildlife-associated recreation) are not mutually exclusive—many of the individuals surveyed report enjoying fish and wildlife in a variety of ways.

The purpose of the FHWAR Survey is to gather information on the number of anglers, hunters, and wildlife-watchers in the United States—the activities they participate in, how often they participate, and how much they spend on these activities. Survey results are presented on a national and a state-by-state basis. Results include participation rates by various demographic characteristics. Total days of participation, trips, and expenditures also are reported, as are the common types of animals hunted, fished, or watched. The results of the 1991, 1996, 2001, and 2006 surveys are available online (http://www.census.gov/prod/www/abs/fishing.html). Because the FHWAR Survey methodology and questions were similar during these four survey periods, estimates and results are comparable.

Human dimensions researchers have used specific data from the FHWAR Survey as a starting point to conduct more detailed analysis of the economic impacts of hunters, anglers, and wildlife-watchers, such as:

- The extent of under-representation by gender, racial, and ethnic groups;
- Changes in participation rates during five-year periods;
- Changes in wildlife recreation expenditures.

These studies are published as separate reports available online from the U.S. Census Bureau (see URL provided above) or from private, for-profit research groups (e.g., Southwick Associates).

The general public may not consider the FHWAR Survey to be of great importance, or even be aware of its existence, but the study provides valuable information to communities and businesses on the potential economic impact that can be realized from local wildlife resources.

THE ROLE OF HUMAN DIMENSIONS IN URBAN WILDLIFE MANAGEMENT

There are many roles for human dimensions in urban wildlife management. For the sake of this discussion, however, four broad categories of assessments will be used; each will be examined in the order listed below:

- Public participation in wildlife-related programs;
- Wildlife values;
- Quality of life issues;
- Conflict resolution.

Public Participation in Wildlife-Associated Recreation

According to data compiled in the 2006 FHWAR Survey (U.S. Department of the Interior 2006), nearly one-third of all U.S. residents (71.1 million) participated in some type of wildlife-watching activities. The popularity of birds and mammals may have a connection to the charismatic association that people have regarding animals with feathers and fur compared to animals with scales and exoskeletons (reptiles and insects).

A quick surf through the television channels provides additional evidence of continued interest in all things wild. Not only are shows about wildlife commonplace, there are now entire networks (e.g., Animal Planet) devoted to wildlife and nature programming.

There are many public wildlife-related programs that use backyard wildlife for appreciation and the betterment of the community. Two examples—bats in bridges and Eagle Days—illustrate how these programs are implemented. Alternatively, there exist some highly questionable public wildlife-related programs that lead to serious negative consequences. Rattlesnake roundups are one

Figure 9.1   Crowds gather on summer evenings in Austin, Texas, to watch the evening emergence of Mexican free-tailed bats from under the Congress Avenue Bridge. (Courtesy Merlin D. Tuttle, Bat Conservation International)

example that represents a form of wildlife exploitation that is difficult to justify by any standard (Adams and Thomas 2008). Several studies have examined the HD consequences of this type of program (Thomas and Adams 1993; Adams et al. 1994; Fitzgerald and Painter 2000). A second example is that of constructed habitats (e.g., wildscaping). These programs, while well intentioned, may have unintended ecological consequences. Few HD assessments of constructed habitat programs have been conducted.

Many urban communities use local wildlife resources to generate revenue for their citizens and cities. As mentioned in Chapter 8, the Mexican free-tailed bat (*Tadarida brasiliensis*) colony under the Congress Avenue Bridge in Austin, Texas, is a perfect example. The number of tourists and the revenue generated from those who visit the bridge to watch the bats emerge during the summer months is enormous (Figure 9.1). The following excerpt from the website of Bat Conservation International illustrates this point:

> As the city came to appreciate its bats, the population under the Congress Avenue Bridge grew to be the largest urban bat colony in North America. With up to 1.5 million bats spiraling into the summer skies, Austin now has one of the most unusual and fascinating tourist attractions anywhere. *The Austin American-Statesman* created the Statesman Bat Observation Center adjacent to the Congress Bridge, giving visitors a dedicated area to view the nightly emergence. It is estimated that more than 100,000 people visit the bridge to witness the bat flight, generating $8 million in tourism revenue annually. (Bat Conservation International website)

Another example of marketing the wildlife in your own backyard is the highly successful Eagle Days program (Figure 9.2), started in Missouri in the late 1970s (Witter, Wilson, and Maupin 1979). Bald eagles (*Haliaeetus leucocephalus*) use the Mississippi, Missouri, and Arkansas Rivers, along with various reservoirs, in their southern migration from Canada. They establish wintering grounds (stopovers) where they feed on fish and rest near urban communities and then move on. Several urban communities (Clarksville, Missouri; Tulsa, Oklahoma; Cassville, Wisconsin; and LaClaire, Iowa) have realized that they can convert a "free" natural phenomenon into a revenue-generating

Figure 9.2 Eagle Days festivals are held annually in Missouri and other states. (Courtesy Harold Selby)

opportunity. Regrettably, there is no hard data on the amount of additional revenue generated by communities that promote Eagle Day events. It is known that what started out as a few hundred observers has grown into the thousands, causing some communities to "gear up" with enhanced goods and services during Eagle Days.

In both of the above cases people—particularly urbanites—are spending time and money to observe wildlife "up close and personal." The wildlife resource is used in a safe and unobtrusive manner that allows each species to go about its business, but under the watchful eyes of curious humans. It is assumed that humans are increasing their awareness, knowledge, and appreciation of wildlife. Whether these transformation conditions translate into an increased stewardship ethic and concern for habitats is unknown. The last statement is rather tragic; the tools used in the human dimensions of wildlife management could be easily used to determine the attitudes, activities, expectations, and knowledge of program participants concerning wildlife and their habitats.

There is a flip side to the story on how humans use the wildlife resources around them; rattle-snake roundups provide an example of exploitation under the guise of recreation and public education (Figure 9.3), and it has been explored in depth in the recently published *Texas Rattlesnake Roundups* (Adams and Thomas 2008). Unlike the previous two examples, rattlesnake roundups have been studied extensively in terms of the human dimensions, educational, ecological, and economic ramifications (Thomas and Adams 1993; Adams et al. 1994; Fitzgerald and Painter 2000, Adams and Thomas 2008). Rattlesnake roundups are events in which rattlesnakes (*Crotalus* sp.) are collected from the wild for public display, sale, and consumption. It would be difficult to identify another public event that uses a wildlife resource in this manner. In the words of the Humane Society of the United States (HSUS),

> Rattlesnake roundups are cruel and ecologically damaging events that have taken place in many parts of the U.S. since the late 1920s. Every year, in the states of Texas, Oklahoma, Kansas, New Mexico, Pennsylvania, Alabama, and Georgia, thousands of rattlesnakes are captured and slaughtered or used in competitive events in ways that violate the most basic principles of wildlife management and humane living. Though they are often promoted as fundraising events for local civic causes, rattlesnake roundups primarily benefit their organizers and corporate sponsors. ["The Truth behind Rattlesnake Roundups" (Humane Society of the United States website)]

Figure 9.3 *(A color version of this figure follows page 158.)* A Texas rattlesnake roundup holding pen for captured rattlesnakes. (Courtesy Clark E. Adams)

In Texas, one of the states in which these events are held, a hunting license is required to pick the snake up off the ground. As long as this requirement is met, there are no additional regulatory statutes that govern the harvest of rattlesnakes. A snake in the bag is considered to be the property of the individual holding the bag—no questions asked (Adams et al. 1994).

Rattlesnake roundups may bring as much as $3 million into a community in one weekend. Civic organizations host the events and use the revenue generated from the sale of snakes to dealers to provide student scholarships and to support other community programs. Roundup spectators number in the thousands. They are local residents, tourists, the media, and members of environmental protest groups. Most are between the ages of twenty-four and fifty-five, reside in large cities (population of 50,000), and are predominantly white males (Thomas and Adams 1993).

The promoters and supporters of rattlesnake roundups tout these events as educational programs that enlighten the public on the natural history of the animals, biological aspects of rattlesnake existence, and how to treat a rattlesnake bite. However, the "educational" programs also include how to butcher a rattlesnake, remove one's body from a sleeping bag occupied by twenty or more rattlesnakes, or safely walk among hundreds of rattlesnakes. Serious consideration needs to be given to public attitudes toward rattlesnakes, and snakes in general, before and after attending one of these events. At least one study has shown that, when spectators leave the educational events, their knowledge of rattlesnakes is similar to those who did not attend the programs (Fitzgerald and Painter 2000).

Some of the most popular educational efforts in urban areas are programs undertaken by both private organizations and government agencies to encourage homeowners to develop wildlife habitat on their properties. The objectives of many of these programs include: (1) increasing awareness and appreciation of wildlife, (2) educating homeowners and community leaders regarding habitat creation and restoration, and (3) increasing the total amount of quality habitat or "green space" in urban and suburban areas. For example, the National Wildlife Federation (NWF), a nonprofit organization, sponsors the well-known "Backyard Habitat Program." Since 1973, the NWF has been providing tools and resources for homeowners to create and restore native wildlife habitats in their backyards. Several groups have similar "wildscaping" programs, and there are many websites on this topic.

Incorporating wildlife habitat into residential areas is becoming increasingly common for homeowners and developers and is touted as a way to reduce some of the impacts of residential development on wildlife populations. A study by Hunter (2002) conducted over a two-year period in Bexar County, Texas, found bird diversity at the wildscaped neighborhood was significantly greater than in traditional neighborhoods or natural areas. Unfortunately, similar studies are rare. Assessment of whether stated objectives have been met often is a missing component in these types of wildscaping programs. In addition to more studies on the impact of wildscaping on species diversity, many other questions need to be asked: Do the programs actually promote more awareness about and appreciation of urban wildlife? Do they increase homeowners' knowledge of wildlife habitats in and outside of urban areas? Do they increase public sensitivity to conservation issues such as loss of biodiversity, exotic species, and pollution?

There are several possible reasons behind the lack of evaluation for public wildlife education and awareness programs. To begin with, program evaluation is usually an afterthought rather than part of the up-front planning of the program in the form of clearly articulated and measurable program objectives. Often the primary goal is simply to get the program "on the street," after which the number of participants is used to "measure" program success. There may be few, if any, personnel associated with the program who have training and experience in program assessment. Lack of professional interest in investigating the questions listed above may be the result of the low representation of urban biologists at the university level. Another explanation may be the fact that dollars drive the direction of research. Organizations may not be eager to find out that a popular program is having ecological repercussions that are not all positive. Finally, program managers may find it difficult to allocate money to do assessments when other programs are waiting for funding.

Wildlife Values

Economic benefit is one way to value wildlife, but there are many others. The terms *value* and *assigned value* are used interchangeably here to mean the level of worth a person gives to one object relative to another (Brown and Manfredo 1987). Researchers have attempted to categorize and measure both values and attitudes toward wildlife; in fact, neither is directly measurable. Using classification models or values scales as measurement tools, the underlying values of individuals are inferred. King (1947) began the process of categorization by defining six overlapping types of values; several decades later Kellert (1980) described ten attitude categories. Both of these typologies are commonly cited in human dimensions publications. Purdy and Decker (1989) developed the wildlife attitude and values scale (WAVS), patterned after King's classification model, to describe noneconomic values of wildlife. A comparison of various wildlife value categorizations can be found in Table 9.3.

Table 9.3 Comparison of Several Wildlife Value Categorizations

King (1947)	Kellert (1980)	Purdy and Decker (1989)	Fulton, Manfredo, and Lipscomb (1996)
Aesthetic	Aesthetic	Problem-Acceptance	Bequest & Existence
Biological	Dominionistic	Societal-Benefits[a]	Fishing/Anti-fishing
Commercial	Ecologistic	Traditional-Conservation	Hunting/Anti-hunting
Educational	Humanistic		Recreational Experience
Recreational	Moralistic		Residential Experience
Social	Naturalistic		Wildlife Education
	Negativistic		Wildlife Rights
	Neutralistic		Wildlife Use
	Scientistic		
	Utilitarian		

[a] Consists of social-significance and ecological-significance components.

Table 9.4   Wildlife Value Typologies

Gilbert and Dodds (2001)	
Monetary values (commercial or economic)	Uses of wildlife that can be exchanged for money or have an economic impact on an individual or society.
Recreational values	The enjoyment people experience from hunting, birding, camping, or wildscaping. There is some overlap between recreational and monetary values when people spend money to pursue these activities.
Scientific/educational values	The roles wildlife species play in advancing human understanding, including the creation and dissemination of knowledge.
Biological, ecological, or ecosystem values	The benefits wildlife species provide by maintaining a functional ecosystem.
Option values	Related to monetary because it is based on the amount a person is willing to pay today to ensure that a resource will be available for his or her use in the future (Krutilla 1967; Randall 1991).
Existence values	The sense of well-being a person experiences from just knowing that wildlife exists in the natural world (Randall 1991).
Additional Wildlife Values (Conover and Conover 2003)	
Historical value	Based on the premise that humans, by their basic nature, are nostalgic and long for the way things used to be, including the presence of certain endemic wildlife species.
Empathetic value	Based on the human ability to empathize with other living beings, such as an animal's fear, distress, or pain.
Character-assessment value	Based on the idea that people who behave ethically or altruistically towards animals are likely to behave similarly toward people; this allows people to ascertain the true character of others.

Other authors have developed their own classification systems for wildlife values. F.F. Gilbert and Dodds (2001) listed six wildlife values generally recognized by biologists. Conover and Conover (2003) argue there are three wildlife values not yet articulated by biologists that people sense at a subconscious level (Table 9.4).

Because people hold a variety of opinions, attitudes, and values concerning natural resources, and because those values are difficult, if not impossible, to measure, wildlife managers face the challenge of trying to make science-based decisions that will also address the needs and expectations of as large a portion of the public as possible. This is not an easy task. Managing wildlife, urban or otherwise, is a relatively simple task compared to managing people.

Quality of Life Issues

Through the ages wildlife has inspired art, music, dance, drama, storytelling, and poetry. Wildlife continues to add to our quality of life. It doesn't seem to matter much whether it's watching visitors to a backyard bird feeder, listening to frogs calling on a summer night, or spotting a white-tailed buck from a hunting blind—millions of North Americans would consider their lives to be less enjoyable without wildlife. In fact, many quality of life indices include wildlife and nature along with good schools, above-average annual income, and cultural opportunities as characteristcs of an appealing community.

The benefits of wildlife and wildlife habitat go well beyond intangibles, as many studies have shown. In Salem, Oregon, urban land adjacent to greenspace was found to be worth approximately $1200 more per acre than urban land 1000 feet away, all other things being equal (Nelson 1986). In the neighborhood of the Cox Arboretum in Dayton, Ohio, the proximity of the park and arboretum accounted for an estimated 5 percent of the average residential selling price, and in the Whetstone Park area of Columbus, Ohio, the nearby park and river were estimated to account for 7.35 percent of selling prices (Kimmel 1985). There has been increased interest in developing urban and suburban wildlife habitats as a way to attract real estate buyers. Housing developers have caught on to the fact that nature sells, and as a result, many have begun to promote the incorporation of "green space" in their development plans.

Studies of groups considered "uncommitted" on attitudes about wildlife issues have found that, while these individuals did not seek out wildlife-related activities, they appreciated wildlife in the context of their other activities (Fleishman-Hillard Research 1994). Access to wildlife may even improve human health. Some researchers have proposed that humans have an innate need to associate with other species, termed *biophilia* by Harvard biologist Edward O. Wilson (1984). Emory University Rollins School of Public Health scientist Howard Frumkin (2001) has presented evidence for the health benefits of four kinds of contact with the natural environment: contact with nonhuman animals, contact with plants, viewing landscapes, and contact with wilderness. Several studies have linked nonhuman animals with human health. Frumkin pointed to research that concludes pet owners have fewer health problems than non-pet owners. Examples include lower blood pressure, improved survival after heart attacks, and enhanced ability to cope with life stresses.

People can be involved with wildlife near their homes in a variety of ways, from watching wild animals to participating in management by providing food, water, shelter, and space for wildlife (Leedy and Adams 1984; Young 1991). As we discussed earlier in this chapter, *wildscaping*—a form of landscaping that relies heavily on native plants to attract birds, butterflies, and other types of wildlife—and other methods for attracting wild animals have become a fulfilling hobby for many Americans (Figure 9.4). State and federal wildlife management agencies have responded to the increased popularity of residential wildlife watching by creating books, pamphlets, and videos, and developing and promoting "watchable wildlife" programs, often in cooperation with local chambers of commerce and tourism departments.

However, urbanization has isolated humans from the natural world. In general, people who live in urban and suburban areas have a lower overall level of knowledge about wildlife compared to rural residents (VanDruff, Bolen, and San Julian 1994). Since the majority of the human population in North America can now be classified as urban, there is a greater need than ever before to educate the public about wildlife. It is hard to enjoy something you don't understand, or even fear (see Perspective Essay 9.1).

Figure 9.4   Wildscaping offers the chance to enjoy nature close to home. (Courtesy John M. Davis)

Figure 9.5 Urbanites have a tendency to misinterpret wildlife behavior, which can lead to a variety of unwanted consequences, including human injury. This man was charged by a small black bear while attempting to take a photo. (National Park Service Historic Photograph Collection, c.1950)

Studies have shown that, while interest in wildlife and the environment may be high, urbanites tend to be unaware of the names of all but the most common species and lack understanding about the way the natural world works (Adams et al. 1987; Penland 1987). One study of high school students (Adams et al. 1987) found that many could not correctly identify common urban wildlife species (e.g., confusing opossums and rats), the relative numbers of selected wild species in Harris County, Texas (e.g., raccoons were considered rare, while cougars were thought to be abundant), the diets of sixteen common mammals, and the effect of human habitation on relative abundance of those mammals.

In addition, urbanites have a tendency to misinterpret wildlife behavior. Examples include thinking that a wild animal that does not maintain a safe escape distance "wants to be friends." This lack of understanding can result in a variety of unwanted consequences, including human injury (e.g., tourists mauled by seemingly tame park bears), removal of animals from the wild (e.g., "rescuing" deer fawns), or contributing to exponential proliferation of certain wildlife populations (e.g., feeding Canada geese [*Branta canadensis*] on city ponds). Lack of public information, education, and awareness of wildlife has been identified as an important wildlife issue (VanDruff, Bolen, and San Julian 1994; Figure 9.5).

People from all walks of life want and need information about wildlife. They include adults and school children, elected officials and business owners, local governments, apartment managers, homeowners, and members of the media. Their needs and reasons for obtaining wildlife information are just as diverse, ranging from a desire to increase opportunities to enjoy wildlife on a daily basis, to avoiding property damage, to improving public health (Young 1991). Unfortunately, when questions about living with wildlife arise, Americans often are unsure whom to contact for help (Reiter, Burnson, and Schmidt 1999).

Natural resource agencies have much at stake in shaping public understanding of wildlife management and conservation (Mankin, Warner, and Anderson 1999). Agencies have been quite successful in identifying and communicating with traditional clienteles (Decker et al. 1987; Witter

1990; Hesselton 1991; Kania and Conover 1991; Gray 1993; Jolma 1994). Some agencies also have attempted to reach out to nonconsumptive users; Texas Parks and Wildlife Department's (TPWD) Nongame and Urban Wildlife Program, for example, has made a major effort to develop programs specifically to appeal to adults excluded by hunting and fishing programs (Adams, Leifester, and Herron 1997). On the whole, however, the wildlife profession has had difficulty communicating effectively with the general public (Decker et al. 1987; Jolma 1994).

Human-Wildlife Conflict Issues

Urban wildlife conflict resolution issues can be placed into one of two subsets: conflicts between humans and wildlife, and conflicts between humans over wildlife resources. Often, although not always, human-wildlife conflicts are the easier of the two to address. Many how-to books and other readily available reference materials provide detailed instructions for solving common problems that arise between humans and wildlife. Rather than repeat those efforts, this section offers an introduction to some of the more prevalent human-wildlife conflict issues natural resource professionals and the public are likely to face.

By the same token, hundreds of books devoted exclusively to solving disputes between humans using well-researched methods of negotiation and conflict resolution are easy to find at almost any bookstore. To repeat those efforts here, even in abbreviated form, would not do justice to the subject. However, the topic is one that is vitally important to wildlife managers in general and urban wildlife managers in particular. Anyone involved in this field is strongly advised to consider taking one or more of the many excellent courses available on this subject. Many reference materials on negotiation and conflict resolution are also available. Discussion here will be limited to an overview of the issues that may arise and the ways in which human dimensions can be applied.

Conflicts between humans and wildlife species have increased as urbanization has increased. This is due in part to greater exploitation of urban and suburban habitats by wildlife. Because people enjoy feeling a connection with nature and wildlife, many homeowners actively work to attract wild animals to their property by offering supplemental food (bird, squirrel, and butterfly feeders), water (bird baths and misters), and shelter (bird, bat, and toad houses). State agencies and nongovernmental organizations (NGOs) such as the National Wildlife Federation and others encourage homeowners to wildscape their yards to provide both cover and food for wildlife. In other cases, people "accidentally" welcome wildlife to their homes and yards by leaving garbage cans and pet food and water bowls outside.

Regardless of a homeowner's intent, often the result of providing supplemental food, water, and shelter is a higher density of animals than would normally be supported by an equal amount of natural habitat. Most members of the public remain largely uninformed about the consequences of attracting wildlife. For example, many do not understand that the number of animals will increase as the amount of food offered increases. This concept might sound quite attractive to the novice wildlife aficionado, but the number of animals expecting to be fed can quickly exceed the homeowner's budget or tolerance level—not to mention that of their neighbors! Furthermore, large congregations of animals promote the transmission of both disease and parasites that can decimate local populations of one or more species—hardly the scenario most wildlife-watchers have in mind when they begin to offer supplemental food.

Research has shown that effective education programs are needed to introduce the public to the realities of wildlife conflict and damage prevention (Adams, Stone, and Thomas 1988; Lowery and Siemer 1999). For example, food may be plentiful in the city, but often natural denning sites are not. This can cause raccoons, opossums, skunks, and squirrels to set up housekeeping in attics, decks, and out-buildings. This, in turn, increases the number of wildlife complaints from homeowners.

Not only is the public uneducated concerning human-wildlife conflicts, relatively few are knowledgeable about which agencies are responsible for managing wildlife. A national study on public attitudes toward wildlife damage and policy found that only 19 percent of survey respondents had ever heard of the U.S. Department of Agriculture's Wildlife Services program, formerly known as Animal Damage Control (Reiter, Burnson, and Schmidt 1999). A study of Texas residents found the majority were unable to identify their state wildlife agency by name (Adams and Thomas 1998). A study in Alabama had similar results (Rossi and Armstrong 1999).

A survey of agency responsiveness to wildlife damage found that most of the organizations were concerned about the issue (Hewitt and Messmer 1997). The majority (75 percent) of responding state and provincial wildlife agencies indicated they had written wildlife damage management policies. However, many did not actively publicize or articulate their policies. Few agencies evaluate their wildlife damage policies; and when evaluations are done, often they are reactive rather than proactive.

An earlier publication on the responsibilities of various agencies for animal damage management found that agency responsibility hierarchies can be confusing (Berryman 1994). In some states, responsibilities for certain wildlife species are vested with the state agriculture agencies, while other species are the responsibility of the department of natural resources. At the federal level, authority over migratory birds is vested within the U.S. Fish and Wildlife Service (Department of Interior), but depredation control of birds is vested in Wildlife Services (Department of Agriculture). Little wonder the public perception of responsibility for wildlife damage management is unclear at best.

People will attempt to find their own solutions to human-wildlife conflicts when they are unable to get professional help (Case Study 9.1). Not all remedies are the ones wildlife professionals would promote or prefer. Some methods are harmless, and often ineffective, but it is not uncommon for homeowners to try bizarre and even dangerous cures. Stories of damage control efforts gone awry are commonplace. One dramatic example involved pouring gasoline down burrow holes and igniting it to exterminate chipmunks (*Tamias* spp.; San Julian 1987).

Wildlife professionals tend to regard human-wildlife conflict management in urban and suburban situations as difficult, due to the perceived resistance of the public to a full range of management options (Decker and Loker 1996). Attitudes among urban and suburban dwellers continue to shift away from utilitarian perspectives and toward moralistic and humanistic perspectives (Kellert 1997; Hadidian et al. 1999), meaning they are less likely to accept lethal management techniques. Control of wildlife in urban environments is more technically complex than control in an agricultural situation, even when the same species is involved. Wildlife management agencies face additional impediments when trying to work with the public on conflict issues, including: (1) agency image and credibility problems, (2) conflicts between recommended solutions and the personal values of a diverse constituency, and (3) public animosity toward regulatory agencies (Lowery and Simer 1999).

Given these obstacles, as well as funding limitations, state and federal wildlife agencies often have left resolution of human-wildlife conflicts to individual initiative or to private nuisance wildlife control operators (NWCOs) (Hadidian et al. 1999). A 1993 national study found that about 63 percent of licensed trappers had been contacted to remove problem animals (International Association of Fish and Wildlife Agencies 1993). There appears to be rapid growth and privatization of the wildlife control field over the past several decades due to the increase in human-wildlife conflicts (Braband 1995; Barnes 1997).

The role of government in the business of wildlife damage control has been to regulate activities through a licensing or permitting process, and to provide extension or educational services. Legal authority for such regulation is vested in federal and state governments, but in reality, authority often is divided among different agencies, such as natural resources, agriculture, and public health (Barnes 1997). Specifics on the nature, scope, and extent of NWCO activities are limited. A 1997 study found that only 36 percent of states required a license for individuals to practice nuisance wildlife control; only fourteen of those states required an annual report of activities (Barnes 1997).

A study of private NWCOs found that most have a high school diploma, but little specific training in wildlife damage management or wildlife management in general; more than half of NWCOs surveyed had not attended a trapper-education course (Barnes 1995a, 1995b). NWCOs come from a diversity of backgrounds, including wildlife biologists, pest control operators, fur trappers, and wildlife rehabilitators (Braband 1995). Barnes (1997) proposed a model for nuisance wildlife control licensing containing three key requirements: education, continuing education, and liability insurance. Schmidt (1998) discussed the importance of training in wildlife identification and wildlife ecology, state and federal wildlife and pesticide laws, parasites and diseases of concern to wildlife and humans, chemical immobilization and euthanasia, and current and emerging technologies in wildlife damage management. Additional studies are needed to determine the extent and volume of human-wildlife conflict resolution activities and the impact these have on wildlife populations.

The effects of wildlife on human health and safety are a subset of human-wildlife conflict, but one that deserves special attention. Examples of this type of conflict include zoonotic diseases (e.g., Lyme disease, hantavirus; see Chapter 13), transportation hazards (e.g., deer-automobile collisions, beaver dams flooding roads), and sanitation problems (e.g., rodents, roosting birds). No national summaries are available on losses in human lives and economic productivity due to these types of human-wildlife conflicts. However, published and unpublished data have been compiled to assess potential cost of wildlife encounters in the United States in terms of human illnesses, injuries, and fatalities (Conover et al. 1995).

One of the more common ways in which wildlife impacts human health and safety is in the area of ground and air transportation, which were discussed in detail in Chapter 8. Aircraft and avian species often compete for the same airspace. When that happens, collisions may occur, resulting in property damage, injuries, and even fatalities. The problem exists nationwide, but airports in the eastern and southeastern United States experienced the greatest number of wildlife-aircraft collisions (Wildlife Services 2001).

Conover et al. (1995) estimated the number of collisions between vehicles and deer (*Odocoileus* spp.) at >726,000 and increasing. Approximately 600 collisions between moose (*Alces alces*) and vehicles reportedly occur in Alaska annually (Romin 1994). The incidence of human injury has been reported as 4 percent nationwide; less than 3 percent of deer-vehicle collisions resulted in a human fatality (Rue 1987). Approximately 29,000 human injuries and 211 human fatalities occurred as a result of collisions with deer (Conover et al. 1995). Unfortunately, in many states there is no formal mechanism to identify loss of personal property or human life to deer-auto collisions. In fact, more often than not, these accidents may be reported as uncategorized. Therefore, the numbers reported in published literature do not always agree.

Of course, deer and moose are not the only wildlife species involved in vehicle collisions. Unfortunately, few if any records are kept on collisions with other mammalian species, such as black bear (*Ursus americanus*), coyote (*Canis latrans*), and elk (*Cervus elaphus*), much less the number of accidents that were caused by drivers who attempted to avoid smaller wildlife species, such as tree squirrels (*Sciurus* spp., *Tamiasciurus* spp.), raccoons (*Procyon lotor*), and birds.

Despite the positive effects of beaver (*Castor canadensis*) on water quality and the creation of wetland habitat, when these animals begin constructing dams they can create public safety hazards by flooding roadways, plugging culverts, and damaging roadbeds and bridges (Harper 2002). In the northern United States, 27 to 40 percent of total beaver damage complaints involved culvert plugging and road flooding (Payne and Peterson 1986). Flooded and washed-out roads can present serious human safety concerns (Jensen et al. 2001).

When wild animals congregate they can create sanitation problems that pose a threat to human health. Large flocks of starlings, grackles, and/or red-winged blackbirds may take up residence in urban neighborhoods, and the resulting accumulation of feces under roost trees can be considerable (Figure 9.6). Accumulation of fecal material also causes conflicts between humans and urban waterfowl, particularly Canada geese. While waterfowl are not a health threat to humans, their droppings are causing concerns about water quality control in municipal lakes and ponds. Mammals also can create sanitation concerns; raccoon latrines may increase the potential for transmission of the para-

Figure 9.6 *(A color version of this figure follows page 158.)* Roosting birds—and their droppings—can become an annoying and costly problem. A one-night accumulation of bird guano leads to a difficult, daily, and costly clean-up process. (Courtesy Clark E. Adams)

site *Baylisascaris procyonis,* while rodent middens and droppings can be a source for hantavirus infection in humans.

A wealth of information is available to the public from governmental and nongovernmental sources concerning disease transmission and prevention and human safety—if they can figure out where to find it. Many studies have been done which identify the need for education on any number of zoonotic diseases. A study conducted at Cornell University's Human Dimensions Research Unit on human perceptions and behaviors associated with Lyme disease included two information transfer objectives: identifying communication mechanisms and networks, and developing an information base and guidelines for land and wildlife managers (Siemer et al. 1992). The California Department of Fish and Game developed a program to educate the public about a behavioral approach for managing predators at the urban-wildlife interface which was reported to have reduced the number of complaints received by that agency (Wehtje 1998).

A 2002 study examined the role of wildlife disease monitoring programs to provide early detection of new and emerging diseases, some of which may have serious zoonotic and economic implications (Mörner et al. 2002). The authors suggested that monitoring programs be integrated within national animal health surveillance infrastructures. These programs should have the capacity to respond promptly to the detection of unusual wildlife mortality and institute epizootiological research.

In 1996, researchers at the Centers for Disease Control and Prevention (CDC) in Georgia surveyed 100 small- or exotic-animal veterinarians in California, considered to be a primary source of information on reptile-associated salmonellosis for reptile owners, to assess the veterinarians' knowledge and information transfer practices (Glynn et al. 2001). Veterinarians have a unique opportunity to inform the public about zoonotic disease prevention strategies, immunization recommendations, and risk assessment following potential exposure, due to their knowledge of the natural history of vertebrate animals (Herbold 2000). In 2002, the Colorado Wildlife Commission published a policy statement on chronic wasting disease that included a mandate for health experts to provide precautionary guidelines to people hunting in game management units. However, a review of the literature did not reveal any research into the existence of formal information transfer structures or evaluation of the effectiveness of agency efforts to educate the public.

Although conflicts have always been a part of human culture, conflicts over wildlife between members of the general citizenry, and between the public and wildlife agencies, began to increase during the late 1960s and early 1970s. At this time, people other than traditional clientele of wildlife management began to express interest in all types of wildlife, not just game species or "nuisance" wildlife. They wanted an equal voice, along with traditional consumptive users, in how wildlife resources were managed. Hikers, campers, and birders wanted nongame species protected, so they could be enjoyed during nonconsumptive recreation. Environmentalists wanted certain species to be protected to ensure ecosystem health (Decker, Brown, and Siemer 2001). By the 1980s, advocates for animal rights had begun to question not only specific wildlife management practices, but whether managing wild animals, especially for the purpose of "harvesting" them, should be done at all.

Conflicts between humans over wildlife usually are the result of different beliefs and attitudes regarding how wildlife resources should or should not be used. In some ways they are quite similar to the kinds of conflicts that arise over other natural resources, including water, air, timber, and minerals.

Stories of conflicts over wildlife can be found almost daily in both local and national newspapers, as well as in other types of media. For example, some New Mexico residents strongly support federal mandates limiting the use of water from the Rio Grande for irrigation in order to protect the endangered silvery minnow, while others, particularly those who depend on irrigated crops for their livelihood, question the fairness of assigning more value to the welfare of fish than the welfare of farmers. Hunters in various western states have pushed wildlife agencies to add a spring hunting season for black bear that, environmentalists argue, will devastate bear populations. In neighborhoods across the country homeowners decide it would be nice to plant a butterfly garden, unaware that they'll soon have an irate neighbor at their door complaining about the damage ravenous caterpillars have caused in a prized vegetable garden.

HD METHODS FOR RESOLVING HUMAN-WILDLIFE CONFLICTS

How can urban wildlife managers hope to solve these types of human-human conflicts? Are solutions even possible? Surely there will be cases in which, despite the best efforts of the manager involved, decisions will need to be made that result in one side "winning" and the other "losing." How can such an outcome be deemed a true solution?

Finding appropriate solutions to urban wildlife management issues requires creativity, skill, and patience. Each situation is unique. In many cases a broad range of conflicting interests and concerns must be addressed, along with the economic, political, and social ramifications of any decision. This can be daunting, in part because techniques have not been fully developed (Carpenter, Decker, and Lipscomb 2000). Classes and books are available that can help managers improve their conflict resolution skills. In addition, Decker, Brown and Siemer (2001) have described some key traits of wildlife managers who have been successful at integrating human dimensions into wildlife management.

First and foremost, managers need to *be receptive* to many different viewpoints expressed in many different forms—telephone calls, office visits, letters to the editor, editorials, news coverage, demonstrations. This type of information is not easily interpreted and must be kept in proper perspective. Minority interests can make a large impact using the media, while majority interests may not be as well organized or vocal.

Second, the authors suggest that managers must *be inquisitive*. Don't expect information to fall into your lap. Think about the many ways in which biological and human dimensions data may be found and gathered: surveys, literature searches, newspaper archives, questioning colleagues, and at public hearings.

Finally, effective managers need to *be diagnostic*. Carefully define the problem and its elements. Develop a plan for approaching the problem. Think about who should be involved in the decision-making process.

Decker, Brown and Siemer (2001) described five general approaches to problem solving using public involvement:

- Expert authority
- Passive-receptive
- Inquisitive
- Transactional
- Comanagerial

Expert authority is a top-down approach typical of agencies in the days when they served a narrow constituency with shared values. Today the public is far less tolerant of a paternalistic "we know what's best" attitude. While still in use, there are very few cases in which it can be used effectively.

The *passive-receptive* approach is a slight variation on the top-down paradigm; managers welcome outside input but don't seek it in any systematic way, and the power to make decisions remains with the manager-expert.

Managers using the *inquisitive* approach are proactive in their attempts to gather information. They recognize that unsolicited input can lead to a biased perspective. Decision-making power is still in the hands of the manager, but decisions are made based on a more thorough understanding of the issues.

The *transactional* approach is distinguished from the other approaches described so far by the fact that managers work with interested parties to find acceptable objectives and actions. Depending on the type of decision, this method may allow interested parties to make binding decisions within bounds set by the wildlife agency.

The *comanagerial* approach is, in some ways, the most radical of the five methods described here. Comanagement involves actually sharing decision-making power by giving local communities greater responsibility for solving local wildlife problems. It requires rethinking the role of state wildlife agencies and wildlife professionals, placing more emphasis on providing biological and human dimensions expertise, training community participants, approving community produced management plans, certifying private consultants and community wildlife managers, and monitoring management activities.

Before managers can choose the best approach for solving a particular problem, they must have a thorough understanding of not only the environmental factors, but the human factors as well. The realization that wildlife management is as much about people as it is about natural resources was a driving force behind the creation of a method that could help wildlife professionals address a wide spectrum of societal needs and values—the stakeholder approach—discussed in Chapter 10.

CASE STUDY 9.1: DUCKS AND TRAFFIC

There is a "no pedestrian crossing" sign to discourage visitors from braving the traffic on Jefferson Street to reach the Buena Vista Lagoon duck-feeding area across the road, but ducks and geese can't read. The sight of dead birds along Jefferson Street after being hit by vehicles prompted a Carlsbad, California, resident to take action in hopes of getting motorists to slow down along that stretch of road.

Robert Feliciano's concern over the situation led him to submit a letter to the city asking officials to post "duck crossing" signs near the entrance to the duck-feeding area. In the letter, Feliciano offers to donate $200 to pay for the signs and supply the labor to install them. "I'd be happy to pay for the signs and put them up," Feliciano said. "I just want to get people to slow down."

Feliciano said the situation could result in an accident, since drivers tend to swerve into an adjacent lane to avoid hitting the birds, Feliciano said. "I love animals, so if I see a duck, I'm going to put on my brakes and let them get by. A lot of people are in such a hurry, they get upset and cut into another lane to get around the animal."

A visit to the duck-feeding area emphasized the conflict between nature and urban activity, when a mother and her eleven-year-old daughter arrived holding a duckling they rescued from the nearby Sears parking lot. On Feliciano's advice, they agreed to take the duckling to an animal shelter. "Obviously, these animals were probably there before the mall was and now the mall has displaced them," said the mother, Mary Lou Jenkinson. Jenkinson said she supported Feliciano's idea.

Finding ducks and geese in the nearby streets is commonplace at this time of the year, said Petey Tade, comanager of the Buena Vista Audubon Nature Center. Tade said the problem was much more prevalent before the feeding area was enclosed by a fence in 1990. She agreed that signs warning motorists are a good idea, especially because of the large number of children visiting the area.

Michael J. Williams, Staff Writer
Excerpt from North County Times, *San Diego, California*
August 21, 2003

PERSPECTIVE ESSAY 9.1   Urbanites' Fear of the Natural World Around Them

I talk to people all of the time about the compulsion to mow tall grass or remove understory. When I get to the heart of the matter, it always involves some type of fear. I hear statements like, "I mow the tall grass to keep snakes away." There it is. Fear. To that person, tall grass equals snakes, which equals fear. Though they have never been bitten or harmed by a snake, they remove all tall grasses to avoid them.

I hear the same fear-based arguments pertaining to understory. I hear people voice opposition to maintaining natural habitat or say that they want the brush and leaf litter "cleaned up" because it will attract rats, or snakes, or criminals, or whatever. It's all fear. And much of it is simply not based on fact or even personal experience on the part of the person experiencing the fear.

I try to explain to them that rodents living in and around our homes are almost always an introduced species. The house mouse (*Mus musculus*), roof rat (*Rattus rattus*), and the Norway rat (*Rattus norvegicus*) are the most common offenders. They are introduced species and prefer to associate with human habitation. They are not the ones that prefer to be in the woods and fields. So a homeowner saying that a woodlot or prairie is causing a rat problem in their home is assigning blame in the wrong place.

However, rats may just be the scapegoat serving as a convenient excuse to get rid of habitat when the real reason the person is opposed to the habitat is fear of some sort. Several years ago I began working with a local park to reestablish native understory to the woodland areas as well as encourage prairie grasses in the openings. About three years into the restoration project when the understory and grasses were progressing nicely, I got a call from a homeowner living across the street from the park. She objected to the restoration project stating that it was causing a rat problem at her home.

I was intrigued and asked her what type of rats she was seeing and after some discussion she admitted that she hadn't actually seen any rats on her property. I asked her if she had seen any rats crossing the street from the park and entering her neighborhood. She admitted she hadn't. I then gently asked her if she had seen any rats actually in the park. She replied, "Oh I don't go over there." Puzzled, I asked her why not and she replied, "I'm afraid of snakes." Fear had shown up again and it wasn't related to rats. I did my best to ease her fears. I can't say that she became a supporter of the project, but she no longer voiced opposition.

Law enforcement officials often oppose maintaining open space in a natural condition. In my experience, they push for understory to be cleared for visibility purposes. They fear that understory hides criminals. I believe this fear, like many others, is blown out of proportion. I recently attended a meeting with a local municipality to determine how a local creek corridor should be maintained by the parks department. The corridor had sections that were highly maintained and other sections that were left natural. At the meeting were representatives from various community interests. Each community group was allowed to present concerns and suggestions as to how the corridor should be managed. The law enforcement officer representing the police department suggested that the understory along the creek be removed for safety concerns. He indicated that natural areas such as this would encourage violent crime. When asked if there had ever been a violent crime in that city associated with that creek corridor he answered, "No." When asked if there had ever been a violent crime associated with natural areas anywhere in the city, he answered, "No."

You see, that city did not have a problem with understory encouraging crime, but that's the fear-based argument that was used against managing the corridor in a natural fashion. This is not to say that crime never occurs in natural areas. There are examples where it has, but according to the Justice Bureau, violent crimes most often occur within our own homes. Should a city have a problem with violent crimes occurring in a particular natural area, then certainly something should be done about it, but to universally eliminate understory out of fear for potential crime seems overreactive to me.

Some people will inevitably argue that removing understory simply "looks better." They'll say their actions aren't based on fear, but aesthetics. I submit that in this situation we determine what "looks good" based indirectly upon fear. That which eases our fear comforts us, and we are attracted to that which we find comforting. Orderly landscapes imply that we have a sense of control and human dominance. This eases our fears.

Natural landscapes do not imply human dominance and, thus, contain a degree of uncertainty for some people. To those who are fearful, a wooded landscape that looks less fearsome is aesthetically pleasing. However, to people who aren't fearful of understory, a woodland with the shrub layer intact is preferred.

Tradition is yet another reason that people will clear understory. It's the "that's the way it's supposed to be" syndrome. I contend that our traditions come from practices that we become accustomed to seeing. Our society has been clearing understory out of fear for so long, that it has now become tradition.

John M. Davis

AN URBAN SPECIES OF SPECIAL CONCERN: URBAN COYOTES (*CANIS LATRANS*)

No wildlife species in North America represents the adversarial relationship between predators and humans better than the coyote. Mark Twain best summarized this struggle:

> The coyote is a living, breathing allegory of Want. He is always hungry. He is always poor, out of luck and friendless. The meanest creatures despise him and even the flea would desert him for a velocipede.

—Mark Twain, *Roughing It* (1926)

Like the raccoon (Chapter 1), and the white-tailed deer and Canada goose (Chapter 14) the coyote (Figure 9.7) is another example of a wildlife species exceptionally well adapted for life in

Figure 9.7 *(A color version of this figure follows page 158.)* Coyote. (Courtesy Sara Ash)

the city. The coyote is a member of the dog family (Canidae), as are domestic dogs, wolves, foxes, and jackals. The coyote stands less than two feet tall, weighs between twenty-five and thirty-five pounds, and is brownish gray in color with a lighter belly. Some distinguishing anatomical features are the erect and pointed ears, thin and lanky body, slender muzzle, yellow eyes, and bushy tail with a black tip (Gehrt 2006).

Urban Wildlife Management would be incomplete without a presentation on how the coyote uses the urban environment to its advantage, interacts with humans and other wildlife species, and is "managed" in urban communities. Unlike other abundant wildlife species in urban environments, the coyote is "the ghost of the cities, occasionally heard but rarely seen" (Gehrt 2006). The coyote vocalization or howl is a distinctive feature that identifies this species in the wild. In fact, the translation of its scientific name (*Canis latrans*) is "barking dog." Human reactions to the coyote howl range from a feeling of peace and tranquility, that is, all is well in the natural order of things, to a reaction of alarm, fear, and foreboding. The latter reactions are largely a product of the film industry (e.g., the old-time cowboy movies) or a lack of human experience with a coyote presence in their community. It is possible that the typical urbanite's only experience with the coyote was largely the result of watching the "Roadrunner and Wile E. Coyote" cartoons. The cartoon series always depicted the coyote as an incompetent predator unable to capture the roadrunner in spite of using every sort of available ACME device. Obviously, the true character of the coyote was lost in the cartoon series. We've done our best to assemble and interpret the vast body of literature on the coyote that has accumulated over the last forty years. In short, the coyote (urban and rural) has been captured in the literature through books (Dobie 1961; Bekoff 1977; Meinzen 1995; and Reid 2004), book chapters (Bekoff 1977; HSUS 1997; Bekoff 1999; Bekoff and Gese 2003), a wildlife monograph (Andelt 1985), and hundreds of manuscripts published in professional and popular outlets. In 2007, there was an entire session at the 12th Wildlife Damage Management Conference devoted to the research and management of the urban coyote.

Because of the potential for human injury, the existence of coyote populations in urban areas can become a volatile issue for political leaders, wildlife managers, and residents.

What Is an Urban Coyote?

What sort of coyote has Los Angeles created? It's a creature that will jump over chain link for a bowl of Alpo. It's an animal that can learn and remember which storm-sewer channels lead to which golf courses ... which dumpsters behind which supermarkets are likely to be overflowing with old vegetables and delightfully rancid fish … It has eaten from the Tree of Forbidden Knowledge, and it recalls fondly the taste of Fifi and Mr. Boots.

—**David Quammen**

There are so many resources available in the print and electronic media on coyote biology and natural history that limited our ability to add anything new on the topic. On the other hand, it was instructive to examine various aspects of the ecology of urban coyotes in order to gain a better understanding of how this species exists in urbanized landscapes. Based on an analysis of radio-telemetry studies of coyotes residing in eight metropolitan communities by Gehrt (2007), and in many other publications, some distinctive features about the ecology of urban coyotes emerged. For example, urban coyotes:

1. Have high survival rates.
2. Have small home ranges.
3. Avoid people and automobile traffic by shifting to night time activity (McClennen et al. 2001; Tigas, Van Vuren, and Sauvajot 2002; Riley et al. 2003). However, there have been significant exceptions to this behavior. In some cities, coyotes will hunt during the day in areas of high human activity (Baker and Timm 1998).
4. Tend to avoid developed areas (e.g., residential) within their home range.
5. Are opportunistic predators and scavengers. Their diet consists primarily of rodents, lagomorphs, and vegetation. Anthropogenic foods (e.g., garbage, pet foods) are consumed only when available within their home range. Urban coyotes are not dependent on pets (e.g., cats) for food. They will consume winter-killed and winter-weakened white-tailed deer and fawns, and resident Canada geese eggs and goslings (Buck and Kitts 2004; Gehrt 2006).
6. Select more "natural" areas for denning and resting sites. Some studies (Riley et al. 2003; Grinder and Krausman 2001) showed that coyotes used forests or habitats with sufficient cover, but also in close proximity to residential or urban parks.
7. Have no natural predators or competitors; the automobile is not a "natural" predator, but it is a significant cause of urban coyote mortality.

Take a moment to imagine the logistics involved in organizing and conducting a credible field study on coyotes in cities, municipalities, and neighborhoods. The information that is now available on urban coyote biology and natural history has been the result of long hours in the field (usually at night), maintaining open lines of communication with public concerns, negotiating diverse urban landscapes, and research technologies (e.g., radiotelemetry) that can only give a partial picture of what is really going on in the city. As a result, scant research exists on coyote behavioral ecology in human-altered landscapes. Much more information is needed on coyote social structure and territorial behavior, foraging activities, birth rates, and survivorship that model population growth, and reliable and cost-effective census techniques that are now missing (Curtis, Bogan, and Batcheller 2007).

Range Expansion

There were several important events, from pre-European arrival to now, that allowed coyotes to expand their range to their current distribution (Figures 9.8a and b). Range expansion was a two-way street that involved either people moving in with the coyotes or vice versa. Coyotes have expanded their range to include eastern Canada and the United States in the past fifty to seventy-five years. One example is Mississippi, where the coyote population increased 7.5-fold over one fifteen-year period (Lovell, Leopold, and Shropshire 1998). This range expansion is considered to be a result of both natural colonization (Gompper 2002) and intentional and accidental introduction by humans (Hill, Sumner, and Wooding 1987). In 1999, a coyote was captured in New York City's Central Park (Martin 1999), and that same year, another coyote was captured after being discovered hiding under

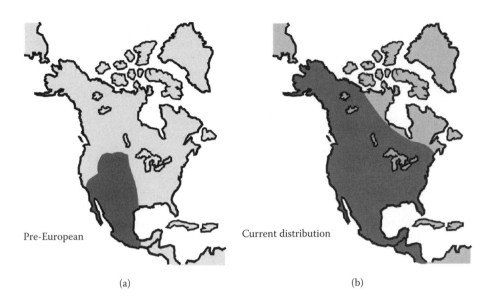

Pre-European

Current distribution

(a)   (b)

Figure 9.8 The historical (a) and contemporary (b) geographical range of the coyote in the western hemi-sphere. (Redrawn from Gehrt 2006 by Linda Causey)

a taxi on Michigan Avenue in Chicago (Kendall 1999). It would be naïve to think coyotes would not exploit cities in the east with the same adaptability and tenacity as they did in the west.

Coping with Coyotes—Management Plans

One of the most comprehensive approaches to urban coyote management was provided by Siemer, Hudenko, and Decker (2007). They laid out a complex plan designed to include many of the critical elements needed to facilitate the best management practices when dealing with urban coyotes. Their plan included consideration of a table of human dimensions information needs (five steps) to support a rational decision-making model for urban/suburban coyote management. Next, they provided a detailed diagram that articulated the ends-means connections required to achieve a successful urban-coyote management outcome—cohabitation—in New York state. However, it would be appropriate to say that "the devil is in the details." For example, a plan may, in theory, have contained all of the essential elements, but be logistically improbable to execute in the real world. The learning curve requiring professionals outside the human dimensions discipline to comprehend, and more impor-tantly, implement the plan in the field would be formidable. At the risk of being faulted as a reduction-ist and over-simplifying a complex process, the urban residents' considerations in most urban wildlife management plans usually focus on time, cost, safety and humaneness, effectiveness, and isolation from the process. For example, urban residents want resolution to the problem immediately at no cost to them personally. They do not want any harm to come to them, their pets, or the coyote; and they want to see results without getting involved. The wildlife biologists have to negotiate an environment of highly variable human values, emotions, and expectations before even considering how to deal with the animal (Timm and Baker 2007). The full scope of this dilemma is discussed below.

Urban coyote management has another dimension that is quite different from management of urban white-tailed deer and resident Canada geese (Chapter 14). Rather than dealing only with over-abundance, another important concern in urban coyote management is dealing with the exceptions—the "problem" coyotes. Problem coyotes are the ones that are guilty of attacks on people or their pets, property destruction, or other forms of negative or aggressive behavior (Orthmeyer et al. 2007). Several major urban areas have experienced human-coyote conflicts, namely attacks on people and pets. For example, Baker and Timm (1998) reported that between 1988 and 1997, coyotes attacked fifty-three people and harassed more than thirty-two people in southern California. A comparison of two eight-year periods, 1990 to 1998 and 1999 to 2006 revealed a 228 percent increase in conflicts between coyotes and pets, and a 300 percent increase between humans and coyotes in California (Orthmeyer et al. 2007). The sequence of coyote behaviors that precede attacks on humans is as follows:

1. An increase in observing coyotes on streets and in yards at night.
2. An increase in coyotes approaching adults and/or taking pets at night.
3. Daylight observance of coyotes chasing or taking pets.
4. Attacking pets on leashes, chasing joggers and cyclists.
5. Presence of coyotes in children's play areas, school grounds, and parks in the middle of the day.
6. Coyotes acting aggressively toward adults in the middle of the day (Timm et al. 2004).

Of note is that all of the above six behaviors exemplify a progressive increase in coyote habituation to humans. Why do some urban coyote populations avoid direct contact with humans while other populations (i.e., southern California cities) harass and prey on humans? Baker and Timm (1998) suggested that coyote attacks are precipitated not by hunger alone, but also by a lack of aggression from humans. Without trapping and shooting control efforts, coyotes lose their fear of humans. Couple this loss of fear with hunger and the need to feed pups in the late summer, and coyote attacks can result. Trapping and shooting not only removes offending individuals, but may modify the behavior of remaining individuals (Baker and Timm 1998). Habituation of wildlife to humans creates wild animals that are potentially more dangerous than their rural counterparts (Geist 2007).

Coyote Economics

The bane of farmers and ranchers since the European settlement of central and western North America, no other wildlife species in North America best represents the adversarial relationship between predators and humans than the coyote. According to Conover (2001), U.S. ranchers lose $71.5 million annually as a result of predation on livestock, the majority of which are taken by coyotes. Consequently, coyotes have been the target of decades of lethal control methods, and continue to be hunted by ranchers and the federal government. The U.S. Department of Agriculture's Wildlife Services killed approximately 90,000 coyotes in 2001 to prevent predation of livestock. Twenty-four percent of U.S. farmers experience damage from coyotes, and the number rises to 40 percent in Great Plains states (Conover 1998). Historically, the conflict between humans and coyotes has been a rural one, but that has changed over the past few decades.

Now there are reported damages due to coyotes living in urban environments. In California, 362 coyote conflicts with pets resulted in $78,232 in damage from 1991 to 1998, compared to 1079 conflicts resulting in $402,540 in damage from 1999 to 2006 (Orthmeyer et al. 2007). However, on the positive side of the ledger, urban coyotes may be saving urban residents money by preying on overabundant white-tailed deer and resident Canada geese (Gehrt 2006).

Coyote Control Considerations

It would be an understatement to characterize the management of urban coyotes at the individual and population levels as a daunting task—coyote control considerations go far beyond deciding how to catch coyotes (Huot and Bergman 2007; Orthmeyer et al. 2007). Urban coyote managers have to negotiate multiple levels of concerns that are all interrelated (Schmidt 2007). For example, the human (e.g., stakeholders) response to urban coyote management can be based on experience (positive vs. negative), fear level, knowledge, safety concerns, legal constraints, control method, cost, and authority ("who is in charge"). In short, the full range of human dimensions issues need to be addressed first and hopefully resolved before a control plan can be developed and implemented.

This management criterion is particularly important because urban coyotes, unlike deer or geese, can become very dangerous animals. This being the case, it is important to give urban residents opportunity to engage in the urban coyote management process.

The involvement of urban residents in the urban coyote management process begins with their acceptance and adherence to a set of guidelines (Sidebar 9.1).

Coyotes are now a common sight in many U.S. cities and suburbs, leading to the development of urban coyote management plans. For example, Glendale, California (Baker and Timm 1998), Austin, Texas (Farrar 2007), and Vancouver, British Columbia (Worcester and Boelens 2007), have well-documented urban coyote management (control) programs. An overarching component of all of these programs centers on public education and focuses nearly exclusively on "problem" coyotes. For example, each of these programs has a "coyote hot line" that provides immediate citizen access to someone who can answer questions about coyotes or take reports of problem coyotes. Websites are another source of public information, for example, CoyoteBytes.org (Timm 2007), which provides a comprehensive collection of information about urban coyotes, links to relevant ancillary organizations and government offices, reported locations of problem coyotes, and factors residents should consider when encountering urban coyotes in their neighborhoods. Other public information venues consist of printed materials, school programs, and community presentations. All programs are strengthened by municipal and state laws prohibiting the feeding of coyotes. Finally, all programs have led to a marked decrease in urban coyote conflicts with pets and humans.

In order to grasp the full range of issues and problems associated with the development and implementation of a residential urban coyote management program, see Sidebar 9.2 by Robert Boelens, an urban wildlife specialist with the Stanley Park Ecological Society.

SIDEBAR 9.1   Ten Proactive Coyote Management Activities

1. Do not feed coyotes, deer or other wildlife!
2. Eliminate sources of water.
3. Position bird feeders so that coyotes can't get the feed. Coyotes are attracted by bread, table scraps, and even seed. They may also be attracted by birds and rodents that come to the feeders. They are also attracted to deer feed.
4. Do not discard edible garbage where coyotes can get to it.
5. Secure garbage containers and eliminate garbage odors.
6. Feed pets indoors whenever possible. Pick up any leftovers if feeding outdoors. Store pet food where it is inaccessible to wildlife.
7. Trim and clean, near ground level, any shrubbery that provides hiding cover for coyotes or prey.
8. Don't leave small children unattended outside if coyotes have been frequenting the area.
9. Don't allow pets to run free. Keep them safely confined and provide secure nighttime housing for them. Walk your dog on a leash and accompany your pet outside, especially at night.
10. If you start seeing coyotes around your home or property, chase them away by shouting, making loud noises, or throwing rocks. Carrying a stick, baseball bat, or golf club will also scare them away. Negative stimuli are commonly suggested as a means for residents to control urban coyotes. However, some experienced wildlife control operators express concerns that such activity on the part of inexperienced residents is "an accident waiting to happen" (Erickson 2007). The concern seems warranted because residents do not know the true condition of the coyote just by its appearance, and the coyote may take retaliatory action on being placed in an adversarial environment (http://www.ci.austin.tx.us/council/bm_urban_coyote_info.htm).

SIDEBAR 9.2   Coexisting with Coyotes in Vancouver, British Columbia

Once Upon a Time … (the coyotes arrive and the program begins)

Coyotes were first spotted in the Vancouver landscape in the late 1980s, having been attracted by the city's green space, access corridors, food supply, and rodent population. With the coyotes came a great deal of surprise, myths, and concern. Coyotes were seen hunting on golf courses, sunbathing in parks and backyards, and trotting along streets and alleys. Soon they began to appear as the topic of local radio talk shows, and café and park conversations throughout the city. Their appearance was unexpected and without invitation, but it soon became clear coyotes were in Vancouver to stay.

Vancouver's coyotes adapted to the urban lifestyle quickly and with ease, leading to an increase in the number of encounters and experiences with people. Some coyotes began to prey on outdoor cats, while others dined on food deliberately left out for them by humans. Public opinion, as on any topic, was divided. There were suggestions of a cull, a trapping and relocation program, a public education campaign, and doing nothing at all. The one constant among all the suggestions and concern was a demand for accurate and consistent coyote information. Something which, at that time, just wasn't there.

Though coyotes had been known to live in regions within 80 to 100 kilometers (50 to 60 miles) of Vancouver since the 1930s, their appearance in city and suburban yards and main streets brought surprise. Residents were shocked to learn that the coyote was *not* the wolf-sized, nocturnal, pack-hunting, carnivore that their first thoughts suggested, but a 9- to 16-kg (20- to 35-lb) master of adaptation, that was perfectly comfortable and amazingly discreet, living in close proximity to active human populations.

Attempts to live trap the first coyotes sighted in Vancouver for relocation proved a failure; the animals would sniff and circle but not one of them would enter the live box-trap. At the dawn of the millennium every neighborhood in urban Vancouver had been, at one time or another, visited by a coyote, with certain areas, usually bordering large natural parks or golf courses, becoming well known for coyote activity. As the 1990s came to an end there were a growing number of reports of coyotes losing their instinctive fear of people, an increased number of outdoor cats reported missing, and, in certain areas, incidents of small dogs being removed directly from their owner's leash. In the year 2000 there were three incidents in which a child was bitten, each incident receiving immense media coverage, causing fragmented panic and clearly demonstrating the need for an organized and effective response. In February 2001, after open consultation meetings with government, environmental and animal welfare agencies, and the public, the not-for-profit Stanley Park Ecology Society, in coopera-tion with the Provincial Ministry of Environment and the City of Vancouver, began the Co-Existing with Coyotes (CwC) public education program. Two subsequent biting incidents in July 2001 provided the CwC program with instant publicity and recognition. The demand for information was greater than ever before.

OUR DEFINITION OF COEXISTING

The program aims to reduce conflict between people, pets, and coyotes by providing information to both targeted and general audi-ences, as well as providing a direct response to individual coyotes that are starting to, or are displaying, behavior of concern. Stanley Park Ecology Society and City of Vancouver wildlife staff track, locate, evaluate, and use nonlethal coyote deterrents with consistent success in neighborhoods throughout Vancouver, but, equally importantly, also recognize that coexistence is not always an option. Program staff worked to identify and help coordinate the removal of any coyote that poses a risk to human safety and supports the Provincial Management Plan that calls for aggressive coyotes to be destroyed. In Vancouver, one such coyote was removed by the Ministry of Environment's con-servation officers in January 2003, and a second in April 2006. In an average year Ecology Society staff visited between twenty-five and thirty neighborhoods to provide a nonlethal response to individual coyotes there, and to train and stimulate area residents to do the same. The nonlethal response that staff provided is simple yet effective, as coyotes consistently and quickly responded to staff displaying loud and aggressive-appearing behavior. The importance and level of the volume and hostility used cannot be overemphasized. Residents who observed staff physically chasing a coyote out of the neighborhood with noisemakers as simple as a cookie tin with a few stones in it or an old broomstick, recognized that the key to deterring a coyote is an aggressive-appearing response and, moreover, having seen it work, one that they themselves could provide. The coyote is pursued as long as its whereabouts are known. If it darts into shrubbery or under a shed, the noisemakers and broomstick are used to chase it out. Once the animal is out of sight, staff spent additional time attempting to locate it, hoping to repeat the treatment. Generally after one to three of these experiences, the sightings in the neighborhood sharply decreased or ceased all together, without appearing or starting in an adjoining area. If the pattern of increased sightings did continue, so did the use and frequency of nonlethal techniques, as well as monitoring the coyote's behavioral changes, for potential removal.

The majority of the program's daily operational work is, however, education based. It was recognized that the majority of conflicts could be prevented if residents became coyote aware, but the question remained, how do you inform them? The average Vancouver resident is not particularly interested in a pause in their day to discuss or learn about urban coyotes, or their management. But that changes when there is a coyote in their yard or on a field at their child's school. Then they are very interested.

WHAT DO I DO? WHO DO I CALL?

The vast majority of the more than 5000 people that have contacted Vancouver's coyote phone line since it started ringing in 2001 wanted two things. The first was to be able to tell someone what they had experienced, and the second was to be told what they should do about it. The coyote phone line provides accurate information (which was not occurring when multiple agencies were answering calls), advice and situation-specific responses, ranging from answering questions about natural history and an appropriate reaction to a backyard coyote, to an immediate on-site response. The phone line receives between 700 and 900 phone calls per year and also serves to monitor the city for areas where individual animals are displaying behavior of concern. Other agencies, including various branches of city and provincial governments, police and fire departments, and animal welfare groups, have all been very happy to refer phone calls about coyotes to a specific, designated line. Printed material is offered to each person who calls the phone line, along with the recommendation that they distribute it to neighbors.

INFORMATION IN YOUR MAILBOX, AT YOUR CHILD'S SCHOOL, VETERINARIAN, ETC.

The Co-Existing with Coyotes brochure is a quick reference point for the general public to learn more about urban coyotes and how coexistence is possible. It informs the reader of the coyote attractants present in their neighborhood, how to prepare for a coyote encounter, to keep pets safe, the dangers of feeding a coyote, and offers contact information for questions and specific concerns. Printed material is sent to each community center, library, golf course, veterinary clinic, pet services business, elementary school and childcare facility in the City of Vancouver. The program distributes 10,000 to 15,000 brochures each year. Notes that report localized pet attacks, coyote feeding, and anony-mous "coyote attractants on your property" are also available. There are also more than one hundred permanent, 60 by 75 cm (24 by 30 inches) metal signs posted in each golf course, off-leash dog park, and area of frequent coyote activity in Vancouver. The signs provide encounter behavior, pet safety tips, coyote natural history and identifying features, as well as providing contact details for additional information.

WHAT ABOUT MY KIDS? WHAT EVERY ELEMENTARY SCHOOL STUDENT SHOULD KNOW

Coyotes 101 is a kindergarten to grade 7, auditorium-style presentation designed to provide audiences with the skills to identify coy-otes (big ears up/bushy tail down/white smile (bib-like patch of white fur), recognize urban coyote attractants in their neighborhood, and be familiar with coyote encounter behavior (Do NOT Run. Appear BIG, MEAN and LOUD). Coyotes 101 emphasizes the above objectives while providing additional natural history information and engaging the students and teachers in an informative, interactive, and entertaining manner. On average, 12,000 elementary school children participate in the program each year.

The Ecology Society also reaches public audiences of several thousands per year by attending community events, and leading interpretive walks through "The Coyote Zone" (any requested neighborhood in Greater Vancouver) highlighting how easy it is for urban coyotes to survive in the urban environment, and how residents can and should respond.

One of the most successful and complimented resources Co-Existing with Coyotes has is the compilation of most of the program materials on its website. It hosts a comprehensive collection of resources, including suggestions for coyote encounter behavior, pet safety tips, reducing neighborhood coyote attractants, blueprints for home-made deterrents, common questions, updated sighting reports, brochures in eleven languages for downloading, identifying coyote features, coyote natural history facts and sounds, coyote conflict statistics, and opportunities to ask specific or incident-related questions, report sightings, share opinions, and leave stories and comments.

COYOTE CONFLICTS IN VANCOUVER—BITERS AND FEEDERS

To date, the Vancouver coyote has yet to bite the hand that feeds it; instead, tragically, approaching children and biting them. Of the six Lower Mainland incidents in which coyotes have bitten a child, four of the animals involved had been deliberately fed by adults, and one was in an area where feeding occurred at the time of the biting. Essentially, the majority of coyotes that had bitten a child had been fed by an adult, a theme that is sadly consistent with coyote bites across the western half of North America. As of September 2006, the last time a child was bitten in the greater Vancouver area was in July 2001.

BYLAWS THAT WORK AND WORKING WITH THOSE WHO ENFORCE THEM

A cooperative effort among involved agencies and government departments is essential, and in Vancouver's case, fortunately, it is the norm. When residents complain about coyotes attracted by the condition and rat habitat of a neighbor's neglected yard, the City Bylaw Office is quick to respond. The Untidy Premises Bylaw makes the property owner responsible to ensure their residence and yard are maintained at a similar level as the rest of the neighborhood. Failure to comply with the bylaw carries a penalty ranging between $50 and $2000. Virtually every urban city or municipality has similar regulations and is prompt to enforce them. On other occasions, when coyote habitat (primarily overgrown or vacant lots) appeared on city owned but undeveloped land, a work order for the removal of the bushes or maintenance of the property was promptly issued once the residents' concern came to the civic department's attention.

Prosecuting individuals who feed wildlife is a problem, though Section 33.1 of the BC Wildlife Act provides a minimum $345 ticket and maximum $50,000 fine and six-month prison sentence for anyone who "with the intent of attracting dangerous wildlife to any land or premises, provides, leaves or places in, or about the land or premises, food, food waste or any other substance that could attract dangerous wildlife to the land or premises." The act includes, of course, the intentional or unintentional feeding of coyotes. The act is logistically difficult to enforce, as the people doing the feeding are discreet and difficult to identify. Thankfully the vast majority of residents recognize the danger that feeding coyotes exposes their communities to, and react to news of feeding in their neighborhoods with shock and anger.

THE BOTTOM LINE (... ALMOST)

Administrative support is provided by Ecology Society staff and operational support is provided by a wide range of public agencies, private companies, and residents. The program operates on a budget of less than Can$50,000 per year and receives core funding from the Government of British Columbia and the City of Vancouver, as well as generating funds from program delivery and material sales.

CO-EXISTING WITH COYOTES TURNS FIVE

The program has begun its sixth year of operation and is recognized as having played a key role in reducing conflict between people and coyotes in Vancouver. The fact there has not been a child bitten by a coyote from August 2001 onwards is a measured result of success, particularly when one considers there were five such incidents between April 2000 and July 2001. The balanced approach that Co-Existing with Coyotes brings has generated and continues to generate support from government and animal welfare agencies, school administrators, and members of the public who have encountered an urban coyote. Most of Vancouver's residents have a coyote story of their own to tell. They have vivid recounts of their coyote encounter, often with varying levels of emotion and opinion, of the time they saw or heard one of this city's coyotes, and many of them now have a story of coexistence to tell as well.

The key elements needed to reduce conflicts between humans and wildlife were all present in the Vancouver plan. First, there was an attempt to understand the human dimension of the conflict. To quote Lampa (2001), "Since it is neither desirable or practical to eliminate the wildlife, it is probably more feasible to change human behavior than the behavior of wildlife." Second, the city did not dismiss the fears of residents regarding the potential risk of a coyote attack. Money was appropriated to fund an officer who dispatches any aggressive coyotes. While the educational program tried to inform residents of the low probability of attack, it educated the residents about how to react if a coyote approaches them. Knowledge generally helps to alleviate fear. Finally, this management plan was culturally sensitive and in line with the philosophy of the majority of residents. We are not suggesting that wildlife populations should be managed based on the court of public opinion, but to disregard cultural and sociological issues is to plan for failure.

Source: From Boelens, 2006. Reprinted with permission from the December 2006 issue of *Public Management (PM)* magazine published by ICMA, the International City/County Management Association, Washington, D.C.

Figure 9.9   American crow. (Courtesy John J. Mosesso/NBII)

SPECIES PROFILE: AMERICAN CROW (*CORVUS BRACHYRHYNCHOS*)

The crow doth sing as sweetly as the lark, When neither is attended

—William Shakespeare

Crows (Figure 9.9) have been and continue to be maligned as dirty and as harbingers of death. Although highly intelligent, crows are one of the least appreciated bird species. Human lack of respect for this bird increases exponentially when crows gather in large roosts. Similar to coyotes, conflicts involving crows previously occurred only in rural areas, but within the past forty to fifty years, crows have been colonizing urban and suburban habitats. A survey of wildlife damage professionals showed that 110 cities in twenty-eight states have active crow roosts (Gorenzel et al. 2000).

Crows are found throughout most of the continental United States and Canada (Stokes and Stokes 1996). They inhabit landscapes that provide open areas for feeding and wooded areas for nesting and roosting (Johnson 1994). Like the coyote, listing the common food items of the crow would be difficult, but in broad terms, approximately one-third of the crow's diet consists of invertebrates, small vertebrates, and eggs. Crows also eat a wide variety of plant matter, including wild foods such as acorns, and cultivated foods such as corn. Crows also eat carrion and garbage (Johnson 1994).

Crows commonly defend feeding territories during the day. In fall and winter they will leave their territories to join large flocks to roost for the night (Stokes and Stokes 1996). Scientists have been trying for years to understand why some bird species gather in these large communal roosts. Two hypotheses suggest foraging behavior influences selection of communal roost sites. The first of these hypotheses is called the information center hypothesis (Ricklefs and Miller 1999). It proposes birds will gather together at night to collect information about foraging success from the other birds. Individual crows assess if other members of the roost were successful at finding food that day. In the morning, unsuccessful foragers from the previous day can then follow successful crows to the food source.

The second hypothesis related to foraging is the diurnal activity center. It asserts a roost of birds is merely a fortuitous aggregation of individuals that have strategically placed their roost site to minimize distance to their supplemental and main food sources (Caccamise and Morrison 1986; Ricklefs and Miller 1999). This hypothesis suggests the foraging site is selected first and then, subsequently, a decision about where to roost is made. Imagine your workplace as your foraging area and your house as your roosting site. Most people look for a house that is in close proximity to their workplace. You don't usually buy a house and then look for a place to work. This hypothesis has not been tested with crows. However, if true, crows would not be faithful to a particular roosting site. Instead, we should see shifts in the memberships of roosts on a regular basis.

Because many birds form communal roosts in late fall and winter, some scientists have hypothesized that individuals minimize heat loss when roosting in large groups. Finally, it has been suggested that communal roosting decreases the risk of predation (Weatherhead 1984).

We may never know why communal roosting behavior originally evolved in some species, but benefits derived from it *can* be experimentally tested. In reality, crows probably gain several benefits from roosting in a large group. We will have to wait for scientists to test these hypotheses before we know the relative importance of each factor.

Another question for which we have no definitive answer is why crows have moved into suburban and urban areas to roost. Are they leaving the country life for the excitement of the city? Is the North American crow population growing so rapidly that there is no room in the country and surplus crows are pushed into city living? Crow experts have hypothesized both of these scenarios may be occurring. Crow populations in the west are growing rapidly, and crows that are not ready to breed disperse into urban areas (Marzluff et al. 2001). In the east, McGowan (2001) asserts the suburban and urban habitats provide new breeding and roosting opportunities for the crows.

Urban and suburban roosts may provide benefits to crows not found in rural habitats. For example, many urban and suburban areas contain large mature trees crows find especially attractive for roosting. Especially in the Midwest, urban and suburban areas may be the only source of large trees. Another factor that may influence urban roost site selection is the relative warmth of cities. Urban areas are usually several degrees warmer than the surrounding countryside. Gorenzel and Salmon (1992) showed roost trees selected by urban crows had higher temperatures than randomly selected trees. The researchers also showed urban crows selected roost sites with street lighting. The lighting may provide extra protection from great-horned owls, the main predator of American crows.

As stated earlier, large numbers of crows in an urban or suburban area can result in problems for human residents. An example of this conflict is currently taking place in Auburn, New York, home of approximately 29,000 human residents (2000 U.S. Census) and the winter home of between 25,000 and 75,000 crows (McGowan 2001). Crows began roosting in the town about fifteen years ago (McGowan 2001). Residents have complained about the feces and noise produced by the roosts. Business owners complain the crows are damaging property with their feces (Fox 2003).

City council members had discussed the crow issue occasionally (City Council Minutes), but no official action was taken until an Animal Nuisance Advisory Committee was formed in November 2002. The city granted the committee $10,000 to deal with urban nuisance animals, including crows and feral cats. The committee has been criticized as largely ineffective because no official attempts have been made to solve the crow problem.

According to media reports (*theithacajournal.com* January 2003), a private group of business owners and citizens organized a crow-shooting event in the late winter of 2000. The goal of the event was to reduce the numbers of crows roosting in the Auburn area. Hunters were asked to shoot crows in the rural areas immediately surrounding Auburn. The event received little attention until the winter of 2003 when the organizers advertised the hunt as a contest in which the four-man team that killed the most crows would receive a monetary prize. Needless to say, the proverbial crow feces hit the fan.

Animal rights activists from several organizations came to Auburn to protest the crow-shooting tournament (Varley 2003). Members from The Fund for Animals threatened to sue the crow-shoot organizers (Fox and Broach 2003). The protest received national attention from several media outlets, including the *New York Times* (February 2, 2003). Approximately 300 crows were killed in the surrounding countryside during the tournament, which accounts for less than one percent of the total population that roosts in Auburn. In the end, the city of Auburn received a lot of bad publicity and was no closer to a solution for dealing with the crow roosts.

Recently, some nonlethal management techniques have been tried. The U.S. Department of Agriculture Wildlife Services Program has teamed up with Auburn city employees to broadcast taped recordings of a crow distress call at all the roosting hotspots in town. The recordings, along with shining red lasers into the trees, usually cause the crows to move along. Pyrotechnics are on hand for any birds that need a little extra convincing. The crows appear to have learned to recognize the USDA truck, and the program appears to be working (www.livingonearth.com, January 20, 2006).

Lastly, some individuals have attempted to make lemonade from lemons by encouraging members of the public to visit Auburn to see this natural phenomenon (www.wildlifewatch.org). Perhaps such an event could turn the Auburn crows into revenue generators for the town, causing business owners and residents to view the birds in a more positive light—or at least providing some additional tax revenue that could be used to help clean up the mess.

CHAPTER ACTIVITIES

1. Access the most recent FHWAR survey data from the U.S. Fish and Wildlife Service website. Provide a brief report on the data collected for your state, including number of hunters, anglers, and wildlife watchers, along with the economic impact of these activities for the state.
2. Provide a list of community-sponsored wildlife-related activities (e.g., "Eagle Days") available in your state. Both positive and negative examples are acceptable.
3. Find out whether your community is conducting an urban coyote (or any other species) management program, who is in charge, how the program is being conducted, and if there are community interest meetings on the program.

LITERATURE CITED

Adams, C.E., J.A. Leifester, and J.S.C. Herron. 1997. Understanding wildlife constituents: Birders and water-fowl hunters. *Wildlife Society Bulletin* 25:653–660.

Adams, C.E., R.A. Stone, and J.K. Thomas. 1988. Conservation education within informational and education divisions of state natural resource agencies. *Wildlife Society Bulletin* 16:333–338.

Adams, C.E. and J.K. Thomas. 1998. *Statewide Survey of the Texas Public Outdoors: A Vision of the Future.* College Station, TX: Texas A&M University, Department of Wildlife and Fisheries Sciences.

Adams, C.E. and J.K. Thomas. 2008. *Texas Rattlesnake Roundups.* College Station, TX: Texas A&M Press.

Adams, C.E., J.K. Thomas, P.-C. Lin, and B. Weiser. 1987. Urban high school students' knowledge of wildlife, in *Integrating Man and Nature in the Metropolitan Environment, ed.* L.W. Adams and D.L. Leedy, 83–86. Columbia, MD: National Institute for Urban Wildlife.

Adams, C.E., J.K. Thomas, K.J. Sternadel, and S.L. Jester. 1994. Texas rattlesnake roundups: Implications of unregulated commercial use of wildlife. *Wildlife Society Bulletin* 22:324–330.

Andelt, W.F. 1985. Behavioral ecology of coyotes in south Texas. *Wildlife Monographs* 94:1–45.

Baker, R.O. and R.M. Timm. 1998. Management of conflicts between urban coyotes and humans in southern California. *Proceedings Vertebrate Pest Conference* 18:200–312.

Barnes, T.G. 1995a. Survey of the nuisance wildlife control industry with notes on their attitudes and opinions, *Proceedings of the Great Plains Wildlife Damage Control Conference* 12:104–108.

Barnes, T.G. 1995b. A survey comparison of pest control and nuisance wildlife control operators in Kentucky, *Proceedings of the Eastern Wildlife Damage Control Conference* 6:39–48.

Barnes, T.G. 1997. State agency oversight of the nuisance wildlife control industry. *Wildlife Society Bulletin* 25:185–188.

Bekoff, M. 1977. *Canis latrans. Mammalian Species* 79:1–9.

Bekoff, M. 1999. Coyote, *Canis latrans*. In *The Smithsonian Book of North American Mammals*, ed. D.E. Wilson, and S. Ruff. Washington, DC: Smithsonian Institution Press.

Bekoff, M. and E.M. Gese. 2003. Coyote (*Canis latrans*). In *Wild Mammals of North America: Biology, Management and Conservation*, 2nd edition, ed. G.A. Feldhamer, B.C. Thompson, and J.A. Chapman, 467–481. Baltimore: The Johns Hopkins University Press.

Berryman, J.H. 1994. Animal damage management: Responsibilities of various agencies and the need for coordination and support. *Proceedings of the Eastern Wildlife Damage Control Conference.* 5:12–14.

Boelens, R. 2006. Co-existing with coyotes in Vancouver (and anywhere else for that matter). *Public Management* 88(11):26–30.

Braband, L.A. 1995. The role of the nuisance wildlife control practitioner in urban wildlife management and conservation. *Proceedings of the Eastern Wildlife Damage Control Conference.* 6:38.

Brown, T.L. and M.J. Manfredo. 1987. Social values defined. In *Valuing Wildlife: Economic and Social Perspectives, ed.* D.J. Decker and G.R. Goff, 12–23. Boulder, CO: Westview Press.

Bryan, H. 1996. The assessment of social impacts. In *Natural Resource Management: The Human Dimension,* ed. A.W. Ewert, 145–166. Boulder, CO: Westview Press.

Buck, W.W. and J.R. Kitts. 2004. Citizen research of Chicago coyotes: A model program. In *Proceedings of the 4th International Symposium on Urban Wildlife Conservation*, ed. W.W. Shaw, L.K. Harris, and L. VanDruff, 186–189. May 1–5, 1999, Tucson, AZ.

Caccamise, D.F. and D.W. Morrison. 1986. Avian communal roosting: Implications of diurnal activity centers. *American Naturalist* 128(2):191–198.

Carpenter, L.H., D.J. Decker, and J.F. Lipscomb. 2000. Stakeholder acceptance capacity in wildlife management. *Human Dimensions of Wildlife* 5(3):5–19.

Conover, M.R. 1997. Wildlife management by metropolitan residents in the United States: Practices, perceptions, costs, and values. *Wildlife Society Bulletin* 25:306–311.

Conover, M.R. 1998. Perceptions of American agricultural producers about wildlife on their farms and ranches. *Wildlife Society Bulletin.* 26(3):597–604.

Conover, M.R. 2001. *Resolving Human–Wildlife Conflicts: The Science of Wildlife Damage Management.* Boca Raton, FL: CRC Press.

Conover, M.R. and D.O. Conover. 2003. Unrecognized values of wildlife and the consequences of ignoring them. *Wildlife Society Bulletin* 31:843–848.

Conover, M.R., W.C. Pitt, K.K. Kessler, T.J. DuBow, and W.A. Sanborn. 1995. Review of human injuries, illnesses, and economic losses caused by wildlife in the United States. *Wildlife Society Bulletin* 23:407–414.

Curtis, P.D., D.A. Bogan, and G. Batcheller. 2007. Suburban coyote management and research needs: A northeast perspective. *Wildlife Damage Management Conference* 12:413-417.

Decker, D.J., T.L. Brown, B.L. Driver, and P.J. Brawn. 1987. Theoretical developments in assessing social values of wildlife: Toward a comprehensive understanding of wildlife recreation involvement. In *Valuing Wildlife: Economic and Social Perspectives, ed.* D.J. Decker and G.R. Goff, 76–95. The Wildlife Society Annual Meeting Symposia on Wildlife Damage, October 1–5.

Decker, D.J., T.L. Brown, and B.A. Knuth. 1996. Human dimensions research: Its importance in natural resource management. In *Natural Resource Management: The Human Dimension,* ed. A.W. Ewert, 29–45. Boulder, CO: Westview Press.

Decker, D.J., T.L. Brown, and W.F. Siemer, eds. 2001. *Human Dimensions of Wildlife Management in North America.* Bethesda, MD: The Wildlife Society.

Decker, D.J. and G.R. Goff. 1987. *Valuing Wildlife: Economic and Social Perspectives.* Boulder, CO: Westview Press.

Decker, D.J. and C.A. Loker. 1996. Human dimensions insights for successful wildlife damage management in suburban environments. Presented at *The Wildlife Society Annual Meeting Symposia on Wildlife Damage,* October 1–5.

Ditton, R.B. 1996. Human dimensions in fisheries. In *Natural Resource Management: The Human Dimension,* ed. A.W. Ewert, 73–90. Boulder, CO: Westview Press.

Dobie, J.F. 1961. *The Voice of the Coyote.* Lincoln: University of Nebraska Press.

Erickson, R.J. 2007. The urban coyote control program. *Wildlife Damage Management Conference* 12:402–404.

Ewert, A.W., ed. 1996. *Natural Resource Management: The Human Dimension.* Boulder, CO: Westview Press.

Farrar, R.O. 2007. Assessing the impact of urban coyote on people and pets in Austin, Travis County, Texas. *Wildlife Damage Management Conference* 12:334–341.

Fitzgerald, L.A. and C.W. Painter. 2000. Rattlesnake commercialization: Long-term trends, issues, and implications for conservation. *Wildlife Society Bulletin* 28:235–253.

Fleishman-Hillard Research. 1994. *Attitudes of the Uncommitted Public toward Wildlife Management.* St. Louis, MO: Fleishman-Hillard.

Fox, C. 2003. City hall fights crows. *The Citizen* online, http://www.auburnpub.com/.

Fox, C. and L.H. Broach. 2003. Crow shoot begins this morning. *The Citizen* online, http://www.auburpub.com/.

Frumkin, H. 2001. Beyond toxicity: Human health and the natural environment. *American Journal of Preventive Medicine* 20:234–240.

Fulton, D.C., M.J. Manfredo, and J. Lipscomb. 1996. Wildlife value orientations: A conceptual and measurement approach. *Human Dimensions of Wildlife* 1(2):24–47.

Gehrt, S.D. 2006. Urban coyote ecology and management. *The Cook County, Illinois, Coyote Project. Bulletin 929*, 1–31.

Gehrt, S.D. 2007. Ecology of coyotes in urban landscapes. *Wildlife Damage Management Conference* 12:303–311.

Geist, V. 2007. How close is too close? Wildlife professionals grapple with habituating wildlife. *Wildlife Professional* 1:34–37.

Gigliotti, L.M. and D.J. Decker. 1992. Human dimensions in wildlife management education: Pre-service opportunities and in-service needs. *Wildlife Society Bulletin* 20:1–14.

Gilbert, D.L. 1964. *Public Relations in Natural Resources Management. Minneapolis, MN:* Burgess Publishing Company.

Gilbert, F. F. and D.G. Dodds. 2001. *The Philosophy and Practice of Wildlife Management,* 3rd edition. Malabar, FL: Krieger Publishing Co.

Glynn, M.K., J.H. Mermin, L.M. Durso, F.J. Angulo, and K.F. Reilly. 2001. Knowledge and practices of California veterinarians concerning the human health threat of reptile-associated salmonellosis (1996). *Journal of Herpetological Medicine and Surgery* 11(2):9–13.

Gompper, M.E. 2002. Top carnivores in the suburbs? Ecological and conservation issues raised by colonization of North-Eastern North America by coyotes. *BioScience* 52(2):185–190.

Gorenzel, W.P. and T.P. Salmon. 1992. Urban crow roosts in California. In *Proceedings 15th Vertebrate Pest Conference,* 97–102.

Gorenzel, W.P., T.P. Salmon, G.D. Simmons, B. Barkhouse, and M.P. Quisenberry. 2000. Urban crow roosts—a nationwide phenomenon? *Proceedings Eastern Wildlife Damage Management Conference* 9:158–170.

Gray, G.G. 1993. *Wildlife and People: The Human Dimensions of Wildlife Ecology.* Urbana, IL: University of Illinois Press.

Grinder, M.I. and P.R. Krausman. 2001. Home range, habitat use, and nocturnal activity of coyotes in an urban environment. *Journal of Wildlife Management* 65(4):887–898.

Hadidian, J., M.R. Childs, R.H. Schmidt, L.J. Simon, and A. Church. 1999. Nuisance-wildlife control practices, policies, and procedures in the United States. In *Wildlife, Land, and People: Priorities for the 21st Century, ed.* R. Field, R.J. Warren, H.K. Okarma, and P.R. Sieverd, 165–168. Proceedings of the 2nd International Wildlife Management Congress. Bethesda, MD: The Wildlife Society.

Harper, S. 2002. VDOT trying kinder method of beaver control. *The Virginian-Pilot,* 22 June.

Herbold, J.R. 2000. What you and your clients need to know about zoonotic diseases: Rabies, Lyme disease, and salmonellosis. *Proceedings of the North American Veterinary Conference* 14:833–834.

Hesselton, W.T. 1991. How governmental wildlife agencies should respond to local governments that pass anti-hunting legislation. *Wildlife Society Bulletin* 19:222–223.

Hewitt, D.G. and T.A. Messmer. 1997. Responsiveness of agencies and organizations to wildlife damage: Policy process implications. *Wildlife Society Bulletin* 25:418–423.

Hill, E.P., P.W. Sumner, and J.B. Wooding. 1987. Human influences on range expansion of coyotes in the southeast. *Wildlife Society Bulletin* 15:521–524.

HSUS (Humane Society of the United States). 1997. *Wild Neighbors.* Golden, CO: Fulcrum Publishing.

Hunter, A.L. 2002. Comparison of avian communities within traditional and wildscaped residential neighborhoods in San Antonio, Texas. Thesis, Southwest Texas State University, San Marcos.

Huot, A.A. and D.L. Bergman. 2007. Suitable and effective coyote control tools for the urban/suburban setting. *Wildlife Damage Management Conference* 12:312–322.

International Association of Fish and Wildlife Agencies. 1993. Ownership and use of traps by trappers in the United States in 1992. Washington, DC: Fur Resources Committee of the International Association of Fish and Wildlife Agencies and the Gallup Organization.

Jensen, P.G., P.D. Curtis, M.E. Lehnert, and D.L. Hamelin. 2001. Habitat and structural factors influencing beaver interference with highway culverts. *Wildlife Society Bulletin* 29:654–664.

Johnson, R.J. 1994. American crows. In *Prevention and Control of Wildlife Damage,* ed. S.E. Hygnstom, R.M. Timm, and G.E. Larson, E-33–E-40. Lincoln: University of Nebraska Cooperative Extension, U.S. Department of Agriculture-Animal Plant Health Inspection Service-Animal Damage Control.

Jolma, D.J. 1994. *Attitudes toward the Outdoors: An Annotated Bibliography of U.S. Surveys and Poll Research Concerning the Environment, Wildlife, and Recreation.* Jefferson, NC: McFarland.

Kania, G.S. and M.R. Conover. 1991. Another opinion on how governmental agencies should respond to local ordinances that limit the right to hunt. *Wildlife Society Bulletin* 19:222–223.

Kellert, S.R. 1997. *The Value of Life: Biological Diversity and Human Society.* Washington, DC: Island Press.

Kellert, S.R. 1980. Contemporary values of wildlife in American society. In *Wildlife Values, ed.* W.W. Shaw and E.H. Zube, 31–60. Fort Collins, CO: U.S. Forest Service, Rocky Mountain Experiment Station.

Kendall, P. 1999. Wounded coyote captured by rush of loop traffic, *Chicago Tribune*, online edition, March 25, 1999.

Kimmel, M.M. 1985. Parks and property values: An empirical study in Dayton and Columbus, Ohio. Thesis, Miami University, Oxford, OH.

King, R.T. 1947. The future of wildlife in forest land use. *Transactions of the North American Wildlife and Natural Resources Conference* 12:454–467.

Krutilla, J.V. 1967. Conservation reconsidered. *American Economics Review* 57:778–786.

Lampa, K. 2001. Coyotes in the city. *Discovery Magazine,* Vancouver Natural History Society, Fall.

Leedy, D.L. and L.W. Adams. 1984. *A Guide to Urban Wildlife Management.* Columbia, MD: National Institute for Urban Wildlife.

Leopold, A. 1933. *Game Management.* Charles Scribner's Sons (Reprinted in 1986 by University of Wisconsin Press, Madison, WI).

Lovell, C.D., B.D. Leopold, and C.S. Shropshire. 1998. Trends in Mississippi predator populations, 1980–1995. *Wildlife Society Bulletin* 26(3):552–556.

Lowery, M.D. and W.F. Siemer. 1999. Resource agencies as effective sources of information on wildlife damage prevention and control: Overcoming the obstacles. *Abstracts of The Wildlife Society Annual Conference* 6:141–142.

Manfredo, M.J., J.J. Vaske, and D.J. Decker. 1995. Human dimensions of wildlife management: Basic concepts. In *Wildlife and Recreationists: Coexistence through Management and Research,* ed. R.L. Knight and K.J. Gutzwiller, 17–31. Washington, DC: Island Press.

Mangun, W.R. 1992. *American Fish and Wildlife Policy: The Human Dimension.* Carbondale, IL: Southern Illinois University Press.

Mankin, P.C., R.E. Warner, and W.L. Anderson. 1999. Wildlife and the Illinois public: A benchmark study of attitudes and perceptions. *Wildlife Society Bulletin* 27:465–472.

Martin, D. 1999. Wild (and unleashed) coyote is captured in Central Park. *New York Times,* 148:51480, 1999, B1.

Marzluff, J.M., K.J. McGowan, R. Donnelly, and R.L. Knight. 2001. Causes and consequences of expanding American crow populations. In *Avian Ecology and Conservation in an Urbanizing World,* ed. J.M. Marzluff, R. Bowman, and R. Donnelly, 331–363. Norwell, MA: Kluwer Academic Press.

McClennen, N., R.R. Wigglesworth, S.H. Anderson, and D.G. Wachob. 2001. The effect of suburban and agricultural development on the activity patterns of coyotes (*Canis latrans*). *American Midland Naturalist* 146(1):27–36.

McGowan, K.J. 2001. Demographic and behavioral comparisons of suburban and rural American crows. In *Avian Ecology and Conservation in an Urbanizing World,* ed. J. M. Marzluff, R. Bowman, and R. Donnelly, 365–381. Norwell, MA: Kluwer Academic Press.

Meinzen, W. 1995. *Coyote.* Lubbock, TX: Texas Tech University Press.

Mörner, T., D.L. Obendorf, M. Artois, and M.H. Woodford. 2002. Surveillance and monitoring of wildlife diseases. *OIE Revue Scientifique et Technique* 21:67–76.

Nelson, A.C. 1986. Using land markets to evaluate urban containment programs. *APA Journal,* Spring, 156–171.

Orthmeyer, D.L., T.A. Cox, J.W. Turman, and J.R. Bennett. 2007. Operational challenges of solving urban coyote problems in southern California. *Wildlife Damage Management Conference* 12:344–357.

Payne, N.F. and R.P. Peterson. 1986. Trends in complaints of beaver damage in Wisconsin. *Wildlife Society Bulletin* 14:303–307.

Penland, S. 1987. Attitudes of urban residents toward avian species and species' attributes. In *Integrating Man and Nature in the Metropolitan Environment, ed.* L.W. Adams and D.L. Leedy, 77–82. Columbia, MD: National Institute for Urban Wildlife.

Purdy, K.G. and D.J. Decker. 1989. Applying wildlife values information in management: The wildlife attitudes and values scale. *Wildlife Society Bulletin* 17:494–500.

Randall, A. 1991. The value of biodiversity. *Ambio* 20:64–68.

Reid, C. 2004. *Coyote: Seeking the Hunter in Our Midst.* Boston, MA: Houghton Mifflin.

Reiter, D.K. M.W. Burnson, and R.H. Schmidt. 1999. Public attitudes toward wildlife damage management and policy. *Wildlife Society Bulletin* 27:746–758.

Ricklefs, R. E. and G.L. Miller. 1999. *Ecology.* New York: W. H. Freeman.

Riley, S.P.D., R.M. Sauvajot, T.K. Fuller, E.C. York, D.A. Kamradt, C. Bromley, and R.K. Wayne. 2003. Effects of urbanization and habitat fragmentation on bobcats and coyotes in southern California. *Conservation Biology* 17(2):566–576.

Romin, L. 1994. Factors associated with mule deer highway mortality at Jordanell Reservoir, Utah. Thesis, Utah State University, Logan.

Rossi, A.N. and J.B. Armstrong. 1999. *Alabamian's Wildlife-Related Activities, Wildlife Management Perceptions, and Hunting Behavior.* Montgomery, AL: Alabama Game and Fish Division.

Rue, L.L. III. 1987. *The Deer of North America.* New York: Crown Publishers.

Salant, P. and D.A. Dillman. 1994. *How to Conduct Your Own Survey.* New York: John Wiley and Sons.

San Julian, G.J. 1987. The future of wildlife damage control in an urban environment. *Proceedings of the Eastern Wildlife Damage Control Conference* 1:229–233.

Schmidt, R.H. 1998. Required knowledge. *Wildlife Control Technology* 5:6–7.

Schmidt, R.H. 2007. Complexities of urban coyote management: Reaching the unreachable, teaching the unteachable, and touching the untouchable. *Wildlife Damage Management Conference* 12:364–370.

Siemer, W.F., H.W. Hudenko, and D.J. Decker. 2007. Coyote management in residential areas: Human dimensions research needs. *Wildlife Damage Management Conference* 12:421–430.

Siemer, W.F., B.A. Knuth, D.J. Decker, and V.A.L. Alden. 1992. *Human Perceptions and Behaviors Associated with Lyme Disease: Implications for Land and Wildlife Management.* HDRU Series 92–8. Ithaca, NY: Department of Natural Resources, Cornell University.

Stokes, D. and L. Stokes. 1996. *Field Guide to Birds: Eastern Region.* Boston: Little, Brown.

Stout, R.J. 1998. Human dimensions of interactions between white-tailed deer and urban dwellers in Missouri. Ph.D. dissertation, Texas A&M University, College Station.

Thomas, J.K. and C.E. Adams. 1993. The social organization of rattlesnake roundups in rural communities. *Sociological Spectrum* 13:433–449.

Tigas, L.A., D.H. Van Vuren, and R.M. Sauvajot. 2002. Behavioral responses of bobcats and coyotes to habitat fragmentation and corridors in an urban environment. *Biological Conservation* 108:299–306.

Timm, R.M. 2007. CoyoteBytes.org: A new educational website. *Wildlife Damage Management Conference* 12:405–412.

Timm, R.M. and R.O. Baker. 2007. A history of urban coyote problems. *Wildlife Damage Management Conference* 12:272–286.

U.S. Department of the Interior, Fish and Wildlife Service and U. S. Department of Commerce, Bureau of the Census. 2005. *National Survey of Fishing, Hunting, and Wildlife-Associated Recreation.* Washington, DC: United States Government Printing Office.

VanDruff, L.W., E.G. Bolen, and G.S. San Julian. (994. Management of urban wildlife. In *Research and Management Techniques for Wildlife and Habitats, ed.* T.A. Bookhout, 507–530. Bethesda, MD: The Wildlife Society.

Varley, R. 2003. From across the state, protestors flock to Auburn, *The Citizen* online, http://www.auburbpub.com/.

Weatherhead, P.J. 1984. Two principal strategies in avian communal roosts. *American Naturalist* 121(2):237–243.

Wehtje, M.E. 1998. Defensible space: A behavioral approach for managing predators at the urban-wildlife interface. *Vertebrate Pest Conference Proceedings* 18:290–292.

Wildlife Services. 2001. Wildlife Services assistance at airports. United States Department of Agriculture, Animal and Plant Health Inspection Service, Washington, DC.

Wilson, E.O. 1984. *Biophilia: The Human Bond with Other Species.* Cambridge, MA: Harvard University Press.

Witter, D.J. 1990. Wildlife management and public sentiment. In *Management of Dynamic Ecosystems, ed.* J.M. Sweeney, 162–172. West Lafayette, IN: The Wildlife Society, North Central Section.

Witter, D.J., J.D. Wilson, and G.T. Maupin. 1979. "Eagle Days" in Missouri: Characteristics and enjoyment ratings of participants. *Wildlife Society Bulletin* 8:64–65.

Worcester, R.E. and R. Boelens. 2007. The Co-Existing with Coyotes program in Vancouver, B.C. *Wildlife Damage Management Conference* 12:393–397.

Young, C. 1991. Fostering residential participation in urban wildlife management: Communication strategies and research needs. In *Wildlife Conservation in Metropolitan Environments, ed.* L.W. Adams and D.L. Leedy, 203–209. Columbia, MD: National Institute for Urban Wildlife.

CHAPTER 10

Working with Urban Stakeholders

We all know we are unique individuals, but we tend to see others as representatives of groups.

—Deborah Tannen, Linguistics Professor

KEY CONCEPTS

1. Stakeholder is defined with examples of potential urban wildlife stakeholders.
2. There are four types of stakes a person might have in the urban wildlife management policy- and decision-making process.
3. There are four stages of the policy life cycle and urban wildife management is just in the recognition stage.
4. The public and private sectors represent different sets of stakeholders in the policy- and decision-making process.
5. There are five levels of government that may be involved in a wildlife management decision.
6. The U.S. Department of the Interior has four different bureaus with responsibility for managing wildlife resources.
7. The relationship between the U.S. government and federally recognized tribal governments contains special considerations.
8. The term "the public" can be described using three different examples.
9. Academic institutions have a unique role to play in the wildlife management decision-making process.

INVITING EVERYONE TO THE TABLE

Because wildlife in North America belongs to the public rather than to individuals, it is a shared resource—a commons. The government, through state and federal agencies, manages the resource for the citizens and in the public's best interest. However, management and decision making can be challenging, as is the case with any common property. The same wildlife resource can be valued for very different reasons. People have diverse and even conflicting wildlife values (see Chapter 9). While the uses and values of wildlife vary between individuals, cultures, and regions, the underlying human dimensions problem is consistent: How do we balance sustainable use and distribution of resources while recognizing the needs and desires of many different interests?

The latter part of the twentieth century saw a change in the focus of wildlife managers. Beginning in the mid-1980s, concepts about who benefits from wildlife management and who should be considered in management decision-making began to broaden. Traditionally, agencies focused on landowners and consumptive users, thought of as clients; but during the 1990s a stakeholder approach gained increasing support.

THE POLICY LIFE CYCLE

According to Wright (2004), environmental policy is developed in a sociopolitical context, usually in response to a problem. Policy development can occur at every level of government. Local policy may focus on zoning for conservation and agriculture. Federal policy generally addresses broader problems, particularly those that transcend political boundaries, such as air pollution.

The development of public policy tends to take a predictable course, referred to as the "policy life cycle" (Figure 10.1). Typically, there are four stages: *recognition, formulation, implementation,* and *control.* Each stage can be thought of as having a certain amount of political importance or "weight."

The recognition stage begins when a problem is first perceived, often, but not always, as a result of scientific research. The problem has little political weight at this point, but once the media starts to spread the word the public becomes involved. Often there are many dissenting viewpoints on what should be done, if anything. If the problem garners enough publicity it will eventually receive some attention from at least one level of government.

As the problem gains political weight it enters the formulation stage. The public begins to look to legislators and policymakers to do something, and debate over exactly what to do occurs in the corridors of power. Political battles ensue over how much regulation is needed, who will be affected, and how much it will cost to take action. Media coverage is probably high at this point. Politicians begin to hear about the issue from their constituents and lobbyists on a regular basis. During this stage, policymakers consider the "Three E's" of environmental policy: *effectiveness* (will the policy accomplish what it intends to do?); *efficiency* (will the objectives be accomplished for the lowest possible cost?); and *equity* (will the financial burden of implementing a solution be fairly distributed among all the stakeholders?).

Once decisions have been made about how to solve the problem the policy has reached the implementation stage. By now, public concern and political weight are starting to ebb. Emphasis shifts to development of specific regulations and their enforcement. Industry begins to adjust to

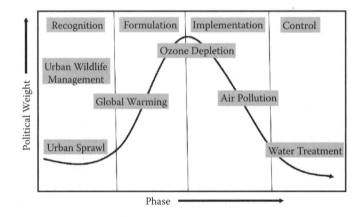

Figure 10.1 Urban wildlife management and other ecological issues and their place in the policy life cycle. (From R.T. Wright. 2004. *Environmental Science: Toward a Sustainable Future.* Upper Saddle River, NJ: Pearson Prentice Hall. With permission.)

the new regulations. Eventually, greater attention may be given to efficiency and equity as various stakeholders gain experience with the policy.

Finally, possibly years after the recognition stage, we reach the control stage. Regulations may become more streamlined but they're considered to be a fact of everyday life (although policies are always vulnerable to political shifts). Policymakers focus on keeping the problem under control and, over time, the public may forget the problem ever existed; or the policy may create new problems, in which case the whole process begins again!

Urban wildlife management policy formulation, as a whole, is still in the recognition stage (Figure 10.1), although certain management issues have advanced to the formulation stage. While it carries little political weight, it does garner some media attention (particularly human-wildlife conflicts) and has been documented as a public service need through scientific research.

Often, it takes a dramatic event to move issues along through the policy life cycle. For example, the crash landing of U.S. Airway's Flight 1549 into the Hudson River that occurred as the manuscript for this book was being prepared, as a result of a collision between the Airbus A320 and a flock of birds, has surely brought greater attention to the issue of birds and air safety.

WHAT (AND WHO) IS A STAKEHOLDER?

A *stakeholder* is any individual who has an interest in a particular issue—in our case, urban wildlife management; in other words, citizens who will be affected by, or who will affect, a management decision or action (Susskind and Cruikshank 1989; Crowfoot and Wondolleck 1991; Decker et al. 1996). A person's stake may be economic, recreational, cultural, social, or related to human health and safety. Typically, stakeholders include individuals and groups who have legal standing, political influence, sufficient moral claims connected to the situation, or who have the power to block implementation of a decision (Susskind and Cruikshank 1989). A partial list of potential urban wildlife stakeholders includes birders and other wildlife watchers, hunters and trappers, homeowners and landlords, businesses, local governments, nongovernmental organizations, and state and federal management agencies.

THE CHANGING FACE OF WILDLIFE STAKEHOLDERS

Based on a model developed by Aldo Leopold (1933), wildlife management agencies have long used a quasi-agricultural approach, in which plans are developed and implemented to allow an annual sustained yield or harvest of specific wildlife species—primarily game animals (Decker, Brown, and Siemer 2001). The stakeholders these management efforts attempted to serve were primarily hunters, trappers, and landowners. As discussed in Chapter 1, consumptive users pay for agency services through license fees, including federal duck stamps, and special taxes imposed on equipment such as firearms and ammunition. While there are millions of acres of publicly owned habitat in the United States, the majority of wildlife habitat is controlled by private land owners. Government wildlife management activities have been designed, in large part, to build and serve a constituency of consumptive users and to return revenues to the agencies (Dunlop 1991).

Hunters and landowners fit the traditional definition of a constituency: a group of people (constituents) who authorize or support the efforts of others (e.g., wildlife professionals) to act on their behalf (Decker, Brown, and Siemer 2001). The traditional wildlife constituency was limited to a few specific, relatively homogeneous groups, and the system functioned well for those interest groups. Managers understood the values of their clientele and, in many cases, shared those values. Often wildlife managers were members of one or more of the constituent groups they served.

During the late 1960s and early 1970s, however, the definition of wildlife management stake-holders was changing. Not all who paid for management (primarily in the form of taxes) used or received services, while not all with an interest in management paid for those services. The concept of "user" was changing too, as nonconsumptive recreational activities, such as birding or even enjoying wildlife while hiking or canoeing, increased in popularity. As the public became more interested in the environment and more distrustful of government, individuals who had never been considered part of the traditional wildlife management constituency began to demand a seat at the table. The public wanted more than to have their opinions heard by agencies; they wanted to play an active role in management decisions.

Also, the people sitting at that table were not as homogeneous as before. Historically, American wildlife stakeholders were overwhelmingly White males. The face of wildlife stakeholders is radically different today, including both genders, diverse races, cultures, and socioeconomic groups, all with wildlife values that are every bit as varied. These individuals have a much wider variety of attitudes and expectations regarding wildlife management than their traditional counterparts. As anyone who has tried to order pizza for a crowd can tell you, this makes the job of making everyone happy far more difficult!

A GUIDE TO MAJOR STAKEHOLDERS

The first step in successfully implementing a stakeholder approach is to evaluate and identify those who will be substantially affected by the decision or action in question. Stakeholders can be grouped into four major categories: government ("the public sector"), nongovernmental organizations (NGOs or "the private sector"), academic institutions, and the public. Keep in mind that an individual stakeholder can belong to more than one of these categories. For example, a U.S. Fish and Wildlife Service employee is also a member of the public, and may have ties to an NGO, particularly some kind of advocacy group. Moreover, the specifics concerning who has a stake in a management decision will vary from case to case.

Government (Public Sector)

There are at least five different levels of government to consider whenever a wildlife management plan is being evaluated: federal, state, tribal, county, and municipal. Jurisdictions may (and do) overlap, for example, state, tribal, and county governments; or municipal and several different county governments. At each level one may find several different departments with varying degrees of interest in the issue at hand. To complicate matters further, in some cases a variety of foreign governments may be affected as well, particularly when the issue involves migratory species of birds and/or mammals.

The following sections will focus on the government agencies and departments most likely to become involved in wildlife management issues, their jurisdiction, mandate, and/or focus. Discussion of how these levels of government overlap and the resulting challenges for wildlife managers can be found in Chapter 11.

Federal

Some federal agencies have direct wildlife management responsibilities while others are simply affected by management decisions. Although it is unlikely that all of the departments and bureaus discussed below will need to be involved in any single management decision, we'll take a look at those that should at least be considered in any stakeholder identification effort. Other federal agencies and departments that may need to be considered when attempting to identify stakeholders include those listed in Table 10.1.

U.S. Department of the Interior (DOI)

Four of the DOI's eight bureaus have some degree of responsibility for managing the nation's wildlife resources. They are: (1) U.S. Fish and Wildlife Service (USFWS); (2) Bureau of Land Management (BLM); (3) National Park Service (NPS); and (4) U.S. Geological Survey (USGS). The DOI manages 500 million acres—approximately one-fifth of all land in the United States—including 548 wildlife refuges, 37 wetland management districts, and vast areas of multiple-use lands (U.S. Department of the Interior 2009).

- *U.S. Fish and Wildlife Service (USFWS):* USFWS is the primary national wildlife resource agency. The agency's mission is "working with others, to conserve, protect and enhance fish, wildlife, and plants and their habitats for the continuing benefit of the American people." The USFWS traces its origins back to 1871 when Congress created the U.S. Commission on Fish and Fisheries (Department of Commerce) and the Division of Economic Ornithology and Mammalogy (Department of Agriculture). In 1939 responsibility for wildlife and fisheries resources was moved to the Department of the Interior and the Fish and Wildlife Service. Among its key functions, USFWS enforces federal wildlife laws, protects endangered species, manages migratory birds, conserves and restores wildlife habitat including wetlands, and helps foreign governments with their international conservation efforts. The department also oversees the Federal Aid in Wildlife Restoration Act (Pittman–Robertson) and the Federal Aid in Sport Fish Restoration Act (Dingell–Johnson) programs that distribute hundreds of millions of dollars in excise taxes on fishing and hunting equipment to state fish and wildlife agencies. Additionally, since the vast majority of fish and wildlife habitat is on lands other than federal lands, USFWS is involved in a number of partnership activities that promote voluntary habitat development on private lands, including Partners in Flight and Partners for Fish and Wildlife (U.S. Fish and Wildlife Service 2009).
- *Bureau of Land Management (BLM):* In 1946, the Grazing Service was merged with the General Land Office to form the Bureau of Land Management. When the BLM was initially created, there were over 2000 unrelated and often conflicting laws for managing the public lands. The BLM had no unified legislative mandate until Congress enacted the Federal Land Policy and Management Act of 1976 (FLPMA). The mission of the BLM is to "sustain the health, diversity and productivity of the public lands for the use and enjoyment of present and future generations." The Bureau is responsible for managing 262 million surface acres of land, most of which are located in the western part of the country. BLM manages a wide variety of resources and uses, including wild horse and burro populations, fish and wildlife habitat, and wilderness areas (U.S. Bureau of Land Management 2009).
- *National Park Service (NPS):* The National Park Service was created in 1916 as a new federal bureau responsible for protecting the forty national parks and monuments then in existence and those yet to be established. The mission of the NPS is "to promote and regulate the use of the … national parks … and to provide for the enjoyment of the same in such manner and by such means as will leave them unimpaired for the enjoyment of future generations" (National Park Service 2009). The national park system has grown to comprise a network of nearly 400 natural, cultural, and recreational sites in forty-nine states, the District of Columbia, American Samoa, Guam, Puerto Rico, Saipan, and the Virgin Islands. In many areas, NPS units represent the last vestiges of once vast undisturbed ecosystems. The national parks serve as outdoor laboratories for the study of physical, biological, and cultural systems and their components (National Park Service 2009).
- *U.S. Geological Survey (USGS):* The USGS was established on March 3, 1879, to study the geological structure and economic resources—including biological resources—of the public domain. The USGS operates sixteen Biological Resource Centers (BRCs), listed in Table 10.2. Although each BRC plays an important role in wildlife management, several are of particular interest.
  - The Center for Biological Informatics (CBI) develops, identifies, and provides access to tools that facilitate collection and use of biological information and data. CBI also cooperates with

Table 10.1   Potential Federal Government Stakeholders

Department of Commerce	Federal Highway Administration
Department of Health and Human Services	Federal Railroad Administration
Department of Housing and Urban Development	National Highway Traffic Safety Administration
Department of Transportation	U.S. Army Corps of Engineers
Environmental Protection Agency	U.S. Customs Service
Federal Aviation Administration	

Table 10.2   U.S. Geological Survey Biological Resource Centers

Alaska Science Center	Northern Prairie Wildlife Research Center
Center for Biological Informatics	National Wildlife Health Center
Columbia Environmental Research Center	National Wetlands Research Center
Florida and Caribbean Science Center	Pacific Island Environmental Research Center
Fort Collins Science Center	Patuxent Wildlife Research Center
Forest and Rangeland Ecosystem Science Center	Upper Midwest Environmental Sciences Center
Great Lakes Science Center	Western Ecological Research Center
Leetown Science Center	Western Fisheries Research Center

others to improve access to existing information and data. The term "biological informatics" refers to the development and use of computer, statistical, and other tools in the collection, organization, dissemination, and use of information to solve problems in the life sciences (U.S. Center for Biological Informatics 2009).

- *The National Wildlife Health Center (NWHC),* established in 1975, is a biomedical laboratory dedicated to assessing the impact of disease on wildlife and identifying the role of various pathogens in contributing to wildlife losses. Located in Madison, Wisconsin, the NWHC monitors and assesses the impact of disease on wildlife populations, defines ecological relationships leading to the occurrence of disease, transfers technology for disease prevention and control, and provides guidance, training, and on-site assistance for reducing wildlife losses when outbreaks occur. The NWHC staff provides expertise regarding animal welfare regulations and their application to wildlife. Technical assistance regarding animal welfare matters is often provided to wildlife biologists and others. Preparation of videotapes, publications, consultations, and training are activities commonly carried out by the NWHC in the animal welfare arena (National Wildlife Health Center 2009).

- *The Patuxent Wildlife Research Center* was established in 1936 as the nation's first wildlife experiment station. It has been a leading international research institute for wildlife and applied environmental research, for transmitting research findings to those responsible for managing the nation's natural resources, and for providing technical assistance in implementing research findings so as to improve natural resource management. Scientists at the Patuxent Wildlife Research Center have been responsible for many advances in natural resource conservation, especially in such areas as migratory birds, wildlife population analysis, waterfowl harvest, habitat management, wetlands, coastal zone and flood plain management, contaminants, endangered species, urban wildlife, ecosystem management, and management of national parks and national wildlife refuges. The Patuxent Wildlife Research Center develops and manages national inventory and monitoring programs and is responsible for the North American Bird Banding Program and leadership of other national bird monitoring programs. The center's scientific and technical assistance publications, wildlife databases, and electronic media are used nationally and worldwide in managing biological resources (Patuxent Wildlife Research Center 2009).

Department of Agriculture (USDA)

President Abraham Lincoln founded the USDA in 1862. In Lincoln's day, nearly half of all Americans were farmers. While that's no longer the case, ties to our agrarian past remain—gardening is one of the most popular hobbies in the United States. Still, it would be stretching the truth to say America is currently an agricultural society.

Several agencies within USDA have some level of responsibility for managing wildlife, including the Forest Service (FS), the Natural Resource Conservation Service (NRCS), and Wildlife Services (part of the Animal and Plant Health Inspection Service). In addition, the Cooperative State Research, Education, and Extension Service (CSREES) is responsible for facilitating some wildlife research and disseminating information on wildlife to the public (U.S. Department of Agriculture 2009).

- *Forest Service (FS):* Established in 1905, the Forest Service manages public lands in national forests and grasslands encompassing 193 million acres of land. The FS has established wildlife programs, including the National Wildlife Program (NWP) and the National Wildlife Ecology Unit (WEU). The NWP covers terrestrial animal species not considered threatened, endangered, or sensitive (TES) and assists field biologists in meeting the Forest Service's wildlife and habitat goals. The WEU facilitates the transfer of technology and technical information from research to field biologist. WEU staff members serve on technical teams addressing terrestrial watershed management issues, including TES species (U.S. Forest Service 2009).
- *Natural Resource Conservation Service (NRCS):* The agency that would eventually become NRCS was started in 1933 as the Soil Erosion Service. Today the NRCS mission is "helping people help the land." Of the agency's various conservation programs, at least five specifically target wildlife and wildlife habitat: Conservation Reserve Program, Conservation Technical Assistance Program, Grazing Lands Conservation Initiative, Wetlands Reserve Program, and Wildlife Habitat Incentives Program (National Resource Conservation Service 2009).
- *Wildlife Services (WS):* Wildlife Services, formerly known as Animal Damage Control, is part of the USDA's Animal and Plant Health Inspection Service (APHIS). WS was granted statutory authority for wildlife damage management by the Animal Damage Control Act of 1931. Its mission is "to provide federal leadership in managing problems caused by wildlife." In spite of a traditional focus on rural and agricultural interests, requests for assistance with urban wildlife issues are becoming more common and WS now employs urban wildlife biologists. Research is conducted to assess the need for wildlife damage control and to develop ways in which to address those needs, including nonlethal methods (Wildlife Services 2009).

National Oceanic and Atmospheric Administration (NOAA)

The ancestor agencies of the National Oceanic and Atmospheric Administration include the U.S. Coast Survey (established in 1807), the U.S. Weather Bureau (1870), and the U.S. Commission of Fish and Fisheries (1871). NOAA protects coasts, bays, estuaries, and the Great Lakes, and is involved in helping to mitigate damage from oil and hazardous material spills. It operates thirteen marine sanctuaries and one marine national monument, and oversees other marine protected areas, such as national seashores, national wildlife refuges, and national estuarine research reserves.

The mission of the National Marine Fisheries Service (NOAA Fisheries) is "the stewardship of living marine resources through science-based conservation and management, and the promotion of healthy ecosystems." In addition to marine fish species, the program protects marine mammals, including whales, dolphins, porpoises, seals, and sea lions, and six endangered species of sea turtles (National Oceanic and Atmospheric Administration 2009).

Tribal Governments

There is a unique relationship between the United States and the 562 federally recognized tribal governments (Bureau of Indian Affairs 2009). These tribes are considered sovereign governments, a fact that hundreds of treaties, federal laws, and court cases have repeatedly affirmed. Sovereignty is an internationally recognized concept, with treaties acting to formalize nation-to-nation relationships. However, while the U.S. Constitution recognizes Indian tribes as distinct governments, it also authorizes Congress to regulate commerce with "foreign nations, among the several state, and with the Indian tribes." State governance is generally not permitted within reservations; only Congress has plenary (overriding) power over Indian Affairs (American Indian Policy Center 2009).

The legal position of Alaska Natives is singular among Native Americans. While the same legal rules apply to Alaskan Natives as to native peoples elsewhere in the United States, the historical development of the relationship between Alaskan Natives and the U.S. government was considerably different. The majority of Alaskan Natives have never lived within reservations. Until recent decades Alaska Natives, many in remote villages, have been affected very little by non-native culture. The application of federal Indian law to Alaska Natives is still evolving, with no clear answers to many basic legal questions (Mertz 1991).

American Indians in the contiguous forty-eight states have jurisdiction over 45 million acres of reserved lands and an additional 10 million in individual allotments; there are another 40 million acres of traditional Native lands in Alaska. Tribal governments tend to place a high priority on preserving these lands and their natural resources. As a result, tribes provide habitat for many vulnerable wildlife species (U.S. Fish and Wildlife Service 2009). Many tribal communities allocate individual use rights to land and specific resources without allowing users to sell the land or the resource rights. The Iroquois and Hopi, for instance, recognize use rights but prohibit sales by individuals. The hunting territories of the Cree are individually allocated and passed down from father to son, but sales are not permitted (Anderson 1995).

Some tribal governments have their own departments of wildlife or natural resource management (e.g., Nez Perce, Navajo, and Southern Utes). Additionally, some federal agencies have created American Indian Liaison Offices specifically to work with tribal governments. The relationship between tribal and state governments can be fraught with challenges and friction can arise between these entities. However, tribes and states are discovering ways to set aside jurisdictional debate in favor of cooperative government-to-government relationships that respect the autonomy of both sides. Tribal governments, state governments, and local governments continue to find innovative ways to work together to carry out their governmental functions. Intergovernmental institutions have been developed in some states, and state-tribal cooperative agreements on a broad range of issues are becoming more commonplace (Johnson et al. 2000). Tribes are now being recognized as prominent fisheries and wildlife managers and, as such, expect full participation in national fisheries and wildlife initiatives.

State and Territorial Governments

Every U.S. state and territorial government has some kind of department to oversee wildlife management. These departments go by many names—Game and Fish, Conservation, Environment—but for simplicity's sake they are usually referred to as departments of natural resources (DNRs). Keep in mind, as when dealing with the federal government, other departments within state government may be affected by wildlife management issues and decisions. Table 10.3 provides a list of generic state departments to consider during any attempt to identify stakeholders.

Local Governments

Local governments include both counties and municipalities, and it is at this level that the overlap of jurisdiction can become incredibly complex. Local governments often have departments with jurisdictions that correlate with and, to some degree, overlap state and federal governments and each other; for example, health, animal control, parks, tourism, and transportation departments are

Table 10.3   Potential State and Territorial Government Stakeholders

Agriculture	Housing/community development
Animal health	Parks
Commerce/economic development	Pest control board
Environmental protection	Public safety
Forestry	Public utilities/public service commissions
Health	Tourism
Historic preservation	Transportation

common at most levels of government. Additionally, there also may be independent townships and incorporated areas within municipal boundaries, each with its own government.

Of course, wild animals recognize none of the governmental boundaries we have discussed in this section, which is one reason the process of identifying governmental stakeholders can be a time consuming and tricky business. Urban wildlife managers should familiarize themselves with the governmental structures within their own area of responsibility. In addition, managers should work proactively to identify the names and phone numbers of relevant department heads *before* a management action or decision needs to be made—and keep the list current.

Nongovernmental Organizations (NGOs)

The public sector works with and depends on the nongovernmental private sector, and together they address most of the society's needs. Nongovernmental organizations (NGOs) include three types of businesses: for-profit, not-for-profit, and nonprofit. A "business" is a commercial or industrial enterprise and the people (paid or volunteer) who constitute it. All three types may become stakeholders in urban wildlife management issues and should be considered during the identification process.

For-profit businesses are probably the most familiar; these are the manufacturers, distributors, wholesalers, retailers, service providers, and any other organization that engages in an activity primarily to generate income and profit for either individuals or the incorporated entity. A variety of for-profit businesses may be directly or indirectly affected by wildlife management decisions, including manufacturers that use natural resource products; importers that ship wildlife and/or wildlife parts overseas as food, pelts and skins, pets, and folk medicine; wholesalers and retailers offering products for consumptive and nonconsumptive wildlife-related recreation; hotels, motels, restaurants, travel agents, guides, and other tourism-related businesses; and other service industries such as wildlife control operators, taxidermists, and public education providers.

Many people don't think of a *nonprofit* company as a business, but it does fit the definition provided above. There are two primary differences between for-profit and nonprofit businesses: (1) nonprofits are tax-exempt organizations that serve the public interest, and (2) no individual or group is supposed to benefit financially from the activities of a nonprofit organization. Nonprofits are created for diverse reasons: to provide a service not addressed by some level of government; to advocate for a particular interest group or industry; to support governmental and academic activities; to conserve, preserve, or restore natural and cultural resources; to educate, inform or publicize; to support research; or to promote a hobby or activity.

The National Wildlife Federation Conservation Directory lists over 4200 nonprofit organizations that may address one or more of the issues listed in Table 10.4. Consideration of these issues and their related organizations is a good place to start during the process of identifying stakeholders.

A *not-for-profit business* is an organization established for charitable, humanitarian, or educational purposes that is exempt from some taxes and in which individuals do not accrue profits or losses. While some would argue there is very little difference between nonprofit and not-for-profit

Table 10.4   Environmental Issues Included in the National Wildlife Federation Conservation Directory

Agriculture and farming	Pollution (general)
Air quality and atmosphere	Population
Climate change	Public health
Development and developing countries	Public lands and greenspace
Ecosystems	Recreation and ecotourism
Energy	Reduce, reuse, and recycle waste
Ethics and environmental justice	Sprawl and urban planning
Executive, legislative, and judicial reform	Transportation
Finance, banking, and trade	Water habitats and quality
Forest and forestry	Wildlife and species
Land issues	Other
Oceans, coasts, and beaches	

organizations, there are subtle distinctions; however, these differences have little effect when it comes to identifying and working with stakeholders.

Academic Institutions

University faculty, researchers, and cooperative extension personnel may play several different roles in wildlife management decision making. Academic personnel may be able to provide insight into public opinion through the development of surveys and focus groups. They may work in cooperation with both the public and private sectors on a variety of wildlife-related projects. Academics may also become stakeholders in their own right, as when agency actions and decisions have the potential to impact ongoing academic research projects. One effective way to identify potential academic consultants and stakeholders is to do an extensive literature search on the topic under consideration and then contact those individuals who have demonstrated their involvement in the form of publications.

The Public

Before we can discuss the public as stakeholders we must first define what is meant by the term. According to the WorldNet Dictionary, *the public* is "a body of people sharing some common interest" or "people in general considered as a whole." Yet this definition is likely to give the false impression that the public is some homogeneous group that acts and reacts as a whole. Nothing could be further from the truth.

The public is the least homogenous, least organized, and least well-defined of all the stakeholder groups we've discussed so far. Members of the public also may be associated with the public sector, the private sector, academic institutions, or some combination of these. For example, a wildlife biologist who works for the U.S. Fish and Wildlife Service (public sector) may be a member of both The Wildlife Society and the Nature Conservancy (private sector) in addition to being a member of the interested public.

The values expressed by the public—both as individuals and as a group—are far more diverse than those of an organization. Well-managed public and private sector organizations develop specific sets of goals and values, often expressed in general terms within the group's mission statement. Employees of both public and private sector organizations may, as individuals, have values that diverge from, or are even in opposition to, those of the organization for which they work. While they are at work, however, they usually are expected to represent the values of the organization rather than their personal opinions, thus creating a consistent organizational persona.

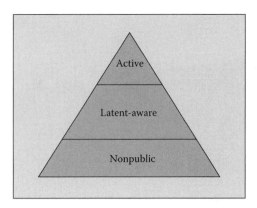

Figure 10.2 Pyramid structure of the number of people in each subset of the public. (From D.J. Decker et al. 2001. *Human Dimensions of Wildlife Management In North America*. Bethesda, MD: The Wildlife Society. With permission.)

When someone becomes a member or a volunteer of an NGO it is usually because they are in agreement with the values and goals of that organization. If either the organization or the member/volunteer has a change of heart, they will simply part company. The result is that the membership tends to be largely consistent in its values and goals. The same cannot be said about the public at large.

Luckily for wildlife managers, not every single member of the public will have a stake in every possible decision that needs to be made. It can be helpful to divide stakeholders into three types of publics: a nonpublic, a latent-aware public, and an active public (Grunig 1983). The *nonpublic* is the subset least aware of (and probably least interested in) the issue at hand. The *latent-aware public* is the subset that is aware of the issue and may even be predisposed to take action, but do not for any number of reasons. The *active public* is the crucial subset—the one that has mobilized around a particular issue.

For any issue, the number of people in each subset of the public will form a kind of pyramid (Figure 10.2). A large portion of the public is minimally aware of any specific wildlife issue, if at all; only a small subset will be motivated enough to actually participate. Keep in mind that a group that is small in proportion to the entire public may still be large in absolute terms.

Identifying active public stakeholders is not an easy task, particularly those individuals who are not also members of the government, NGO, or academic categories. Communication and research can help to reveal stakeholder groups and individuals. Mass media also can be used to solicit participation as well as to educate the public at large about a particular issue.

CASE STUDY 10.1: STAKEHOLDERS DISAGREE ON BEST APPROACH FOR MANAGING FALLOW DEER

They are easily spotted from the road here, lounging in fields and munching grass with little fear of predators. Introduced for hunting six decades ago, fallow and axis deer are popular with tourists eager to see wildlife at Point Reyes National Seashore (Figure 10.3). But park rangers see them as an invasive species that threaten native deer and elk, devour excessive amounts of vegetation, hurt agriculture, and possibly spread disease.

"Invasive species"—organisms introduced into an environment by humans deliberately or by accident—are a problem nationwide. From kudzu in the South to zebra mussels in the Great Lakes, hearty, fast-multiplying invaders can overwhelm an ecosystem, displacing native species and hurting local economies.

Figure 10.3 Fallow deer are an introduced species at Point Reyes National Seashore. (Courtesy NPS)

Now, Point Reyes officials want to eliminate more than 1000 nonnative deer—using shotguns and contraception—from the 71,000-acre national park about forty miles north of San Francisco.

The park's draft plan calls for eradicating the two exotic species by 2020 by sterilizing about one-quarter of them and shooting the rest. After the sixty-day public comment period ends on April 8, the park is expected to issue a final plan late this year or early next year, and start eliminating the deer in 2006.

But, as might be expected in liberal-minded Northern California, the idea of killing such attractive creatures has prompted intense debate among wildlife biologists, conservationists, dairy ranchers, and animal activists.

Park scientists and some environmentalists say the invaders must be eliminated to protect the native ecosystem. "Protecting native species and biological diversity should be the park's prime priority," said Gordon Bennett, chairman of the Sierra Club's Marin County group. "These exotic species are basically agents of genocide. They displace and occupy the habitat of native species."

But many nearby residents and animal rights advocates question whether the deer really threaten the environment, and argue that the animals shouldn't be killed just because they aren't natives. "I think it is being both cruel and insensitive to say just because they're not native we should kill them," said Elliot Katz, president of In Defense of Animals, based in Marin County.

Dozens of residents voiced their opposition to shooting the deer at a recent meeting at Point Reyes, where park officials presented their preferred plan, one of five options being considered. "I don't agree with killing the deer at all. I think it should not be an option," said Ilka Hartmann, who lives in the coastal community of Bolinas. "These are beautiful, majestic animals that were brought here against their will to be hunted."

Fallow deer, originally from Asia Minor and the Mediterranean, come in a range of different colors, from white to brown; males grow large antlers. Axis deer, which come from India, are spotted. About three dozen fallow and axis deer, purchased by a landowner from the San Francisco Zoo, were brought to Point Reyes to be hunted in the 1940s, before the region became a national park in 1962.

Park rangers had been allowed to use rifles to cull both herds to keep each population steady at about 350 individuals. The culling stopped about ten years ago because of public discomfort with shooting the deer, and both populations have grown steadily with no natural predators to keep them in check. The fallow population of 850 is rising 11 percent per year, while the axis population of 250 is increasing 20 percent annually. The nonnative deer compete for food and habitat with the park's two native species: black-tailed deer and tule elk. Known for their aggressive behavior, fallow deer have been seen chasing away tule elk, which are three times larger. The nonnatives can also carry Johne's disease, a virus that usually doesn't sicken the deer but can be deadly for tule elk.

Farmers complain that the deer dig up their beets and carrots, and ranchers say they harass cows, eat hay and livestock feed, damage fences, and devour vegetation in their pastures. Under the National Park Service proposal, expected to cost about $4.5 million, meat from the killed animals would be donated to charities, and an experimental form of contraception could be tested.

Eliminating the deer through contraception alone wouldn't work, because it's virtually impossible to sterilize every animal, experts said. It's also expensive—about $3,000 per animal, compared with $300 to shoot them.

"This is a tough decision because it's very emotional," said park superintendent Don Neubacher. "This is a very complicated issue. In the end I don't think there will be a solution that pleases everybody."

Terence Chea
March 29, 2005
Excerpted from an Associated Press article published at Merced Sun-Star.com

SPECIES PROFILE: BLACK BEAR (*URSUS AMERICANUS*)

We are told that the black bear is innocent, but I should not like to trust myself with him.

—Samuel Johnson, Author

Similar to coyotes, the black bear (Figure 10.4) has had a tenuous relationship with humans in the rural landscapes of North American since European settlement. On one hand, bears are revered as

Figure 10.4 Black bear. (Courtesy Mike Bender/USFWS)

iconic figures in several cultures. Teddy bears, Winnie the Pooh, and Paddington Bear all represent comfort and friendship to millions of children and adults. Native Americans appreciated bears for their abilities to hibernate for months (Wilson and Ruff 1999). However, bears are also perceived as dangerous by many people. Additionally, black bears cause economic damage to forestry operations, apiaries, and livestock, although estimates of the extent of damage are unknown (Conover 2001).

Black bears are widely distributed throughout Canada, the Pacific Northwest, along the Rocky Mountains south into Mexico, and within forested landscapes in the eastern United States (Wilson and Ruff 1999). They inhabit diverse habitats, including urban developments. Densities of black bears vary according to habitat type and food distribution. For example, Beckmann and Berger (2003b) found black bear density at an urban-wildland interface was 120 bears/100 km^2, and in the wildland area, density was 3.2 bears/100 km^2.

The black bear is a generalist forager able to consume various types of food. Most studies indicate black bears eat primarily vegetation, fruit, and nuts (Wilson and Ruff 1999), with animal matter constituting a small percentage of their diets (Utah Division of Wildlife Resources 2000). Because of their dietary plasticity, black bears are able to exploit urban food sources. Black bears are probably more suited to urban areas than are coyotes, because they rely less heavily on animal matter.

While research on urban black bears is extremely limited, two very valuable studies (Beckmann and Berger 2003a, 2003b) have contributed greatly to our understanding of how urban sources of food affect both the ecology and behavior of this species. Beckmann and Berger (2003a) found, in comparison to bears living in wild areas, urban bears that ate garbage were active for fewer hours each day, more likely to be active at night, entered hibernation later in the year and, consequently, stayed in dens for a shorter duration. Additionally, the researchers (2003b) found black bears in urban areas were approximately 30 percent heavier, had significantly smaller home ranges (90 percent and 70 percent reduction for males and females, respectively), and gave birth to triple the number of offspring compared to their wildland counterparts. Interestingly, the mortality rate of dispersing cubs in urban areas was 83 percent, while wildland cubs exhibited 100 percent survival during dispersal. All urban cub deaths could be attributed to anthropogenic sources such as vehicles.

In addition to differences between the two groups at the individual level, population characteristics such as density and sex ratios also differed. As stated previously, the urban bear density was 120 bears/100 km^2; wildland bear density was 3.2 bears/100 km^2. The urban bear population had a 6.8:1 male to female ratio, while the wildland population exhibited a 1.6:1 male to female ratio (Beckmann and Berger 2003b). Based on their analysis, the authors made an interesting argument regarding the dispersion of bears across the urban and wildland landscapes. When they compared the current population density and distribution (urban vs. wildland) to data from a previous study conducted ten years earlier from the same region, they found the population size had not changed appreciably. However, no urban bear population existed ten years earlier and the wildland population density was less than the historic wildland density (20 to 41 bears/100 km^2). The authors argued that, based on these findings, the establishment of an urban bear population and the increases in nuisance bear complaints were more likely a result of a *redistribution* of the bear population across the landscape instead of an actual population increase.

The authors conjectured that urban sources of anthropogenic food were basically luring bears out of the wildland areas and into urban areas. This conclusion has interesting management implications for urban wildlife. When human-wildlife interactions increase, most laypeople and some scientists assume that occurs because the offending wildlife population is increasing or experiencing a population "explosion." While this assumption may be true in certain circumstances, it is important to understand that interspecific interactions (such as human-bear interactions) are also dependent on the behavior of each species and the dispersion of resources for each species across the habitat.

For example, one interesting human behavior is the deposition of extra resources (garbage) at landfills. This rich resource (one man's garbage is another bear's dinner) is highly clumped in both time and space. Consequently, bears will abandon typical territorial behavior and feed in large groups at these landfills (Rogers 1989). The two species' behaviors in this circumstance can increase the probability of contact between them. Abiotic factors such as climate can also affect the likelihood of human-bear interactions. Zack et al. (2003) studied the effect of the El Niño-Southern Oscillation (ENSO) on the probability of human-bear conflicts in New Mexico between 1982 and 2001. The La Niña phase of ENSO is characterized by dry winters and springs, while the El Niño phase of ENSO is characterized by wet winters and springs. The authors hypothesized that, during dry periods caused by La Niña, bears must travel longer distances in search of food. They predicted that the increased movements would increase the likelihood that bears would come into contact with humans. The results of their analysis showed that high rates of human-bear encounters were 4.7 times more likely to occur in La Niña years (dry phase) than in El Niño years (wet phase), thereby supporting their hypothesis.

Whatever the driving factor, several communities across North America have experienced increasing numbers of human-bear conflicts. Often solutions to these conflicts are slow in coming and fraught with controversy. As with the coyote example (Chapter 9), changes in human behavior, not bear behavior, have the most valuable impact but are typically very difficult to produce. Peine (2001) compared bear management strategies used by four different urbanizing communities, including the following: Juneau, Alaska; Mammoth Lakes, California; West Yellowstone, Montana; and Gatlinburg, Tennessee.

In his analysis, Peine (2001) showed that so-called nuisance bears from each community were typically "unnatural-food-conditioned" bears. Sometimes this conditioning of bears to unnatural foods was exploited for monetary gain. For example, "rumors" have circulated within Gatlinburg for many years that some business owners encourage or participate in feeding bears to attract tourists. Hunters found the best opportunity to kill a bear in the county was in and around Gatlinburg, partially because of the bears' dependence on garbage. Consequently, there was some general resistance to implementing measures to reduce bear activity in the city. However, factors other than unnatural food sources contributed to some nuisance bear activity in Gatlinburg. For example, the bear population in Great Smoky Mountains National Park, adjacent to Gatlinburg, grew rapidly in the 1990s because of a combination of abundant natural foods and a reduction in poaching practices. However, in 1997, a late frost and a summer drought precipitated the movement of park bears into Gatlinburg in search of food.

All four communities described in the article developed somewhat similar solutions for their bear problems, although Peine (2001) described in great detail the complexity of the decision-making process and how the factors driving the process for each community were very different. The common theme among the solutions was the creation of ordinances about garbage storage and collection, although the level of enforcement of these ordinances varied across communities.

We should expect more human-bear conflicts in the future for various reasons, including those discussed above. First, some populations of bears may be increasing in density. Second, some bear populations may be redistributing themselves according to available resources such as garbage in suburban and urban areas. Third, as we have stated previously, the expansion of human habitation into wild or rural areas is showing no signs of deceleration. So in simple terms, bears and humans are moving toward the urban fringe (albeit in opposite directions) and it is inevitable that their paths will cross.

CHAPTER ACTIVITIES

1. Look for a local urban wildlife management issue in your community, and create a list of potential stakeholders in both the public and private sectors.
2. Management plans for threatened/endangered species often are controversial and draw the attention of many different stakeholder groups. Visit your state wildlife management agency and access your

state T/E species list. Choose a species that interests you from that list, request a copy of the management plan, and write a report on the concerns various stakeholders could potentially have.
3. Visit the Bureau of Indian Affairs website and find out which tribes, if any, hold lands within your state. Then attempt to learn whether or not these tribes have wildlife management offices and staff.

LITERATURE CITED

American Indian Policy Center. 2009. Website. http://www.airpi.org/pubs/indinsov.html.

Anderson, T.L. 1995. *Sovereign Nations or Reservations? An Economic History of American Indians.* San Francisco, CA: Pacific Research Institute for Public Policy.

Beckmann, J.P. and J. Berger. 2003a. Rapid ecological and behavioural changes in carnivores: The responses of black bears (*Ursus americanus*) to altered food. *Journal of Zoology London* 261:207–212.

Beckmann, J.P. and J. Berger. 2003b. Using black bears to test ideal-free distribution models experimentally. *Journal of Mammalogy* 84(2):594–606.

Brown, G. 1993. *Great Bear Almanac.* Lyons Press.

Bureau of Indian Affairs. 2009. Website. http://www.doi.gov/bia.

Conover, M.R. 2001. *Resolving Human–Wildlife Conflicts: The Science of Wildlife Damage Management.* Boca Raton, FL: CRC Press.

Crowfoot, J.E. and J.M. Wondolleck, eds. 1991. *Environmental Disputes: Community Involvement in Conflict Resolution.* Washington, DC: Island Press.

Decker, D.J., T.L. Brown, and W.F. Siemer, eds. 2001. *Human Dimensions of Wildlife Management in North America.* Bethesda, MD: The Wildlife Society.

Decker, D.J., C.C. Krueger, R.A. Baer, Jr., B.A. Knuth, and M.E. Richmond. 1996. From clients to stakeholders: A philosophical shift for fish and wildlife management. *Human Dimensions of Wildlife* 1(1):70–82.

Dunlop, T.R. 1991. *Saving America's Wildlife: Ecology and the American Mind, 1850–1990.* Princeton, NJ: Princeton University Press.

Grunig, J.E. 1983. *Communication Behaviors and Attitudes of Environmental Publics: Two Studies.* Columbia, SC: Association for Education in Journalism and Mass Communication.

Johnson, S., J. Kaufmann, J. Dossett, and S. Hicks. 2000. *Government to Government: Understanding State and Tribal Governments.* Washington, DC: National Conference of State Legislatures.

Leopold, A. 1933. *Game Management.* Madison: University of Wisconsin Press.

Mertz, D.K. 1991. A primer on Alaska Native sovereignty. Alaska.Net website. http://www.alaska.net/~dkmertz/natlaw.htm.

National Oceanic and Atmospheric Administration. 2009. Website. http://www.noaa.gov.

National Park Service. 2009. Website. http://www.nps.gov.

National Resource Conservation Service. 2009. Website. http://www.nrcs.usda.gov.

National Wildlife Health Center. 2009. Website. http://www.nwhc.usgs.gov.

Patuxent Wildlife Research Center. 2009. Website. http://www.pwrc.usgs.gov.

Peine, J.D. 2001. Nuisance bears in communities: Strategies to reduce conflict. *Human Dimensions of Wildlife* 6(3):223–237.

Rogers, L.L. 1989. Black bears, people, and garbage dumps in Minnesota: Bear-people conflicts. In *Proceedings of a Symposium on Management Strategies,* Northwest Territories Department of Natural Resources.

Susskind, L. and J. Cruikshank. 1989. *Breaking the Impasse: Consensual Approaches to Resolving Public Disputes.* New York: Basic Books.

U.S. Bureau of Land Management. 2009. Website. http://www.blm.gov.

U.S. Center for Biological Informatics. 2009. Website. http://biology.usgs.gov.

U.S. Department of Agriculture. 2009. Website. http://www.usda.gov.

U.S. Department of the Interior. 2009. Website. http://www.doi.gov.

U.S. Fish and Wildlife Service. 2009. Website. http://www.fws.gov.

U.S. Forest Service. 2009. Website. http://www.fs.fed.us.

Utah Division of Wildlife Resources. 2000. *Utah Black Bear Management Plan.* Publication No, 00–23. Salt Lake City, UT: Utah Division of Wildlife Resources.

Wildlife Services. 2009. Website. http://www.aphis.usda.gov/ wildlife_damage.

Wilson, D.E. and S. Ruff. 1999. *The Smithsonian Book of North American Mammals.* Washington, DC: Smithsonian Institution Press.

Wright, R.T. 2004. *Environmental Science: Toward a Sustainable Future.* Upper Saddle River, NJ: Pearson Prentice Hall.

Zack, C.S., B.T. Milne, and W.C. Dunn. 2003. Southern oscillation index as an indicator of encounters between humans and black bears in New Mexico. *Wildlife Society Bulletin* 31(2):517–520.

CHAPTER 11

Legal Aspects of Urban Wildlife Management

The problems of pre-empted, overlapping, and concurrent jurisdiction over American wildlife become even more complex when we consider multiple uses and users.

—Susan J. Buck, Political Scientist

KEY CONCEPTS

1. There are four federal acts that have major impacts on urban wildlife management:
   a. Lacey Act of 1900,
   b. Migratory Bird Treaty Act of 1918,
   c. Animal Damage Control Act of 1931, and
   d. Endangered Species Act of 1973
2. Wildlife law has a layered characteristic which creates challenges for wildlife managers.
3. Municipal weed ordinances prevent urban residents from using alternative landscape designs, for example, the *City of New Berlin v Hagar* case.
4. There are two versions of modified weed laws.
5. There are four urban wildlife damage management principles.
6. Live-trap and release strategies for managing human-wildlife conflicts are, for the most part, unsuccessful.

WILDLIFE LAW 101

Laws govern just about every aspect of our lives, and urban wildlife is no exception. Many people are unaware of the hundreds, if not thousands, of federal, state, county, and municipal laws related to wildlife, and no one could be expected to know every pertinent subdivision covenant and deed restriction. As the human population grows and encroaches on natural wildlife habitat, more wild animals exploit urban landscapes, and more human-wildlife conflicts arise. Society attempts to address these issues through legislation and statutes. Wildlife professions need to have a basic understanding of wildlife laws and regulations, why they were created, and how they impact the lives of both humans and wild animals.

As discussed in Chapter 10, there are at least five different levels of government to consider whenever a wildlife management decision is being evaluated: federal, state, tribal, county, and municipal. In each case the agencies and jurisdictions will change and may even overlap (Figure 11.1). It is the layered characteristic of wildlife law that can create challenges for wildlife managers. Simply

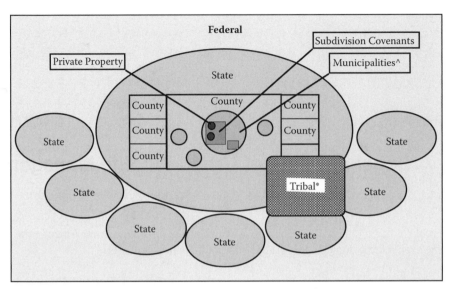

* Sovereign tribal lands may overlap state, county, and municipal boundaries
^ Municipalities may contain townships or other incorporated areas within their boundaries.

Figure 11.1   Government jurisdiction overlap.

establishing which entity at which governmental level has jurisdiction can be a major undertaking. An important first step in this process is an understanding of the major federal laws most likely to apply to urban wildlife management.

Federal Laws

There are numerous federal wildlife laws, but four are of particular importance in the management of urban wildlife. They are, in order of passage: the Lacey Act; the Migratory Bird Treaty Act; the Animal Damage Control Act; and the Endangered Species Act.

The Lacey Act of 1900

The Lacey Act of 1900 was the first federal law to address wildlife protection on a national scale. This act was passed to help curtail the *interstate* (between states) trafficking of illegally taken wildlife. With the passenger pigeon (*Ectopistes migratorius*) hunted to extinction and other bird and mammal species in declining numbers, legislation was needed to protect native wildlife. In its original form, the act focused on the introduction of harmful exotic avian species, the growing scarcity of many native birds, international trade in bird feathers, and interstate commerce in illegally killed and transported wildlife. Its original language authorized limits on avian importation, initiated programs aimed at protecting native bird species, and targeted illegal wildlife trade by strengthening state laws and requiring accurate labeling of wildlife shipments (Anderson 1995).

As a result of revisions over the past 100+ years, the Lacey Act has been expanded to include rare plant species and fish. As it reads today, the act prohibits interstate and foreign commerce (import, export, transport, sale, receipt, acquisition, or purchase) of fish, wildlife, or plants that were taken, possessed, transported, or sold in violation of federal, tribal, state, or foreign laws. The act is enforced by requiring accurate labeling of wildlife shipments, and by criminalizing most types of trafficking in fish, wildlife, and plants (Figure 11.2). Violators face fines (up to $250,000

Figure 11.2 *(A color version of this figure follows page 158.)* Some of the items confiscated under the Lacey Act. (Courtesy Steve Hillebrand/USFWS)

for individuals and $500,000 for organizations), forfeiture of wildlife and equipment, and criminal penalties (up to five years in prison).

U.S. agencies that investigate violations of federal wildlife laws include the Fish and Wildlife Service, the National Park Service, the National Marine Fisheries Service, the Bureau of Land Management, the Forest Service, and the Animal and Plant Health Inspection Service. State wildlife officers investigate violations of state wildlife laws, and sometimes they develop into federal cases. Partly because of the increasing number and complexity of such cases, the Wildlife and Marine Resources Section (WMRS) was created within the Department of Justice's Environment and Natural Resources Division in 1979. Prosecutors from the WMRS consult with federal investigators and assistant U.S. attorneys who are involved in criminal wildlife investigations or litigation; in some circumstances they take the lead prosecutorial role in complex or unusual cases (Anderson 1995).

The Lacey Act occupies a central position within the legal framework of the United States for three reasons: (1) it applies to a wider array of wildlife, fish, and plants than any other single wildlife protection law; (2) it provides for a longer potential term of incarceration than most other wildlife laws containing felony provisions; and (3) the scope of acts it prohibits is broader than most other wildlife laws (Anderson 1995).

In 1988 the U.S. Congress made several changes to the Lacey Act. It added a subsection explicitly defining a sale of wildlife to include the provision or purchase of guiding or outfitting services for the illegal acquisition of wildlife. Another change involved falsification of documents related to shipments of fish, wildlife, or plants. Prior to 1988, the section outlining this offense had applied only to shipments that were actually imported, exported, or transported; in contrast, the 1988 version applied to documents related to wildlife, fish, or plants intended for import, export, or transport. Other amendments authorized federal wildlife enforcement officers to make warrantless arrests for any federal offense committed in their presence and for any felony based on reasonable belief that a felony has been or is being committed. The 1988 language also authorized wildlife officers to make warrantless searches and seizures (Anderson 1995).

The Lacey Act is administered by the Departments of the Interior, Commerce, and Agriculture through their respective agencies—U.S. Fish and Wildlife Service, National Marine Fisheries Service, and APHIS.

The Migratory Bird Treaty Act of 1918

Following on the heels of the Lacey Act, the framers of the Migratory Bird Treaty Act (MBTA) were determined to put an end to the commercial trade in birds and feathers that had wreaked havoc on many native bird populations by the early twentieth century. The MBTA represented one of the first major federal legislative attempts to protect a particular type of wildlife. The MBTA was originally created as a treaty between the United States and Great Britain (for Canada). Since the first treaty, the United States has entered into treaties with Japan, Mexico, and the former Soviet Union (Musgrave et al. 1998).

The original treaty between the United States and Great Britain was created because of concern over the decreasing population of migratory birds in general and waterfowl (geese, ducks, coots, etc.) in particular. At the time, these species were being hunted to the point of extinction. During this same period, several species of birds were on the verge of extinction as a result of a fashion trend for using bird feathers, parts, and even whole birds as adornments for women's hats and clothing (Figure 11.3). However, the MBTA created tension between federal and state management authority that continues to shape federal involvement in the regulation of wildlife (Musgrave et al. 1998).

As the law reads today, it is illegal, except if permitted by regulation (i.e., legal hunting), to take, capture, hunt, or kill; attempt to take, capture, or kill; possess, offer to sell, purchase, barter, deliver, or ship, export, import, or transport any migratory bird, part, nest, egg, or product. Essentially, what this means is that almost all wild birds, eggs, and nests are protected by law. The MBTA and a

Figure 11.3 The popularity of women's hats decorated with wild bird feathers and body parts threatened to push many species to extinction. (Courtesy National Archives and Record Administration/ USFWS)

complete listing of birds protected under this act can be found at the U.S. Fish and Wildlife Service website (http://www.fws.gov).

The MBTA authorizes certain Department of the Interior employees to: arrest a person violating the act in the employee's presence or view without a warrant; execute a warrant or other process issued by an officer or court; and seize and hold all guns, traps, nets, vessels, vehicles, and other equipment used in pursuing, hunting, taking, trapping, ensnaring, capturing, or killing migratory birds (see Case Study 11.1).

The Animal Damage Control Act of 1931

In contrast to statutes protecting wildlife, the Animal Damage Control Act was passed in 1931, in part to assist with eradication of wildlife that threatens livestock grazing and agriculture on federal and private lands. Federal involvement in predator control actually dates back to the late 1800s and was originally authorized by the U.S. Congress in 1915. Originally, management of animal damage control was vested in the Department of the Interior, but was transferred to the Department of Agriculture (USDA) in 1985 (Musgrave et al. 1998).

The act gives the Secretary of Agriculture broad authority to conduct investigations, experiments, and tests to determine the best methods of eradication, suppression, and control of a variety of wildlife species. One purpose of these activities is to protect livestock and other domestic animals through the suppression of rabies and tularemia in wild populations. The secretary may cooperate with states, individuals, agencies, and organizations to carry out the act (Musgrave et al. 1998).

Wildlife Services, an agency within USDA, has been given the mission of providing "federal leadership in managing conflicts between humans and wildlife" (Figure 11.4). The agency's traditional focus was rural and agricultural interests, but it has become increasingly involved in urban wildlife issues. The primary role of government in overseeing private wildlife control operations has been to regulate activities through a licensing or permitting process—at least in some states—and to provide extension and/or educational services. Legal authority for such regulation is vested in federal and state governments, but often is divided among different agencies, including natural resources, agriculture, and public health (Hadidian et al. 1999; Lindsey 2007).

Figure 11.4   One activity of Wildlife Services is to conduct experiments to determine the best methods of wildlife control. (Courtesy APHIS)

Figure 11.5   Black-footed ferrets are one of the species protected under The Endangered Species Conservation Act. (Courtesy Tami S. Black/USFWS)

The Endangered Species Act of 1973

The Endangered Species Act has its origins in the 1960s, with the Endangered Species Preservation Act (ESPA) of 1966. The ESPA required a listing of threatened and endangered species and also required that monies from the Land and Conservation Fund be used to purchase suitable habitat for these species. However, the ESPA did not provide for regulations against the killing or trading of threatened and endangered species.

In 1969 The Endangered Species Conservation Act (ESCA) provided additional protection for threatened and endangered species (Figure 11.5). ESCA created two lists of animals (one for foreign species and one for native species), prohibited foreign species from being brought into the United States, and made the purchase or sale of species on these lists illegal.

These two acts evolved into the single Endangered Species Act of 1973 (ESA), which replaced the earlier, weaker legislation. The ESA is considered "the most comprehensive legislation for the preservation of endangered species ever enacted by any nation" (Musgrave et al. 1998). ESA not only requires federal consultation before major federal action impacting threatened or endangered species is undertaken, but it outlaws the taking of such species and provides for acquisition of habitat to protect threatened and endangered (T/E) species. Federal support also is provided to states that enter into cooperative agreements for conservation of listed species (Musgrave et al. 1998).

The ESA provides broad protection for species of fish, wildlife, and plants listed as threatened or endangered in the United States or elsewhere. The ESA requires the Secretary of the Interior to list species as T/E due to a variety of factors, including habitat destruction, overuse, disease or predation, inadequacy of regulatory mechanisms, or other natural or man-made factors. In the case of marine plants, fish, or wildlife, the Secretary of Commerce determines whether the Secretary of the Interior will list a species or change the status of a species from T/E (see Chapter 10 for additional information on the federal departments involved in management of wildlife). Listing determinations must be made solely on the basis of the best scientific and commercial data available. Any person may file a civil action to compel the secretary to apply emergency listing procedures or protective measures against the taking of a resident T/E species within a state, or alleging a failure to determine a species as T/E or no longer so, if the determination is not discretionary with the secretary (Musgrave et al. 1998).

Violations of the ESA may result in criminal penalties of up to $50,000, imprisonment for one year, or both, and civil penalties of up to $25,000 per violation. District courts have jurisdiction to enforce the act's provisions and regulations, or to order the secretary to perform an act or duty.

For more information on the Endangered Species Act and a complete listing of threatened and endangered plants and animals, readers can visit the U.S. Fish and Wildlife Service website (http://www.fws.gov).

State Laws

Generally speaking—and it is beyond the scope of this book to do otherwise—the federal government is responsible for migratory birds and federally threatened and endangered species; the states are responsible for all other native wildlife within their borders, and each state has its own set of laws and regulations. Many states have similar laws regarding furbearers, nuisance wildlife, state-listed threatened and endangered species, and game animals, but there can be variation in how each state chooses to implement these laws as long as the state law is in compliance with federal law.

In some states, wildlife management agencies share responsibility for establishing and enforcing wildlife regulations with other governmental entities. These include state agricultural, health, and animal health departments/agencies (see Chapter 10). Texas offers an appropriate example. A law was passed in 1996, as a result of a canine rabies outbreak in the southern part of the state, granting authority to the Texas State Health Department to declare statewide rabies quarantines on raccoons, coyotes, and species of foxes indigenous to North America. In the case of such a quarantine, the Texas State Health Department had the legal authority to forbid individuals from transporting these animals to, from, or within the state.

County and Municipal Laws

Wildlife laws created by county and municipal governments also must be considered when dealing with urban wildlife. Some city and county laws may not directly address wildlife, but the way in which a wildlife problem is dealt with may be affected by the law. For instance, in order to use pyrotechnics to disperse birds from an urban roosting site, one must comply with the laws and ordinances pertaining to the use of fireworks and/or discharging a firearm within the city limits. Permission from the police department, fire department, or city leaders may be required as a result of the pyrotechnics' loud noise violating local noise level laws and being potential fire hazards.

When investigating which laws may be relevant in any particular urban wildlife management situation, don't stop at the state and federal levels. Local governments cannot enact laws that negate state and federal laws, but they can enact laws with more rigorous standards. For example, states establish legal hunting seasons for mourning doves (and many other wildlife species), but even during these periods it is illegal to hunt doves within the boundaries of a subdivision or city if ordinances have been passed prohibiting the discharge of a firearm.

Local Ordinances

Some municipalities and subdivisions inadvertently become de facto wildlife sanctuaries. Because of the relative abundance of food and protection from natural predators, urban areas may be particularly attractive to a number of wildlife species (see Chapter 7). Conversely, many local governments and even neighborhood associations generate regulations that can result in prevention of human-wildlife conflicts. For example, some neighborhood covenants may have regulations against feeding deer and other wildlife. This not only discourages deer from eating landscaping plants, but it may also decrease the possibility of a deer-vehicle collision (Davis 2003).

Tree preservation ordinances provide an example of regulations that can benefit wildlife. The understory is an important layer of the forest ecosystem for many wildlife species. Unfortunately, in many urban and suburban areas, the rich diversity of native tree species is replaced by relatively few species of exotic trees. For any significant change in this trend to occur, city planners, developers, homeowner's associations, and private landowners need to understand the benefits of landscaping with native species, which includes reduced expense for fertilizers, pesticides, and watering. Well-written tree ordinances address the protection of native trees, their root zones, and understory, and ensure that replacement procedures increase the diversity of native species.

Weed ordinances often have the opposite effect of tree ordinances. The concept of a "weed" is the product of our previously agrarian society. Weeds—any unwanted plants—are the bane of farmers because they compete with crops for nutrients and moisture. Early urban weed laws were enacted by communities in the 1940s to protect the public from neglectful landowners (Figure 11.6). One downside of weed laws is that they protect and encourage proliferation of exotic mono-turf lawns (e.g., zoysia grass). A Chicago weed law is a typical example; this regulation flatly outlaws "any weeds in excess of an average height of 10 inches" (USEPA 2004).

As a result of weed ordinances, those charged with enforcement often penalize homeowners who have natural landscapes. "Overgrown" yards are still commonly thought to attract rats and mosquitoes or present a fire hazard. This idea was soundly refuted in the case of *The City of New Berlin v Hagar*. New Berlin, Wisconsin, sued Donald Hagar for violating its weed law by practicing natural landscaping on a meadow consisting of several acres. After hearing expert witnesses refute

Figure 11.6 Weed ordinances are enacted to protect public health and property values. (Drawing by Linda Causey)

the city's claims that Hagar's property was a health hazard, the court struck down New Berlin's weed ordinance (see Sidebar 11.1).

Ecologically sensitive weed ordinances do exist. Madison, Wisconsin was the first city of its size to recognize the legitimacy of natural landscapes by enacting an ordinance validating them (Rappaport 1992). Many cities have since enacted ordinances modeled after Madison's. This type of ordinance should be written so as to encourage natural landscaping while protecting society from noxious, invasive exotic plants. Well-written weed ordinances do not contain phrases such as "weeds" and "grass." With the possible exceptions of poison ivy and poison oak, native plans should not be included on an official "weed" list. The term "grass" should be defined either as "turf grass" or by listing species names, in order to allow maximum turf height to be regulated while exempting ornamental grasses.

There are two versions of modified weed laws: (1) the setback ordinance, and (2) the natural landscape exception ordinance. Setback ordinances generally require an area measured from either the front or the perimeter of the lot, in which the vegetation may not exceed a certain height (e.g., ten or twelve inches), exclusive of trees and bushes. The vegetation behind the setback and within the yard is unregulated. Setback distances depend on the type of community and size of the typical lot (Rappaport 1992). Setback laws have several advantages and represent a workable compromise between the sometimes diverse interests of the municipality, natural landscapers, and neighbors. Setback ordinances are also easy to understand and enforce. A setback solves the practical problems caused by large plants and grasses lopping over into neighbor yards or across sidewalks. Neighbor complaints are generally satisfied by such compromise, and living in a community makes compromise essential (Rappaport and Horn 1998).

The second type of modified weed laws are those limiting the blanket weed ordinances with broadly worded exceptions for environmentally beneficial landscapes (Rappaport and Horn 1998). These exceptions include:

1. Native plantings—the use of native plant species for aesthetic and/or wildlife reasons;
2. Wildlife plantings—the use of native and/or introduced plant species to attract and aid wildlife;
3. Erosion control—to offset and control any soil loss problems occurring and/or predicted;
4. Soil fertility building—the enrichment and eventual stabilization of soil fertility through the use of various plant species;
5. Governmental programs—plantings carried out under federal, state, or local programs that require the unimpaired growth of plants during a majority or all of the growing season;
6. Educational programs—plantings designated for educational purposes;
7. Cultivation—any plant species or group of plant species, native or introduced, and grown for consumption, pleasure, or business reasons;
8. Biological control—the planting of a particular species or groups of species that control or replace a noxious or troublesome species;
9. Parks and open space—any and all public parks and open space lands maintained by federal, state, or local agencies, including private conservation organizations;
10. Wooded areas—all areas that are predominantly woods.

These types of modified weed laws are easy to understand and adequately balance the interests of natural landscapers and neighbors. Additionally, exceptions can be added or deleted from this list to tailor the weed ordinance to the needs of the community and the bioregion in which the community is located.

A fourth and most recent concept in weed ordinances is simple promotion. These ordinances do not "regulate" vegetation. Rather, they promote (or require) natural landscaping. According to Rappaport and Horn (1998), Long Grove, Illinois, has no law regulating vegetation height. Their ordinance actually requires developers to include 100-foot scenic easements between homes and major streets in their subdivisions. The easements are planted with native plants, wildflowers, and grasses (Figure 11.7).

Figure 11.7 Volunteers use native plants to modify the banks of a creek in Texas. (Courtesy John M. Davis)

WHO IS IN CHARGE HERE?

By now, it should be apparent there are many laws and regulations that must be considered when dealing with urban wildlife issues. A search of city ordinances and deed restrictions can assist on the local level. State wildlife management agencies (i.e., departments of natural resources, health agency, and animal health commission) often can assist with general questions. The U.S. Fish and Wildlife Service is a good source for information on federal laws. Information on the laws regarding wildlife conflicts can be addressed by APHIS-Wildlife Services, as well as by state wildlife damage control departments and even certain nongovernmental organizations.

PROTECTING THE HEALTH AND SAFETY OF ALL

Wildlife laws are in place not only for the protection of the animals, but also for the protection of human health and safety. In the United States, wild animals are a commons, owned by all and managed for the public good by government agencies, but most wildlife habitat is privately owned and managed (Conover 2002). Any species of wildlife can come into conflict with people. Often the problem is caused, inadvertently, by homeowners and businesses. Dumpsters, trash cans, and pet food are a virtual "Welcome" mat for wildlife. The North Carolina Division of Wildlife Management has created a guide with wildlife damage control principles that all urban residents should have (see Sidebar 11.2). Other agencies have developed similar guidelines.

A popular and seemingly humane solution many people choose for solving wildlife conflicts is to use a live-trap to capture the offending animal, transport it elsewhere, and then release it. However, this "solution" is not as humane as it may first appear because there are a number of problems associated with relocation of wildlife.

First, in many states an individual cannot legally remove an animal and relocate it without permission from the state wildlife agency and/or permission from the individual who owns the land identified as a release site, regardless of whether it is private or public property. Accessing someone's property without first gaining permission may be considered trespassing. Second, studies suggest most relocated animals die shortly after release. One study found approximately 75 percent of relocated raccoons die within the first year after release (Craven, Barnes, and Kania 1998). Many factors, including stress, unsuitable habitat, disorientation, inability to locate food, water, and shelter, and the territorial behavior of animals already living in the area contribute to this mortality rate. Third, relocating wildlife increases the chance of spreading zoonoses (see Chapter 13) such as rabies, Lyme disease, and plague. When a wild animal is relocated, there's always a chance it may be carrying a disease or parasite into a previously unaffected population.

A fine example of how wildlife relocation can spread disease is the raccoon rabies outbreak along the East Coast. Raccoon rabies was first reported in Florida in the 1950s and spread by raccoon-to-raccoon transmission from Florida into Georgia. Between 1977 and 1981, raccoons from Florida were relocated to Virginia to restock raccoon populations; in these shipments were animals that were later confirmed as having rabies. Since 1977, when the first report of raccoon rabies was confirmed in the state, 45,000 raccoons have died of the disease, along with about 4200 cats and 3000 dogs (Centers for Disease Control 2004).

For the reasons cited above, the best solution for most human-wildlife conflicts is a thorough understanding of the underlying factors contributing to the problem, including both human and wildlife behavior. Often conflicts can be resolved or avoided completely by making relatively simple changes. Additionally, this type of approach usually provides a more long-term solution than trapping and relocating, which simply provides an opportunity for a new animal to move in to the now vacant territory that proved so attractive in the first place.

SIDEBAR 11.1   New Berlin v Hagar

The Hagar decision marks a significant watershed in the natural landscaping movement. It is the first, and best, judicial recognition of the practice and the irrational assumptions that underlie the use of weed laws to prosecute natural landscapers. The City of New Berlin, Wisconsin, elected not to appeal Judge Gramling's decision, and as a result, the opinion is unpublished.

In April 1976, New Berlin sued Donald Hagar for violating its weed law by practicing natural landscaping and cultivating a several-acre meadow. Hagar, a wildlife biologist, fought back. He brought in experts to refute the city's claims that his landscape was a health hazard. The testimony was convincing.

Forest Stearns of the University of Wisconsin demonstrated that the Norway rat does not inhabit or find food in a natural landscape. U.S. Forest Service fire expert, David Seaberg, testified that Dr. Hagar's prairie did not create a fire hazard. Philip Whitford, a botanist from the University of Wisconsin, testified that a prairie fire, unlike a forest fire, does not create large and persistent embers that can be carried by the wind. David Kopitzke, a Milwaukee County Public Museum botanist, established that wildflowers and natural landscapes do not create a pollen problem. Exotic plants, like Kentucky bluegrass, and trees, like oaks, create more allergenic pollen than a native prairie.

After three months of deliberation, on April 21, 1976, Judge William Gramling issued his decision. The court found nothing in the testimony to justify the fire and pollen hazard claims that the city cited to support the weed ordinance. He found that natural landscaping did not negatively affect neighbor property values. The court struck down the ordinance as violative of the Equal Protection Clause, because the factual underpinning for the law was too thin to be rational. Following his victory, Mr. Hagar continued his natural landscaping, and New Berlin has not bothered any natural landscapers since.

Excerpted from the *John Marshall Law Review*, Vol. 26, No. 4, Summer 1993.

SIDEBAR 11.2   Urban Wildlife Damage Principles

1. Identify the wildlife species involved (e.g., deer, skunks, bats, squirrels).
2. Be certain the problem is severe and persistent enough to warrant action.
3. Consider alternatives in the following order:
   - Remove food sources that attract wildlife.
   - Remove overhanging tree limbs or other means of access to buildings.
   - Establish protective structures or barriers to prevent wildlife from entering or damaging property.
   - Humanely remove wildlife from buildings and grounds (check first to see if a wildlife depredation permit is needed to trap, transport, or kill wildlife).
   - Permanently repair buildings to prevent reentry.
4. Monitor buildings and grounds periodically for recurring problems, taking appropriate, immediate action to control and prevent damage.

Reprinted from the North Carolina Wildlife Damage Control Agent Program, 2000
North Carolina Division of Wildlife Management

CASE STUDY 11.1: OPERATION REMOVE EXCREMENT

Before dawn on July 23, 1998, the city of Carrollton, Texas, bulldozed a rookery for egrets, herons, and other avian species near a municipal park, saying it had received numerous complaints about odor and noise (Figure 11.8). City officials said they had monitored the rookery and related problems since the fall of 1997, consulted experts, and considered many options before deciding to clear the trees. What the city failed to do was apply for and receive a permit to destroy the breeding grounds from the U.S. Fish and Wildlife Service, as required by the federal Migratory Bird Act of 1918.

Residents at the site said they were outraged by the action, dubbed "Operation Remove Excrement" by city staffers. Jack Laivins said he saw the trees being cleared from about a one-acre site as he drove to work at 4:30 a.m. "They had big lights on. The thing that struck me was that there were thousands of birds circling overhead in the light. At first, I thought it was smoke. But it was the birds."

Irene Simpson, a 20-year Carrollton resident, said, "This is the worst thing I've seen. It makes me sick."

Laivins, Simpson, and other passersby helped wildlife rescue volunteers find injured baby birds trapped in large piles of brush. Many birds had neck or spine injuries, and one had its left leg nearly severed.

Carrollton Mayor Milburn Graveley said later that day he was not aware of anyone in Carrollton who opposed the city's action. "This is, purely and simply a health issue," Mr. Graveley said, adding he believed the city did not act too hastily and needed no permission from state or federal authorities.

A day later, facing a federal investigation and growing tension between neighbors, the city officials who ordered the rookery bulldozed said they regretted the decision and acting without consulting the law. "I'm very disappointed to learn at this point that there are federal regulations that we did not consult," City Manager Gary Jackson said. "My expectation is that we would have known what was permissible. We would have done things differently."

If officials had simply waited a few weeks until the nesting season ended and the egrets and other birds had left for the winter, the city could have acted safely and legally without ruffling the feathers of either the avian or human residents. And, ironically, Dallas County health officials said that razing the site, authorized on the basis of concern for human health and safety, may have temporarily increased residents' risk of contracting an illness. In stirring up accumulated bird droppings at the site, the bulldozers may have spread fungus spores that can cause histoplasmosis, Dr. Karine Lancaster, medical director of the Dallas County Health and Human Services Department, explained.

The decision to bulldoze the rookery cost the community dearly, both financially and in terms of public relations. City Council members agreed to pay $126,800 to the Rogers Wildlife Rehabilitation Center in Mesquite to rehabilitate more than 300 wounded and orphaned birds. Following a criminal investigation, the City Council agreed to pay an additional $70,000 to the U.S. Fish and Wildlife Service for violating one of the nation's oldest conservation statutes. The violation is a misdemeanor and could have led to criminal indictments, but no city staff members were prosecuted.

By the following year the birds had come back home. "I think there are more birds now than last year," said Bob Lanier, chairman of a wildlife and environmental advisory committee that Carrollton created after the rookery was destroyed. The egrets now live in a wooded area just a few dozen yards from the old rookery. Residents who live across the street from the wildlife sanctuary said they no longer mind the egrets. Ana Machado, who lives near the park, said, "I love them. They're my friends."

"The only problem I have with them is that they had to cancel the fireworks," said Belinda Burns. The City Council scrapped Carrollton's Fourth of July fireworks display this year after fireworks testing spooked some egrets.

Mr. Lanier said the city—its residents, as well as its officials—had made progress from an environmental perspective. "I really believe that Carrollton will make no environmental or wildlife move without checking with the right people."

Excerpted from articles published in the *Dallas Morning News*, July 24, 1998, through July 11, 1999.

SPECIES PROFILE: CATTLE EGRET (*BUBULCUS IBIS*)

… Suddenly his neck bends in three different directions,
a rope climbing the air. Houdini, what a trick!

—David Posner, Poet

The cattle egret has seen the world, and it's been a whirlwind tour. This species has undergone the most rapid natural expansion of any bird (Telfair 2006). Originally a native of southern Spain and Portugal, tropical and subtropical Africa and Asia, it began expanding its range in the late nineteenth century. It was first spotted in the Americas in 1977 in Guiana and Suriname after, apparently, crossing the Atlantic Ocean (Krebs, Riven-Ramsey, and Hunte 1994). By the 1930s the species was established in that area. The first cattle egrets arrived in North America in 1941, but were dismissed as escapees or accidentals. Breeding was first observed in Florida in 1953 and, spreading rapidly, in Canada in 1962. In the latter part of the twentieth century, cattle egrets began to expand through the Iberian Peninsula and colonize other parts of Europe, including France, Italy, and the United Kingdom (Martinez-Vilatta and Motis 1992). The species was deliberately introduced to Hawaii in 1959 (Lever 1987).

Range expansion can be tied directly to humans and their domesticated livestock. Cattle egrets are adapted to a commensal relationship with large grazing mammals and it seems to make little

Figure 11.8   Adult cattle egrets. (Courtesy Clark E. Adams)

difference to the birds if that mammal is a water buffalo or Holstein cow. Travel through any pastoral landscape in the continental United States and you're likely to see these small, stocky birds spectacularly adorned in white (and with buffy-orange plumes on the crown, back, and neck when wearing breeding plumage; National Geographic 1999). Their diet includes insects and other small creatures flushed by the large grazing animals they pal around with, as well as crustaceans, mollusks, and earthworms (which don't generally flush). Pairs bond via stereotyped displays, including neck stretches, side-to-side swaying, bill clapping, and mutual preening. Male and female are both involved in chick-rearing.

At first glance, it would appear that people would have an easy time getting along with this bird. But cattle egrets are communal nesters, and they often locate their rookeries near human habitation. While out in the fields they are relatively quiet, but when they return home after a hard day they have a lot to say. The noise and mess associated with rookeries make them a less than desirable neighbor (see Case Study 11.1 above). Moreover, the open grassy areas found at airports offer a seemingly perfect opportunity for cattle egrets to forage, which creates hazards for egrets, airplanes, and the people in them (Fitzwater 1988) (see Chapter 8). Cattle egrets may also cause financial losses at aquaculture facilities and hatcheries, but they do not appear to be as great a problem as other wading birds, cormorants (*Phalacrocorax* spp.) and anhingas (*Anhinga anhinga*).

CHAPTER ACTIVITIES

1. Write a brief report on the existence of a weed ordinance in your town and whether or not it includes modifications such as those discussed in the local ordinances section of Chapter 11.
2. Choose one of the four federal acts discussed in Chapter 11 and then search the archives of your local newspaper to find examples of how this act has been applied in your community. Write a brief report on your findings.
3. Find a member of your community who has wildscaped his or her property (the National Wildlife Federation may prove to be a good resource). Ask the property owner if he or she would be willing to be interviewed by you regarding the reactions of their neighbors to their landscaping approach, and how they responded.

LITERATURE CITED

Anderson, R.S. 1995. The Lacey Act: America's premier weapon in the fight against unlawful wildlife trafficking. *Public Land Law Review* 16:27.
Centers for Disease Control. 2004. Department of Health and Human Services website. http://www.cdc.gov/ncidod/dvrd/rabies (accessed February 17, 2004).
Conover, M.R. 2002. *Resolving Human-Wildlife Conflicts: The Science of Wildlife Damage Management.* Boca Raton, FL: Lewis Publishers.
Craven, S.R., T.G. Barnes, and G. Kania. 1998. Toward a professional position on the translocation of problem wildlife. *Wildlife Society Bulletin* 26(1):171–177.
Davis, J.M. 2003. Urban systems. In *Texas Master Naturalist Statewide Curriculum,* 1st edition, ed. M.M. Haggerty. College Station, TX: Texas Parks and Wildlife Department.
Ehrlich, P.R., D.S. Dobkin, and D. Wheye. 1988. *The Birder's Handbook: A Field Guide to the Natural History of North American Birds.* New York: Simon and Schuster.
Fitzwater, W.D. 1988. Solutions to urban bird problems. *Proceedings of the Vertebrate Pest Conference* 13:254–259.
Hadidian, J., M.R. Childs, R.H. Schmidt, L.J. Simon, and A. Church. 1999. Nuisance-wildlife control practices, policies, and procedures in the United States. In *Wildlife, Land, and People: Priorities for the 21st Century,* ed. R. Field, R.J. Warren, H.K. Okarma, and P.R. Sieverd, 165–168. *Proceedings of the 2nd International Wildlife Management Congress.* Bethesda, MD: The Wildlife Society.

Krebs, E.A., D. Riven-Ramsey, and W. Hunte. 1994. The colonization of Barbados by cattle egrets (*Bubulcus ibis*) 1956–1990. *Colonial Waterbirds* 17(1):86–90.

Lever, C. 1987. *Naturalised Birds of the World*. Essex, UK: Longman Scientific & Technical.

Lindsey, K.J. 2007. Privatization and regulatory oversight of commercial nuisance wildlife control activities in the United States. Dissertation, Texas A&M University, College Station.

Martinez-Vilatta, A. and A. Motis. 1992. Family Ardeidae (herons). In *Handbook of the Birds of the World, Vol. 1: Ostrich to Ducks*, ed. J. del Hoyo, A. Elliot, and J. Sargatal. Barcelona, Spain: Lynx Edicions.

Musgrave, R.S., J.A. Flynn-O'Brien, P.A. Lambert, A.A. Smith, and Y.D. Marinakis. 1998. *Federal Wildlife Laws Handbook with Related Laws*. Rockville, MD: Government Institutes.

National Geographic. 1999. *Field Guide to the Birds of North America*. Washington, DC: National Geographic.

Rappaport, B. 1992. Weed laws. A historical review and recommendations. *Natural Areas Journal* 12(4):216–217.

Rappaport, B. and B. Horn. 1998. Weeding out bad vegetation control ordinances. *Restoration and Management Notes* 16(1):51–58.

Telfair, R.C. II. 2006. Cattle egret (*Bubulcus ibis*). A. Poole, Ed. *The Birds of North America*. Ithaca, NY: Cornell Laboratory of Ornithology.

USEPA. 2004. Green Landscaping with Native Plants. United States Environmental Protection Agency website. http://www.epa.gov/glnpo/greenacres/weedlaws/.

PART V

Special Management Considerations

CHAPTER 12

The Ecology and Management Considerations of Selected Species

KEY CONCEPTS

1. Most urban wildlife management problems are the result of interrelated ecological and sociological factors that operate in concert.
2. Endangered, introduced, and feral species present different urban wildlife management challenges.
3. Endangered species in urban habitats do not seem to fit the definition normally used to list a species as endangered.
4. Introduced species were brought into the United States as pets, escaped into the wild, and now have become a common feature among urban fauna.
5. Feral species have divested themselves from domestication and have become exceptionally well established in urban areas.

INTRODUCTION

The previous chapters in this book have demonstrated the complex factors that operate in urban areas with respect to wildlife issues. Circumstances that arise as a result of wildlife populations inhabiting urban and suburban areas will likely contain nuances specific to the particular geographic area and wildlife species involved. However, it is helpful to analyze each situation with the following conceptual model: Ecological and sociological factors work in concert to create an urban wildlife management dilemma. The ecology and behavior of the species explain the presence and abundance of the population (why the species is here and in what numbers). Sociological factors such as economics, politics, and culture set the framework for how humans respond to that wildlife presence and abundance.

This chapter is not just about direct *conflict* between humans and wildlife species, although that is the most common theme. We have selected three categories with relevance to the urban/suburban ecosystem. These include management of endangered species, introduced species, and feral species. Of course, if we were to include every species that could arguably be placed in these categories we'd end up with an encyclopedia, so we have chosen representative examples of each scenario in the hope that the reader will be able to better appreciate the unique challenges inherent in each. We collected information from several sources, including scientific journal articles, newspaper and other media reports, interviews, government documents, and personal experience. In addition, we invited colleagues to contribute case studies on Florida key deer (*Odocoileus virginianus clavium*) and feral hogs (*Sus scrofa*).

ENDANGERED SPECIES

As described in previous chapters, the process of urbanization often results in the simplification of habitats. Consequently, the number of wildlife species decreases (lower biodiversity) while the number of animals in an urban population is often larger than a rural counterpart. Species that remain or are able to colonize urban habitats tend to be generalists, capable of exploiting the urban environment and adapting to human presence. The generalists are typically very widely distributed, and their population numbers are very large. However, it should be noted that specialists, including some threatened or endangered species, can be found in urban environments when they happen to need the conditions found there.

Endangered species are found in urban or suburban habitats as a result of two mechanisms. First, endangered species are sometimes introduced into urban areas. The peregrine falcon (*Falco peregrinus*), discussed in Chapter 6, is one example. Second, and more importantly, endangered species are found in urban and suburban areas as a result of rapid human development. As development encroaches into the species range, the populations become concentrated in smaller, more fragmented areas. These fragments of remaining habitat become surrounded by development. Essentially, the original habitats become islands within the developed matrix, and individual animals become isolated inhabitants of those islands. The populations must survive on small parcels of remaining habitat and, therefore, become more vulnerable to extinction. Sometimes the species can partially adapt to the urban areas and use them for part of their biological needs.

For example, one subpopulation of endangered bighorn sheep (*Ovis canadensis*) commonly utilizes urban sources of food and water (USFWS 2000; Rubin et al. 2002). Bighorn sheep in southern California were officially listed as endangered in 1998 because of habitat loss and fragmentation, resulting, in part, from urbanization. Urbanization not only has reduced and fragmented the sheep's habitat, but also has eliminated access routes between the small subpopulations of sheep (USFWS 2000).

The goal of the Endangered Species Act of 1973, as mentioned in Chapter 11, is to provide protection for ecosystems upon which endangered species depend for survival and reproduction (Meffe and Carroll 1997). Reaching this goal in an urban or suburban area can be problematic for a number of reasons. First, ownership of important habitat for the species is often split between multiple parties. Managing these fragmented parcels of private land for a common goal is difficult at best. Second, natural areas that border existing urban locations have high commercial value and are usually sought after by a variety of interests for development. If endangered species inhabit these areas or need them for survival, economic gains by private developers and cities are thwarted.

Following is a review of information available for three endangered species that can be found in urban and suburban habitats. The San Joaquin kit fox (*Vulpes macrotis mutica*) was chosen as a case study because ... well, frankly, they are just so darned cute, which may seem to be a trite reason for including them, but their physical characteristics make them very appealing to the urban public. According to San Joaquin kit fox experts, the management of the urban fox population has caused very little conflict with humans. The Houston toad (*Bufo houstonensis*) was chosen precisely because it is not particularly charismatic, and it's usually a lot harder to make a case for saving animals that aren't cute and furry. Last, we selected Key deer (*Odocoileus virginianus clavium*) because one of our colleagues has professional experience in the management of this species and accepted an invitation to write about his experiences. Second, this species is unique in that, while it is highly endangered, it maintains a locally abundant population in a developed area in south Florida.

Houston Toad (*Bufo houstonensis*)

If God had wanted us to be concerned for the plight of the toads, he would have made them cute and furry.

—**Dave Barry, Humorist**

Figure 12.1   Houston toad. (Courtesy Robert Thomas/USFWS)

In 1970, the Houston toad became the first amphibian ever granted protection under the Endangered Species Act (Gottschalk 1970). Critical habitat was designated in Bastrop and Burleson counties in Texas in 1978 (Peterson et al. 2004). At the same time, the Houston Zoo initiated a captive breeding program to help supplement remaining populations or establish new ones in protected areas (Maruska 1986; Pritchard 1995). From 1982 to 1986, 62 adults, 6985 newly metamorphosed toadlets, and over 400,000 eggs were distributed to ten sites within the Attwater Prairie Chicken National Wildlife Refuge (Houston Zoo website), but efforts to establish a viable population at that site were unsuccessful.

Ironically, the Houston toad was eradicated from the Houston area as a result of development during the 1960s. Officially, it is a year-round resident in nine Texas counties—Austin, Bastrop, Burleson, Colorado, Lavaca, Lee, Leon, Milam, and Robertson (Dixon 2000), but in 2006, the species was heard calling in only two counties, Bastrop and Leon; in Leon only a single male was located. The largest population is found in Bastrop County but studies also suggest this population is in decline as well (Dixon 2000).

Houston toads are about 2 to 3.5 inches long when fully grown; females are larger and bulkier than males. Coloration is generally brown and speckled but may vary from light brown to gray or even black (Figure 12.1). Two dark bands extend down from each eye to the mouth. A white stripe is usually found down the middle of the back, but may be completely absent. The underside is pale and may have small, dark spots. Males have a dark throat that can look bluish when distended while the animal is calling.

Most breeding activity takes place in the early spring months of February and March, and is stimulated by warm evenings and high humidity. Males call from in or near shallow water or small mounds of soil or grass surrounded by water, attracting females with a high, clear trill (Swannack 2007). The female lays her eggs as long strings in the water that are fertilized by the male as they are laid (Hillis, Hillis, and Martin 1984; Jacobson 1989). The eggs hatch within seven days. Tadpoles become toadlets in anywhere from 15 to 100 days, depending on the water temperature.

Young toadlets are about one-half inch long when they complete metamorphosis and leave the pond. Males breed when they are about one year old, but females may not breed until they are two years of age (Quinn and Mengden 1984). Annual survivorship of males has been estimated between 15 and 43 percent (Swannack 2007); female survivorship is about 20 percent (Hatfield et al. 2004).

Habitat loss is the most serious threat facing the Houston toad. Conversion of ephemeral and permanent natural wetlands for urban and agricultural uses eliminates breeding sites. Clearing native vegetation from uplands and near breeding ponds reduces the quality of breeding, foraging, and resting habitat, and increases the chances of predation and hybridization. Conversion of native grassland and woodland savannah to sod-forming grasses found in urban habitats, such as Bermuda grass and Bahia grass, pose a problem for the Houston toad because these grasses are usually too dense for the toad to move freely and inhibits burrowing (Campbell 1996). In addition to loss of habitat through development, toads often find roads and other linear features (e.g., pipelines and transmission lines) to be an insurmountable barrier to movement between foraging, hibernating, and breeding sites.

Long-term fire suppression has caused loss of habitat in a significant part of the toad's range, as the savannah grasslands that formed the bulk of the species' habitat grow into brush thickets devoid of herbaceous vegetation these toads need for cover and foraging habitat. The Houston toad is believed to be adapted to fire regimes, but ironically, prescribed burning for habitat management may result in toad mortality. Frequent and/or severe burns may be harmful, particularly for small, fragmented toad populations. However, increased fuel loads may result in hotter wildfires. Additional research is needed to determine the effects of prescribed burning programs (Wildlife Fact Sheets, Texas Parks and Wildlife Department website).

The red imported fire ant (*Solenopsis invicta*) makes it harder to ensure the long-term survival of the Houston toad (Peterson et al. 2004). Fire ants kill young toadlets moving out of the breeding pond into the surrounding land habitat. Current research shows that fire ants have a devastating impact on local arthropod communities, and thus may also limit the toad's food supply.

Human-toad conflicts are primarily related to the fact that both find the same acreage attractive. The *Austin Chronicle* reported several years ago (Proctor May 2005) that a desirable development zone east of Austin in Bastrop County was named one of 595 "centers of imminent extinction" worldwide thanks to the Houston toad—not exactly the kind of panache one hopes for when building a dream home. The study, published in the *Proceedings of the National Academy of Science*, predicted that without dramatic conservation efforts there would be a sad end to a three-decade effort to protect the endangered Houston toad. Finding a rare amphibian on your property can be a thrill … and a headache. The Austin Council of the Boy Scouts of America spent $1 million and six years trying to develop a plan to preserve the Houston toad after a population was discovered on a ranch donated to the Scouts for a camp (Kamb 2009).

San Joaquin Kit Fox (*Vulpes macrotis mutica*)

"Men have forgotten this truth," said the fox. "But you must not forget it. You become responsible, forever, for what you have tamed."

—Antoine de Saint-Exupery, Author

The San Joaquin kit fox (Figure 12.2) is a highly endangered subspecies of kit fox currently distributed in the southern San Joaquin Valley and Salinas Valley in southern California (Cypher and Spencer 1998). According to Brian L. Cypher, research ecologist for the Endangered Species Recovery Program in Bakersfield, California, kit foxes weigh between 3.3 to 6.6 lbs (1.5 and 3.0 kg), less than the average weight of a healthy house cat (personal communication). Kit foxes prey primarily on small mammals, including mice and ground squirrels. Other prey items include rabbits, hares, birds, eggs, and insects (USFWS 1998).

Like other canid species, the kit fox has been able to adapt to urban areas. Urban kit foxes use habitats such as golf courses, city parks, undeveloped lands, open spaces, school grounds, canal, railroad, and powerline right-of-ways, and drainage basins (Cypher personal communication). Studies of urban kit foxes show they eat some sources of anthropogenic food (Cypher and Warrick

Figure 12.2 San Joaquin kit fox. (Courtesy B. Peterson/USFWS)

1993), although the importance of these foods in maintaining the population is not understood. We do not know if kit foxes *must* have these food sources to survive the urban landscape, or if it's simply a matter of preference and convenience. Cypher and Frost (1999) found juvenile urban kit foxes weighed more than their rural counterparts. The researchers hypothesized the consistent availability of their urban food sources likely contributed to this difference.

Kit foxes use underground den sites as shelter from environmental conditions, to escape from predators, and for reproduction (USFWS 1998). Urban den sites include culverts, abandoned pipelines, and banks in fenced drainage basins (Speigel et al., cited in USFWS 1998; Cypher personal communication). According to the USFWS Recovery Plan (1998), urban kit foxes exhibit different behavioral characteristics from their rural counterparts. Many urban foxes are reported to be relatively tame. Much like urban coyotes, they are active during the day and scavenge from human food sources such as garbage.

As a result of encroaching development, the San Joaquin kit fox population is highly fragmented, with subpopulations scattered throughout the San Joaquin Valley. In addition to subpopulations in natural lands, kit foxes have been detected in and near towns such as Tulare, Visalia, Porterville, Maricopa, Taft, and McKittrick. Additionally, a relatively large population (approximately 200) occurs in the metropolitan Bakersfield area (USFWS 1998).

Kit foxes are not generally recognized as a threat to human and/or pet safety, probably because of their petite size (Cypher personal communication). A five-pound kit fox would have to be psychotic, self-delusional, or rabid to attempt an attack on most pet dogs. According to Cypher, most residents probably are not aware that kit foxes live in Bakersfield. Those residents who are aware of the foxes are sometimes protective with regard to the ecologist's activities (e.g., setting traps for foxes). So the typical nuisance issues related to many other wildlife species have not been relevant to the urban kit fox. However, development in the Bakersfield area creates the potential for conflict between the endangered kit fox and economic interests. Yet, according to Cypher, there has been relatively little conflict because of the Habitat Conservation Plan written by the city of Bakersfield.

The Endangered Species Act (ESA) prohibits the "take" of any federally endangered plant or animal species. This law applies to private landowners and has been a point of contention since its enactment. In an effort to resolve this issue, the ESA was amended in 1982 to authorize "incidental take" by private landowners if they developed and implemented a Habitat Conservation Plan. The objectives of these plans are to address the potential impacts of a given activity (development) and to outline mitigation procedures to offset those impacts on the endangered species (Nelson 1999).

The Metropolitan Bakersfield Habitat Conservation Plan (MBHCP) was written in 1994 by the city of Bakersfield and Kern County in an attempt to help developers and landowners reach compliance with the laws of the ESA. By writing the MBHCP to include the entire metropolitan area, a separate permit application is not required for each proposed activity by each developer or landowner. When a landowner or developer applies for a building permit with the city, they pay a mitigation fee to offset the potential negative impacts, as outline by the MBHCP. Money collected by the city is used to purchase habitat for the endangered species listed within the plan, including the kit fox, Tipton kangaroo rat (*Dipodomys nitratoides nitratoides*), blunt-nosed leopard lizard (*Gambelia silus*), and Bakersfield cactus (*Opuntia treleasei*). The plan allows developers to be compliant with state and federal laws without paying for biological assessments and mitigation for each project.

Florida Key Deer (*Odocoileus virginianus clavium*)*

A Key deer outside the Florida Keys is no longer a Key deer.

—Anne Morkill, Manager Key Deer National Wildlife Refuge

In this section, we will discuss the uniqueness of the endangered Florida Key deer (Figure 12.3), the impact of urban development on the species in the last fifty years, and some of the current management challenges for the population. First, we will provide some background and a historical perspective to the Key deer. Next, we will review some of the changes in habitat due to urban development in the last fifty years and the effect this development has had on the deer population. We will focus on the history and trends of Key deer that reside on two islands, Big Pine (6296 acres) and No Name (1139 acres) Keys, because these two islands support approximately 75 percent of the entire deer population (Lopez et al. 2003a) and have been the center of environmental controversy for the most part. Finally, we will conclude with some of the current management challenges for this endangered species due to the urbanization of its habitat.

The endangered Key deer, the smallest subspecies of white-tailed deer in the United States, are endemic to the Florida Keys on the southern end of peninsular Florida (Hardin, Klimstra, and Silvy 1984). In comparison to other white-tailed deer, Key deer are smaller, with an average shoulder height between 24 and 31 inches, and average weights of 63 lbs (maximum 100 lbs) and 83 lbs (maximum 145 lbs) for females and males, respectively (Lopez 2001). Female Key deer less than one year old are not reproductively active; female yearlings and adults average one to two fawns per year (Hardin 1974).

Since 1960, urban development and habitat fragmentation in the Lower Florida Keys have threatened the Key deer population (Lopez 2001; Lopez et al. 2004). In addition to a loss of habitat, an increase in urban development is of particular concern, because highway mortality accounts for the majority of the total deer mortality. Over half of the deer-vehicle collisions occur on U.S. Highway 1, the only highway linking the Keys to the mainland (Lopez et al. 2003b). Approximately 70 percent of the total Key deer mortality is due to human-related causes (Lopez et al. 2003b).

The Florida Keys are a series of islands located off the southern coast of peninsular Florida. It is believed that Key deer were isolated from the mainland after the melting of the Wisconsin Glacier following the last ice age nearly 4000 years ago (Maffei, Klimstra, and Wilmers 1988; Folk and Klimstra 1991). The Florida Keys stretch 100 miles from end to end, which forms an effective barrier from the mainland to gene flow and has allowed the Key deer to adapt to its unique habitat

---

* The authors thank Dr. Roel Lopez of the Wildlife and Fisheries Sciences Department at Texas A&M University for this section.

Figure 12.3 Key deer. (Courtesy Roel Lopez)

(Maffei, Klimstra, and Wilmers 1988; Folk 1991). For example, fresh water is a limiting factor for the Key deer, but they are able to drink brackish water (water approximately as salty as sea water; Folk 1991). The small population and lack of gene flow with mainland white-tailed deer also has promoted morphological and behavioral divergence in the subspecies (Hardin, Klimstra, and Silvy 1984). For example, shorter snout lengths and wider skulls give a "toy deer" appearance to Key deer (Folk 1991).

The former range of the Key deer is hypothesized to have extended from Key Vaca to Key West but never beyond the Lower Florida Keys (Folk 1991). Currently, Key deer occupy twenty to twenty-five islands from Sugarloaf to the Johnson keys (Lopez 2001). Within this range, deer use all habitat types but show preference for upland areas (i.e., hammocks and pinelands; Lopez et al. 2004). Few reliable estimates of the original population size have been made; however, by the 1930s it was believed that less than fifty Key deer were in existence throughout their range (Dickson 1955). An early cartoon by Jay "Ding" Darling, cartoonist, conservationist, and chief of the U.S. Bureau of Biological Survey (predecessor to the U.S. Fish and Wildlife Service), drew national attention to the Key deer's plight and was instrumental in the drafting of early legislation to protect the deer. Early hunting practices used by poachers included burning the vegetation on islands and driving deer into the water where they were either clubbed or shot to death (this was the image depicted in Darling's cartoon that was later published in the *Washington Post*). Increased law enforcement and the estab-lishment of the National Key Deer Refuge in 1957 provided protection for the deer and its habitat. Consequently, the deer population grew to an estimated 300 to 400 animals by 1974 (Klimstra et al. 1974). Currently, the Key deer population is estimated to be 600 to 700 deer throughout their range (Lopez 2001).

Until the 1950s, the Florida Keys had little human population growth (Lopez et al. 2004). This changed when reliable supplies of fresh water and electricity became available, in addition to improved transportation along the now contiguous stretch of U.S. 1 (Lopez et al. 2004). Between 1950 and 1980, a dramatic increase in the human population occurred, causing the State of Florida to declare the Keys an "Area of Critical State Concern" in 1975; in that same year, Monroe County adopted a Land Use Plan and policy for habitat preservation and reduced human population growth

(Peterson et al. 2002). Despite these regulations, since the mid-1970s the human population on Big Pine and No Name Keys (the core of Key deer habitat) has continued to increase and is currently estimated at 5000 people for these two islands (Lopez 2001; Lopez et al. 2004). As previously mentioned, rapid urbanization is the primary threat to Key deer in the form of highway mortality and other anthropogenic causes of mortality (Lopez et al. 2003b). This rapid urbanization also has been detrimental to other endangered species in the Lower Florida Keys, including the silver rice rat (*Oryzomys palustris natator*), Lower Keys marsh rabbit (*Sylvilagus palustris hefneri*), and the eastern indigo snake (*Drymarchon corais couperi*). It is the Key deer, though, that has drawn the majority of the local and national attention with regards to urban development (Peterson et al. 2002).

Native flora of the Lower Keys is primarily of West Indian origin (Dickson 1955); vegetation types include pinelands, hardwood hammocks, freshwater marshes, buttonwood forests, mangrove forests, and urban areas (Lopez et al. 2004). Vegetation varies by elevation with red (*Rhizophora mangle*), black (*Avicennia germinans*), and white (*Laguncularia racemosa*) mangroves, and buttonwood (*Conocarpus erecta*) forests occurring near sea level (maritime zones). As elevation increases inland, maritime zones transition into hardwood (e.g., gumbo limbo [*Bursera simaruba*], Jamaican dogwood [*Piscidia piscipula*]) and pineland (e.g., slash pine [*Pinus elliottii*], saw palmetto [*Serenoa repens*]) upland forests with vegetation intolerant of salt water (Dickson 1955; Folk 1991). Key deer occupy many islands in the Lower Florida Keys; however, islands with fresh water and upland habitats that provide important plants for the Key deer's diet are preferred (Lopez et al. 2004).

Since the arrival of settlers, approximately 1583 acres (21 percent of total area) on Big Pine (1507 acres, 24 percent) and No Name (76 acres, 7 percent) Keys have been developed. The largest land conversion, clearing for large subdivisions, occurred prior to 1985, while the majority of home construction occurred in later years (Lopez et al. 2004). In the last fifty years, Big Pine and No Name Keys have seen a significant amount of urban development. During this same time period Key deer numbers have increased from less than 100 in the 1950s (Frank et al. 2003) to approximately 400 to 500 in 2000 (Lopez et al. 2003a). As we mentioned earlier, previous research suggests that rapid urbanization is the primary threat to Key deer, in the form of highway mortality and other anthropogenic mortality factors (Lopez et al. 2003b). Why then has the relationship between urban development and deer densities been *positive* rather than *negative*? Is urban development really a threat to the Key deer? The answers lie in reviewing where and how development occurred.

Historically, urban development occurred in tidal areas (e.g., mangrove, buttonwood). That was attributed to market demand to build home sites with water access. Since the enactment of federal and state laws such as the Florida Coastal Management Act of 1985 (Title XXVII, Chapter 380, Part II, F.S.), development of these vegetation types has been restricted (Gallagher 1991). The result was an increase in the development of upland areas (e.g., pinelands, hammocks) or the infilling (i.e., building on existing scarified lots) of existing subdivisions with scarified lots. Lopez (2001) hypothesized that islands with high deer densities were those with a substantial upland component, while islands that were mostly tidal (e.g., Summerland, Ramrod) supported fewer deer than similar-sized islands with more upland area (e.g., No Name Key, Little Pine Key).

Empirical data supports this idea, and the change in the amount of "usable space" (upland area that is readily available to Key deer; Guthery 1997; Lopez et al. 2004) might explain the increase in Key deer numbers in the last fifty years (Lopez et al. 2003b). Early urban development occurred in tidal areas and included the clearing of large tracts of land for subdivisions. Early practices used by developers in subdivision construction included (1) the clearing of native vegetation in primarily tidal areas (remember, people want to be close to water), and (2) the digging of canals to allow "backyard" water access for each of these homes. Thus, most subdivisions in the Florida Keys have a system of "alleys" that serve as direct water access for human residents. The clearing and increase in elevation from canal fill, in essence, resulted in an increase of "upland" areas for Key deer, which provided a variety of additional food resources (i.e., planted ornamentals, pet food, etc.) for the deer!

With changes in federal and state laws in the mid-1980s prohibiting the development of tidal or wetland areas, urban development pressure shifted toward upland habitats. Recent development (post-1985) includes the infilling of subdivisions and the conversion of pinelands and hammocks to urban areas. Infilling during this period actually was greater than the conversion of native upland areas due to the previously mentioned Monroe County Land Use Plan (1975; Peterson et al. 2002), thus, the net loss of upland habitats was in fact negative. Assuming that upland areas (e.g., pinelands, hammocks, and urban areas to a certain degree) have equal Key deer numbers, we would predict the conceptual model between Key deer density and urban development would be bell-shaped (Lopez et al. 2004). If current development trends continue, the amount of useable uplands will likely decrease, which in turn will decrease Key deer numbers and increase secondary impacts such as road mortality and fence entanglement (Lopez et al. 2003b; Lopez et al. 2004).

The story of the Key deer is interesting in that it offers some unique challenges for wildlife managers. The role of urban development in this story is critical to understanding many of the historic and current management issues, namely (1) deer overabundance/domestication, (2) habitat fragmentation and anthropogenic mortality factors, and (3) social conflict. We will conclude this section by offering some insight into these challenges.

Like other subspecies of white-tailed deer, the response of Key deer to urban development has been favorable (McShea, Underwood, and Rappole 1997). However, we would caution the reader about believing that urban development is equivalent in value to pinelands and hammocks. Urban development does not necessarily provide for all the Key deer's life history requirements (e.g., fawning areas) as compared to pinelands and hammocks (Folk 1991). In the last fifty years, Key deer numbers have increased due to the previously mentioned habitat changes. In fact, current deer estimates suggest the population is at or near carrying capacity (Lopez et al. 2003a; Nettles et al. 2002) and experiencing problems with density-dependent diseases, domestication, and habitat damage.

Ironically, due to the endangered status of Key deer, many commonly used deer management practices are prohibited, both legally and philosophically. Wildlife managers are faced with some difficult questions: Why is the Key deer "endangered" if population numbers are "overabundant"? Can you legally control deer numbers, for example, with the use of a contraceptive that aborts a fetus? Would that abortion be a "take" under the Endangered Species Act?

Obviously these questions do not have simple answers; however, part of the answer stems from the Endangered Species Act itself, which emphasizes population *numbers* rather than other population factors (Lopez et al. 2001). Future Key deer management will require a radical change in the current management paradigm that should include changes in recovery criteria that emphasize other parameters such as habitat quantity/quality and herd health (Lopez 2001). Another issue with deer overabundance is the "taming" of deer with increasing human-deer interactions. Most deer disease issues are related to the domestication or "ganging" behavior currently observed in Key deer (Folk and Klimstra 1991; Nettles et al. 2002). For example, highly concentrated deer groups increase the likelihood of disease in urban areas. Furthermore, the carrying capacity for Key deer is artificially raised with illegal feeding and watering of deer in urban areas (Folk and Klimstra 1991; Lopez et al. 2004). Attempting to reduce Key deer-human interactions will be a future challenge for wildlife managers.

Approximately 65 percent of Big Pine and No Name Keys are in public ownership. Currently, approximately 1700 lots are privately owned, undeveloped vacant lots (Lopez 2001). The average lot size for privately owned property is one-third acre (Lopez 2001). Monroe County has the highest cost of living and the fifth highest per capita income in Florida (Peterson et al. 2002), which means that land value is high due to demand. For example, a scarified quarter-acre lot in a subdivision is approximately $50,000 to $75,000 (2004 prices).

High land value and small, mixed ownership patterns in the Florida Keys offer yet another suite of headaches for wildlife managers. First, land acquisition programs are limited due to the high cost of land purchases. The mosaic of property ownership also is problematic in the implementation of many land management practices. For example, the use of prescribed fire in native pinelands is necessary for their continued maintenance; however, the use of prescribed fire is limited in areas that are a "checkerboard" of both public and private lands.

A high demand to build by developers/landowners and restrictions in obtaining building permits invariably has led to conflict among stakeholder groups (Lopez et al. 2001; Peterson et al. 2002). Conversely, environmental groups also have great expectations of wildlife managers to fulfill their mission statement from their respective natural resource agencies. As the debate continues between these two groups, we find that the Key deer is normally at the center of the debate. "We would want to build a school for our children on the island" or "We want to preserve the 'small town' atmosphere of Big Pine and No Name Keys [these islands have fewer residents than other islands]" are both legitimate, reasonable requests. As a manager of the Key deer, you have to be prepared to offer acceptable solutions to both of these questions.

Undoubtedly, the coexistence of a unique wildlife species with humans will continue to offer challenges for managers, biologists, and community leaders in the management of this endangered deer population. Conflict resolution will require an attempt to balance the needs of the Key deer and offer regulatory relief to residents of these two islands. In this process, decisions will need to be made based on the current biology of the Key deer in addition to incorporating other social values. Management of an endangered, urban wildlife species requires such an approach in the successful recovery of the species.

INTRODUCED SPECIES

A species—plant or animal—is defined as introduced when it is found outside its native distributional range and it arrived in this location through human activity. The difference between introduced and feral (which usually refers to animals alone, rather than both plants and animals) is that the original stock for the population was not domesticated. However, both introduced and feral species can have a deleterious effect on native wildlife, agriculture, human uses of natural resources, and on humans and their livestock and companion animals. Some examples of introduced vertebrate fish and wildlife species include common carp (*Cyprinus carpio*), sea lamprey (*Petromyzon marinus*), brown anole (*Anolis sagrei*), Mediterranean gecko (*Hemidactylus turcicus*), European starlings (*Sturnus vulgaris*), house sparrow (*Passer domesticus*), mute swans (*Cygnus olor*), house mouse (*Mus musculus*), sika deer (*Cervus nippon*), and nutria (*Myocastor coypus*).

Antos et al. (2006) examined the richness and relative abundance of introduced bird species in small, medium, and large remnants of native vegetation within an urban matrix and found a significant relationship between introduced species richness and remnant size, with larger remnants supporting more introduced species. Introduced species, as a proportion of the relative abundance of both native and introduced species, did not vary significantly between remnants of differing sizes. All remnants were equally susceptible to invasion.

House Sparrow (*Passer domesticus*)

> The sparrows chirped as if they still were proud,
> Their race in Holy Writ should mentioned be.

> **—Henry Wadsworth Longfellow**

The familiar house sparrow is an Old World sparrow (Passeridae) that was deliberately introduced and established in the New World of North America (Figure 12.4). The species is sexually dimorphic. Males in breeding plumage have a gray-to-buff belly, darker gray crown, chestnut nape, white chin, and a black eye mask, bill, and bib. Females have a streaked back, buffy eye stripe, and a solid gray-to-buff belly (National Geographic 1999).

Figure 12.4   Female house sparrow. (Courtesy Thomas G. Barnes/USFWS)

House sparrows are known by several common names: English sparrow, city sparrow, town sparrow, and others. These birds settled down to life in the company of humans about the same time humans in the Middle East began to adopt an agricultural life. They are believed to have been migratory at one time, but this behavior trait has either been lost or can be ignored when there's a reliable, year-round food supplier (i.e., humans). There were several attempts to introduce house sparrows to North America, in part because people found them appealing, but also because it was hoped they would help control insect pests. The first successful introduction occurred in Brooklyn, New York, around 1850. The species expanded its North American range rapidly, primarily through nature expansion, and in only fifty years they were found to occupy habitats across the entire country. House sparrows have differentiated geographically so that populations vary in color and body size depending on region (Erlich, Dobkin, and Wheye 1988).

Sparrows are social birds that tend to flock through much of the year. A flock's range covers about one to two square miles, on average (Eno 1996). This ubiquitous bird can be found in habitats from hard-core urban to agricultural lands to woodlots and everything in between. The adult diet is heavily dependent on seeds, but also includes insects, fruits, and blossoms; young birds are fed mostly insects. As with so many successful urban species, adaptability is key; food discarded by humans—a French fry tossed to birds hanging out at an outdoor café or tossed into the garbage dumpster out back—has become a common food source. House sparrow numbers were at their highest in the early 1900s, decreasing somewhat as cars replaced horses and reduced the ready supply of grain-filled manure. Still, they remain an abundant and common feature of daily life in the United States (Erlich, Dobkin, and Wheye 1988).

House sparrows are primarily cavity nesters—and the cavity could be in a tree or a commercial business sign. In true adaptable urban species fashion, though, the fork of a tree branch will serve if housing options are limited. They will aggressively take over bluebird and swallow nests, even destroying the existing eggs and nestlings (Erlich, Dobkin, and Wheye 1988), and this trait has not endeared them to birders and others concerned with native bird populations. The female lays four to six white, greenish, or bluish eggs that may also be mottled with gray or brown. Eggs are incubated for ten to thirteen days and the chicks hatch in an altricial state. Both the male and female are involved in chick-rearing, and fledglings leave the nest fourteen to seventeen days after hatch.

House sparrows don't cause widespread damage in the urban landscape, but they are opportunists and can create a nuisance. Any nook or cranny will be exploited for nesting (Eno 1996)—gaps in the roof soffit, dryer vents, billboards, brush piles, or between boxes stored on high shelves in a warehouse. Complaints related to direct human-wildlife conflict generally cover one (or more) of three topics—droppings, nesting, or noise (personal observation). However, people do tend to feel personally involved in conflicts between house sparrows and native birds. House sparrows dominate native species at feeders, and they invade bluebird (*Sialia* spp.) and purple martin (*Progne subis*) nest boxes, frustrating backyard birders. These issues, which are not technically human-wildlife conflicts per se, tend to be viewed as such.

At least one study has investigated the potential of house sparrows to serve as an early warning of a St. Louis encephalitis (SLE) outbreak (McLean et al.1983). Birds are the primary vector for SLE virus in most of North America. The increased prevalence of antibody in house sparrows has been related to human cases, making them useful as a sentinel species.

Interestingly, house sparrow populations have declined sharply in Western Europe over several decades. The reasons have not yet been identified in spite of public interest and concern. One study that combined field experimentation, genetic analysis, and demographic data suggests a reduction in agriculturally based winter food supply may be the principal explanation for local extinctions in southern England (Hole et al. 2002).

Nutria (*Myocastor coypus*)

[Nutria] destroy coastal marsh the old-fashioned way—they eat it.

—**USGS Publication**

The nutria (Figure 12.5) is a large, semiaquatic rodent endemic to South America that belongs to the same family as the beaver (Dunlap and Theis 2002). Smaller than a beaver (*Castor canadensis*) but larger than a muskrat (*Ondatra zibethicus*), nutria are similar to both morphologically although the tail is round and slightly furred (Nowack 1999). An average-sized adult weights about 12 lbs (5.4 kg); males are slightly larger than females.

Figure 12.5   Nutria. (Courtesy Dan and Lin Dzurisin)

Nutria are generalist herbivores that forage primarily on the basal portion of plant stems, as well as roots and rhizomes, consuming approximately 25 percent of their weight daily (Johnson and Foote 1997).

Nutria breed throughout the year and, like many rodents, are quite prolific. Females can breed again a day after giving birth. Gestation lasts for 130 days. On average, litters consist of four to five young, but may be as small as one or as large as thirteen. Males are sexually mature at between four and nine months, females between three and nine months.

The nutria's original range included Argentina, Brazil, Bolivia, Chile, Paraguay, and Uruguay. Originally, nutria were imported into California to establish a fur farm. The Gulf Coast population originated in Louisiana in 1937 when E.A. McIlhenny of Tabasco sauce fame imported thirteen animals into Avery Island, Iberia Parish, Louisiana. Trapping nutria for their pelts was a key component of the Louisiana trapping industry from the 1960s until the early 1980s when prices for furs on the world market and in Louisiana fell drastically (Baroch and Hafner 2002). The population expanded its range through natural expansion from Louisiana, as well as through the sale of nutria for aquatic weed control and stocking efforts by state and federal wildlife agencies (Evans 1970).

Nutria are commonly mentioned as a source of human-wildlife conflict. From a habitat management standpoint, they are problematic because their foraging activity creates large unvegetated patches of marsh while overturning the upper peat layer (Kinler, Linscombe, and Ramsey 1987). This disrupts and weakens the root-soil matrix, creating a vulnerability to erosion that may lead to additional marsh fragmentation and loss (Johnson and Foote 1997). Moreover, native marsh vegetation plays a crucial role in buffering storm surges (Baroch and Hafner 2002), an issue that has gained more attention since Hurricanes Katrina and Rita in 2005. Nutria will construct burrows in levees, dikes, and embankments, weakening these structures.

Nutria have been found to live in rice fields year-round if not controlled. Damage to rice in southwest Louisiana consists of grazing, which retards or prevents the production of mature grain, and burrowing into levees, which interferes with water management at various stages of cultivation that are vital to rice production. The burrowing issue causes additional problems, as when cattle grazing on large levees step into nutria burrows and become injured (Evans 1970).

Disease transmission is another concern. Giardiasis is a diarrheal illness caused by a microscopic parasite. Once the host has been infected, the parasite lives in the intestine and is passed in feces. Because the parasite is protected by an outer shell, it can survive outside the body and in the environment for long periods of time (i.e., months). During the past two decades, *Giardia* infection has become recognized as a common cause of waterborne disease in humans in the United States (Centers for Disease Control website).

Beaver are a recognized reservoir of *Giardia* sp. but few studies have investigated the prevalence of infected nutria. Dunlap and Thies (2002) found that *Giardia* sp. is a common and abundant intestinal parasite in nutria, at least in southeast Texas. Over 73 percent of nutria tested positive for *Giardia* sp. The species' semiaquatic habitat, burrowing activities, and coprophagy make nutria highly susceptible to *Giardia* sp. infection. Other studies have suggested nutria may be a vector for other diseases as well, including *Toxoplasma gondii*, *Chiamydia psittaci*, and *Leptospira* spp. (Howerth et al. 1994).

The U.S. Congress passed the Coastal Wetlands Planning, Protection, and Restoration Act, also known as the Breaux Act, in 2000. This appropriation provided funds for a number of research studies on nutria, as well as for coastal restoration and conservation (Steyer and Llewellyn 2000). In 2002, a final report on nutria control methods was completed by Genesis Laboratories (Baroch and Hafner 2002) under contract by the Louisiana Department of Natural Resources. After reviewing a number of possible methods to reduce nutria, the report concludes that the incentive payment program is the best option for coast-wide control. The report confirms the method advocated by the Louisiana Department of Wildlife and Fisheries. This program was put in place when the trapping season opened in November 2002.

FERAL SPECIES

A feral species is defined as a domestic species that has escaped from direct human control and consequently established free-ranging populations. Examples of feral organisms that are common both in rural and urban areas in the United States include cats, hogs, and pigeons. These populations are of concern to wildlife managers because of their potential impacts on native wildlife populations. While not all urbanites recognize and empathize with different species of native wildlife, nearly all of them will recognize and have some level of emotional response to dogs, cats, and hogs. Consequently, it becomes especially difficult for wildlife managers to identify control measures that will be acceptable to urbanites.

Pigeons (*Columba livia*)

[Feral pigeons] aren't wild animals and they aren't domestics. They hover around humanity like a guilty memory, flighty but ineradicable.

—David Quammen, Author

The feral pigeon—also known as the rock dove—may be the most recognized bird on earth. Not only are they a common feature of urban and, increasingly, suburban life, humans and pigeons have a long history together, as in thousands of years. Researchers estimate the birds were first domesticated in the eastern Mediterranean between five and ten thousand years ago (Quammen 1998). At some point after we began putting pigeons into cages they began to escape from those cages and, in the fullness of time, developed the unique characteristics we recognize today (Johnston and Janiga 1995; Figure 12.6).

Humans seem to have an affinity for the rare, so it's not often you'll hear anyone sing the praises of pigeons. They are often thought to be dirty or even dangerous pests. While they are certainly involved in their fair share of human-wildlife conflicts, as we'll discuss later in this section, they are also a phenomenal adaptation success story. Feral pigeons are not simply domestic pigeons that have rejected domesticity. It may be fair to call them superdoves. Compared to other pigeons they fly faster, have a higher annual reproductive potential and greater genetic variability. They are able to eat a more diverse diet, hear high-frequency sound, and see ultraviolet light. Research has even suggested they are capable of symbolic reasoning (Quammen 1998).

Figure 12.6   Pigeons. (Courtesy Clark E. Adams)

An adult feral pigeon is about 12.5 inches (32 cm) in length, with highly variable, multicolored plumage. Those that resemble their wild ancestors have a head and neck that is darker than the back, with black bars on the inner wing, a white rump, and a black band at the end of the tail (National Geographic 1999).

Found in rural areas as well as almost every city and suburb, pigeons are always found near humans. They will nest and roost almost anywhere—on high-rise window sills, under bridges, on streetlights. Pigeons are gregarious and prefer to feed, roost, and nest as a flock. Their diet may include grain, leaves, and some invertebrates, as well as a variety of castoff human foods. Males court females with an inflated neck, tail spread wide, and much bowing and circling (Ehrlich, Dobkin, and Wheye 1988).

The nest may be a shallow assemblage of stems and leaves, or nothing much at all. The female will lay one or two white, unmarked eggs and incubate for sixteen to nineteen days. The young, referred to as squabs, are altricial at hatch. Both adults are involved in chick-rearing, feeding a regurgitated secretion referred to as "crop milk," although it's not actually milk; rather, it is "produced by a sloughing of fluid-filled cells from the lining of the crop" that is quite nutritious (Ehrlich, Dobkin, and Wheye 1988). Pigeon milk contains more protein and fat than either cow or human milk. Squabs fledge at twenty-five to twenty-six days.

There is a potentially dark side to this master of adaptation. To start, Haag-Wackernagel and Moch (2004) documented 176 transmissions of illness from feral pigeons to humans between 1941 and 2003. Feral pigeons were found to harbor sixty different human pathogenic organisms. In spite of their worldwide distribution and frequent contact with people, zoonotic disease transmission to healthy humans is infrequent. That said, individuals who have regular, close contact with pigeons and their nesting sites should be aware of the risks and protect themselves appropriately (Haag-Wackernagel and Moch 2004).

Additionally, a bird that is willing to nest or roost on any surface is going to make a mess. Complaints tend to center on excrement—on buildings, cars, statues, and even people. Pimentel et al. (2002) reported that the control costs for pigeons are at least $9 per pigeon per year. This is not a problem that's isolated to North America and Europe. For example, the species was introduced to Sri Lanka by the British as a pet, and has become a serious problem in many parts of the country.

But now it appears feral pigeons may have a competitor: the Eurasian collared dove (*Streptopelia decaocto*). Both species prefer the same type of habitat. Collared doves have spread from the Bahamas, where some individuals were released in 1974, to Florida, then across the United States. They reproduce even more prolifically than rock doves, but it is not likely they will displace the common city pigeon (Johnson and Donaldson-Fortier, 2009).

There is at least one benefit of having a species that lives in such close proximity to humans— feral pigeons can serve as a coal mine canary, providing an early warning system for a variety of environmental toxins. Studies have been conducted with pigeons to monitor the presence of lead and cadmium (Nam and Lee 2006), manganese (Loranger et al. 1994), iron (Janiga et al. 1990), and even *p,p*-DDT (Sidra and Walker 1980).

Pigeons have become a natural part of everyday life in cities and suburbs and, for the most part, the rock doves and the human residents are able to live and let live. Sure we complain about them, just like we complain about the weather. But you can't really do much about either one.

Free-Ranging Domestic Cats (*Felis catus*)*

A pygmy lion who loves mice, hates dogs, and patronizes human beings.

—Oliver Herford, Author and Illustrator

* The authors would like to thank Sara Ash, Department Chair and Assistant Professor of Biology at the University of the Cumberlands, for her contributions to this section.

Over the course of human history, domestic cats have been viewed as everything from gods to representatives of the devil, depending on the culture. In the United States, the relationship has morphed from primarily utilitarian (barn cats kept to manage rodent populations) to pampered companion animal. The 2007 edition of the *U.S. Pet Ownership & Demographic Sourcebook* (published by the American Veterinary Medical Association), reports 32.4 percent of U.S. households include at least one cat, for a total of over 81 million feline pets. Consequently, where you find humans, you will undoubtedly find cats as well. But not all domestic cats in the United States are cherished members of the family; large populations of free-ranging feral cats have become established in both rural and urban landscapes.

Cat rescue groups have claimed between 10 and 50 million unowned cats live in the United States, but accurate population estimates are understandably hard to come by (Mahlow and Slater 1996; Patronek 1998). The presence of feral cats in urban areas has created conflict with humans because of nuisance behaviors such as caterwauling, fighting, spraying, defecating, and the production of numerous litters of kittens. In addition, some have cited concern about flea infestations and diseases such as ringworm, cat scratch fever, toxoplasmosis, toxocariasis, rabies, and salmonellosis (Ablett 1981; Passanisi and Macdonald 1990). Wildlife professionals also object to the presence of free-ranging cats because of their potential negative impact on native wildlife populations, particularly small mammals and birds.

The domestic cat has a wide distribution and occupies a variety of habitats (Kerby and Macdonald 1988). Dependence on humans for food and shelter varies among cat populations. Some populations exhibit behaviors similar to wild species of cats, and in these populations, individual cats may occupy exclusive territories in "wild" habitats and subsist entirely on prey such as small mammals (Genovesi, Besa, and Toso 1995). Studies have linked predation by domestic cats to the decline of several wildlife species in Australia (Dickman 1996). Many authors (Errington 1936; McMurray and Sperry 1941; Hubbs 1951; Parmalee 1953; Eberhard 1954; Toner 1956; George 1974, 1978; Fitzgerald 1988; Soulé et al. 1988; Coleman and Temple 1996; Lepczyk, Mertig, and Liu 2004) in the United States have written about the impacts of domestic cat predation on wildlife.

Cats in urban and suburban areas usually live in groups and obtain their food directly from humans or by scavenging (Dards 1978; Natoli 1985; Yumane, Emoto, and Ota 1997). Even though cats living in developed areas benefit from feeding by humans, they commonly prey on wildlife (Churcher and Lawton 1987; Clout and Gillies 2003; Woods, Macdonald, and Harris 2003). What effect this predation has on native wildlife populations has not been experimentally demonstrated. We do not know if cats are killing surplus individuals that would have died anyway or if the cats are limiting and/or regulating the populations of their prey.

American college campuses are ideal locations to find populations of feral cats (Figure 12.7), for two reasons. First, the transient nature of students is believed to result in a high incidence of pet loss and/or abandonment, thereby establishing and maintaining cat populations on campus and in nearby neighborhoods. Some support is given to this assumption by Oppenheimer (1980) who found a positive correlation between cat density and the proportion of renters in Baltimore neighborhoods. Second the typical campus environment meets the habitat requirements for feral cats, including ample cover (e.g., basements, maintenance sheds, dense landscaping) and food (e.g., direct feeding, garbage, and prey). Conditions on campuses are so ideal for feral cats that control measures are often necessary to reduce the populations.

The traditional method of control in U.S. cities (including college campuses) is eradication. Domestic animal shelters records are not always available to the public so shelter-based estimates of the number of cats euthanized annually are imprecise and vary widely—from approximately 5 million to as high as 9.5 million in the United States alone (Rochlitz 2007). However, a growing number of people are supporting the use of the TTVAR method in which cats are trapped, tested for, and vaccinated against infectious diseases, altered (sterilized), and returned to the capture site, where volunteers feed and monitor the cats daily (Zaunbrecher and Smith 1993). This

Figure 12.7 Feral cat, commonly found in urban neighborhoods, university campuses, and shopping malls throughout the United States. (Courtesy Sara Ash)

approach was first promoted as a method for reducing numbers of unwanted cats in Europe over twenty years ago (Remfry 1996). Volunteer organizations on several U.S. college campuses have implemented TTVAR for the management of their campus cat populations. Stanford University was the first to promote its use of TTVAR (Johnson 1995), but employees of other campuses like the University of Washington had been using the approach for several years prior to Stanford's public endorsement.

Our colleague, Sara Ash of University of the Cumberlands, undertook a comparison of the TTVAR programs at the University of Washington (UW), University of Texas at Austin (UT), and Texas A&M University (TAMU), which is summarized here. For each university, pertinent stakeholders and their agendas are identified and described, along with political, cultural, economic, and ecological ramifications of the TTVAR program and future considerations.

On each campus, female staff members were responsible for initiating the use of TTVAR. Sharon, a UW staff member, started trapping cats on her campus in 1988. Sharon is a self-described cat lover with experience volunteering for animal shelters and providing foster care for abandoned animals. She first got involved with the feral cat issue when she began taking campus cats home with her. She quickly realized her home could not support the entire campus population, which she estimated at over 100 cats. She began secretly feeding and trapping the cats on campus and found veterinarians who would test, vaccinate, and sterilize the cats at a reduced rate. Her primary objectives were to improve the cats' quality of life and reduce the population. In 1994, after providing several reluctant interviews with the local media about her activities, she and another staff member formed the organization Friends of Campus Cats (FCC). The university sanctioned their organization and TTVAR activities.

Two University of Texas at Austin staff members and a representative from the Austin Society for the Prevention of Cruelty to Animals (ASPCA) formed the volunteer organization Campus Cat Coalition (CCC) in 1995 as a reaction to attempts at feral cat eradication at UT. According to one of the cofounders of CCC, during the 1994 Christmas break, the UT Office of Environmental Health and Safety hired a pest control company to remove the feral cat population from campus. Approximately fourteen cats were trapped and euthanized. When the campus community returned after break, many staff members were outraged by the university's actions. Three staff members were asked to form CCC, which would assume responsibility for the campus cats and implement a TTVAR program to control the population. CCC emphasizes the welfare of the individual cats and eventual reduction in the population.

Dawn, a former staff member of the College of Veterinary Medicine at TAMU, first became interested in the feral cat issue when she witnessed the euthanasia of kittens that were trapped by university pest control. She believed destroying healthy kittens was unnecessary and a waste, so she searched for the campus policy regarding the issue of cat control and was unable to find one. Dawn submitted a proposal to the College of Veterinary Medicine outlining a plan to use the TTVAR approach for controlling the campus cat population, referred to as the Aggie Feral Cat Alliance of Texas (AFCAT). By using this approach, the college offered veterinary students the opportunity for additional surgeries during their third- and fourth-year rotations. Although Dawn believed that TTVAR is the most humane and effective method of controlling the numbers of feral cats, her main objective in promoting its use was to create additional educational opportunities within the College of Veterinary Medicine.

Each cat rescue group expected their respective university administrations would have similar concerns about the use of TTVAR on campus, including costs, liability, and public reaction. According to Sharon, the UW administration's only request was that she not conduct FCC activities on university time and not use her employee Internet account for FCC use. The UT administration's reason for initiating the eradication attempt included concerns about human safety and complaints about offensive odors associated with feeding the cats and the cats' spraying behaviors. However, because the campus community reacted negatively to the eradication, the administration was open to suggestions that would appease the disgruntled staff. The TAMU administration's primary concern was costs of the program. Dawn and her faculty advisor procured external funding for the program, resulting in support and approval from the College of Veterinary Medicine.

Vocal opposition to the TTVAR program was expressed by a UW faculty member in the Zoology department who was concerned about the impact of the cats on local wildlife. Similar wildlife concerns were expressed in a letter to the editor of a local newspaper by two TAMU faculty members of the Wildlife and Fisheries Sciences Department (personal observation). Other opposition included TAMU staff members of the Physical Plant and Athletics Department who were concerned about human health and safety. Cats commonly lived in the football stadium and could sometimes cause "problems." For example, during the opening kickoff at a TAMU-Nebraska football game, a feral kitten ran out onto the field. A staff member was bitten when trying to capture the cat. However, none of the expressions of opposition on the campuses resulted in change in activities by the cat rescue groups; rather, they indicated a lack of support of the program by a small number of the campus community.

All three TTVAR programs garnered attention from the media. The AFCAT program received international attention from the Discovery Channel-Canada in September 1998. The report aired in early March 1999. Positive media attention reflects well on each university, thus leading to improved public relations.

Positive cultural consequences of the TTVAR programs included the increased morale of staff members who are invested in the cats' welfare. Several UT and TAMU staff members had been feeding campus cats for many years prior to the TTVAR program, resulting in cherished relationships. Unlike eradications, the TTVAR approach protects these bonds. Cat feeders are notoriously committed to their animals, and they will care for a colony of cats for several years (Sara Ash, personal observation).

Negative cultural consequences of the TTVAR programs can arise if objectives and intentions of the programs are not clearly defined. TTVAR is not designed to be a solution to the cat overpopulation problem, but rather an approach to stabilize and potentially reduce an existing local population of cats. In addition, it may take several years to substantially reduce the numbers of cats. For example, although Sharon (UW) had been using the TTVAR approach for eleven years at the time of our interview, five colonies of more than seven cats each still existed on campus. The campus community must understand that TTVAR practices alone are not enough to reduce the numbers of feral cats on campus. Abandonment and pet loss must be reduced simultaneously to stem emigration from pet to feral populations. Some cat rescue groups (not those described above) have proposed artificially establishing colonies with unwanted or feral cats from animal shelters (Patronek 1998), a scenario that would alarm most wildlife professionals.

Zaunbrecher and Smith (1993) showed that long-term costs for eradication exceed those needed to support a TTVAR program. TTVAR activities on the UW and UT campus are funded by private donations and fund-raisers, while the program at TAMU is supported by a private grant. None of these three universities, therefore, incurs any costs, although they would have to pay for any eradication program. By using campus cats for educational purposes, the TAMU College of Veterinary Medicine saves money on purchases of laboratory animals for vet student surgeries.

All three cat rescue groups described here claimed their programs may minimize the potential negative impacts of cats on wildlife by providing a reliable food source for the cats. They also stated that neutering may reduce roaming, which would bring cats into contact with more wildlife. However, no studies have shown that TTVAR programs minimize the impact of cats on wildlife or that sterilized adult cats will reduce their roaming behaviors. In fact, Ash (2001) showed that TTVAR-managed cats continued to roam large distances. Across the country, the use of TTVAR management has become a volatile issue between the cat rescue movement and wildlife managers (The Wildlife Society 2001). Let's examine the scientific reasoning behind this opposition.

Studies have shown that cats hunt regardless of hunger state (Adamec 1976; Churcher and Lawton 1987) and that the presence of cats can lead to a change in the local biodiversity (Hawkins 1998). Additionally, it is important to note that cats behave differently according to environmental factors (Liberg and Sandell 1988). Urban cats kill more birds than their rural counterparts (Churcher and Lawton 1987). However, the diet of urban cats is primarily garbage and direct handouts from humans (Fitzgerald 1988). Cats in rural and suburban areas are more likely than urban cats to come into contact with desirable wildlife species such as native small mammals, game birds, and predators. However, as we learned in the chapter on urban habitats, green spaces within highly urbanized locations can support several species of important wildlife. Therefore, it is important to consider the location of a cat colony before deciding to use TTVAR. Both UW and UT are located in highly urbanized areas; therefore, cats are not expected to have a substantial impact on native wildlife. However, TAMU is located in a suburban/rural landscape, and cats on the perimeter of campus will come into contact with several species of native wildlife, potentially creating a conflict of interest among cat rescuers and wildlife interest groups.

Human populations and, consequently, cat populations are not limited in range to highly urbanized areas. As demonstrated by earlier chapters in this book, urban and suburban development of agricultural and ecologically important land is expected to continue. With this progression we expect to see a continued juxtaposition of humans, their domestic animals, and wildlife. Developed areas with cats, dogs, and humans are currently positioned very close to natural habitats. If TTVAR management is used in close proximity to these areas, we can expect to see the following:

1. Native animals such as raccoons, skunks, and coyotes will be attracted to the feeding stations (Ash 2001).
2. Managed cat colonies will prey on native animals, including resident species such as small mammals and migratory species such as songbirds.
3. Humans will dump their unwanted cats in these TTVAR-managed areas (Castillo and Clarke 2003).

Figure 12.8   Feral hog. (Courtesy Texas Agrilife Extension)

Free-Ranging Hogs (*Sus scrofa*)*

The wild hog is the only animal that gives birth to 6 young and 8 survive to adulthood.

—**Billy Higginbotham, circa 2001**

Rarely can the origin of a wildlife issue be dated and blame placed on an individual or group, but such is not the case with wild hogs (*Sus scrofa*) in the United States. Not to be confused with the javelina or collared peccary (*Tayassu tajacu*) that are native to the American southwest, wild hogs (Figure 12.8) first came to the continental United States with the Coronado and de Soto expeditions in 1540 and 1541, respectively (Towne and Wentworth 1950). Brought along as a reproducing food source, the hogs were driven along the expedition route. Some escaped and became feral, while others were stolen by Native Americans and released to establish wild populations (Towne and Wentworth 1950).

In the more than 450 years to follow, North American wild hog populations have been supplemented through traditional agricultural practices of free-ranging livestock, where some animals escape recapture, and through intentional releases to establish huntable populations (Mayer and Brisbin 1991). Wild hogs have long been a sought-after game species prized for their meat and the thrill of the hunt. The Wild Boar Conservation Association and various local groups have developed breeding programs to supply European wild boars for stocking purposes (Myers 1993). In Texas, hunters have happily paid from $25 to $1,000 (average $169) for the opportunity to participate in a wild hog hunt (Rollins 1993).

By the late 1970s, wild hogs were concentrated in the southeast portion of the United States stretching from Florida west to Texas, and into California (Wood and Barrett 1979). Ten years later, hogs had expanded their range to include nineteen states, but were still concentrated in the southeastern states and California (Mayer and Brisbin 1991). By the early 1990s, Miller (1993) reported twenty-three states with resident hog populations and that populations were increasing rapidly. By 2003, Mayer (J.J. Mayer, Westinghouse Savannah River Company, personal communication) reported that thirty-two states and four Canadian provinces had resident wild hog populations.

---

* The authors would like to thank Rob Denkhaus and Suzanne Tuttle of the Fort Worth Nature Center and Refuge for their contributions to this section.

Obviously, wild hog populations have been on the increase, and they are expanding their range. The question then is, how were they doing it?

Gipson, Hlavachick, and Berger (1998), in a survey of wildlife professionals, found the two most common explanations for the burgeoning hog populations were that wild hogs from established populations have been captured and relocated to new areas to create additional hunting opportunities, and that wild hogs have escaped from "confined" populations created and maintained by hunting clubs. Of course, some wild hogs dispersed from established populations, and moved into new areas without any human assistance.

Wild hogs include free-ranging domestic (i.e., feral) swine, Eurasian wild boar, and hybrids between the two. Only one population of pure Eurasian wild boar is thought to survive in the United States, and that is found in Corbin's Park, New Hampshire (Mayer and Brisbin 1991). However, many of the releases in and around areas such as Great Smoky Mountains National Park were reportedly of pure wild stock (Tate 1984). Wild hogs from the Great Smoky Mountains were then used to establish populations in California and other areas (Baber and Coblentz 1986; Mayer and Brisbin 1991). The three types (feral, European stock, and hybrids) exhibit slight differences in morphology (Mayer and Brisbin 1991), behavior (Jones 1959), and habitat preference.

Regardless of the explanations for how and why wild hog populations have increased and which type is found in an area, it is apparent this exotic species is quickly invading available habitat throughout the continent. And, like other species, as available habitat gives way to urbanization, hogs find themselves faced with the opportunity to move into urban areas where their needs can be met but where the problems caused by their existence are magnified.

Wild hogs have much in common with their barnyard cousins, since they are the same species. However, a life spent foraging for food while exposed to the rigors of the weather causes some modifications in the basic pig template. The best adjective to describe wild hogs is "variable," because they can be found in all shapes, sizes, and colors.

Wild hogs tend to be longer legged and leaner than their captive brethren. While hunters often claim to have bagged hogs weighing over 600 lbs (272 kg) and Nowak (1991) reports domestic hogs weighing as much as 1000 lbs (450 kg), adult males, known as boars, average 130 lbs (59 kg), and average adult females, or sows, weigh 110 lbs (50 kg) (Stevens 1996). Boars and sows can, and do, frequently attain weights of up to 300 lbs (136 kg) and 200 lbs (91 kg), respectively. Maximum shoulder height for wild hogs is around 3 ft (0.9 m) (Stevens 1996).

Wild hogs tend to have longer, thicker coats than their domestic counterparts. The Eurasian hogs exhibit a brown to black coat with a lighter shade on the tips of the bristles that give a grizzled appearance. Hogs that flaunt their domestic ancestry can be solid black, brown, blond, white, or red, spotted, or belted. A belted hog has a white band across the shoulder and forelegs. A striped pattern is sometimes found in piglets less than six months old, although this pattern is never maintained to adulthood (Mayer and Brisbin 1991).

Boars have two pairs of continually growing tusks (canine teeth) that protrude from the sides of the mouth. The upper and lower tusks wear against each other to maintain a high degree of sharpness and can reach several inches in length unless they are broken or worn from use. Tusks are used for defense and to establish dominance during the breeding season. As protection from a rival's tusks, boars develop a thick, tough layer of cartilage and scar tissue (frequently referred to as a shield) across the shoulder area. Development of the shield continues as the boar ages and engages in combat with other boars. Sows have smaller tusks and do not develop a shield.

Wild hogs are omnivorous. They will consume virtually anything that they can find, and their preferences change with the seasons as new foods become available (Wood and Roark 1980). Hogs feed by rooting, which involves churning up the soil with their snouts (Figure 12.9). A feeding area can resemble a plowed field with furrows sometimes a foot or more deep, depending on soil type. Surprisingly, given their mode of feeding, Beyer, Connor, and Gerould (1994) found that only 2.3 percent of a wild hog's diet is ingested soil.

Figure 12.9 *(A color version of this figure follows page 158.)* Damage by feral hogs to homeowner's lawn. (Courtesy Linda Tschirhardt)

Hog diets are primarily herbaceous, with fruits and seeds making up the bulk of the diet (Baber and Coblentz 1987). When available, acorns are an extremely important dietary item (Henry and Conley 1972; Wood and Roark 1980; Baber and Coblentz 1987). Grasses and sedges are important foods in the spring (Wood and Roark 1980) and less important in the fall (Henry and Conley 1972). Roots, particularly those of plants on hydric, or wet, sites are preferred in the summer, while fungi are fed upon throughout the year (Wood and Roark 1980). Woody material is rarely found during stomach content analysis, but researchers (Wahlenberg 1946; Wakely 1954) described extensive damage to longleaf pine seedlings, and Wood and Brenneman (1977) theorized that hogs chew the root, swallow the sap and starches, and reject the woody tissue. They reported finding balls of chewed woody tissue in areas where hogs had been rooting among woody plants.

While animal matter does not normally comprise a large percentage of a hog's total diet, both invertebrates and vertebrates are frequently consumed. Wood and Roark (1980) found invertebrates in at least 62 percent of the hog stomachs examined during all seasons of the year. Invertebrates found included grubs, earthworms, centipedes, clams, and mussels. Howe, Singer, and Ackerman (1981) found animal matter in 94 percent of the hog stomachs examined, yet it accounted for very little of the total volume. Wild hogs are known to be opportunistic predators on ground-nesting birds (Rollins and Carroll 2001), fawns (*Odocoileus* spp.; Springer 1975; Hellgren 1993), small mammals (Wood and Roark 1980), reptiles and amphibians (Springer 1975; Wood and Roark 1980; Howe, Singer, and Ackerman 1981), and domestic livestock (Springer 1975), and have been observed scavenging fish carcasses (Baron 1982).

The wild hog's search for preferred foods through the seasons leads it on a tour of available habitats. Hogs can be found virtually anywhere that has permanent water and sufficient cover (Sweeney and Sweeney 1982). Water is used for drinking and thermoregulation. Hogs lack sweat glands and control their body temperature by wallowing in mud holes. The resulting coat of mud also serves to protect the hog from biting insects. The need for water makes swamps, marshes, and hardwood bottomlands very attractive to wild hogs (Wood and Brenneman 1980; Baber and Coblentz 1986; Ilse and Hellgren 1995). As other foods become available, hogs move into upland areas, especially oak forests.

Wild hogs have the highest reproductive potential of any large mammal in North America (Wood and Barrett 1979; Hellgren 1999). Female hogs reach sexual maturity at approximately ten months, while male maturity occurs between five and seven months (Sweeney, Sweeney, and Provost 1979). Wild hogs are polygynous, meaning that one male will breed with many females. Breeding can occur at any time of the year, although Taylor et al. (1998) observed most breeding in midwinter with a secondary peak in the spring. The gestation period is approximately 115 days (Henry 1968).

Litter size in wild hogs is usually less than that observed in domestic pigs (Baber and Coblentz 1986) and more than is reported in pure Eurasian stock (Taylor et al. 1998). Wild hog litter size varies widely depending on region, climate, and food availability but generally falls between four and seven young per litter (Sweeney, Sweeney, and Provost 1979; Baber and Coblentz 1986; Taylor et al. 1998). Although they generally only produce one litter per year, two litters in one year have been reported (Baber and Coblentz 1986; Taylor et al. 1998).

Sows and their piglets generally form loosely associated groups of eight or less, with rarely more than three adults per group (Sweeney and Sweeney 1982). Boars form small bachelor herds or are solitary except when a sow is in estrous. A sow in estrous will entice boars into the area, where battles for dominance and breeding rights may occur.

With the continuing conversion of undeveloped lands into residential areas, wild hogs are being found closer and closer to urban and suburban areas (Brown 1985), where they have increased contact with humans. Wild hogs are capable of inflicting injury upon humans, especially when cornered or when piglets are involved (Lucas 1977; Tinsley 2002). They have also been known to injure pets and even prey upon livestock (Beach 1993). In Texas, wild hogs are a significant predator of domestic goats and sheep and have been known to prey upon calves (Littauer 1993).

The most common source of conflict surrounding wild hogs is the property damage that they cause. In their never-ending search for food, wild hogs often destroy large amounts of row crops. Much of the total damage is from the actual plants being rooted up, but the amount lost due to trampling can also be considerable. Even tree seedlings, such as those on pine plantations, are susceptible to hog depredation (Whitehouse 1999). Fences are often used to prevent hogs from gaining access to agricultural crops, but this frequently results in damaged fences along with the lost crops (Beach 1993).

As the lines that divide hog habitat from human habitat blur, hogs are even found rooting up lawns and golf courses in search of earthworms and other foods. In addition to the destruction of the turf, costly sprinkler systems are sometimes destroyed. This type of damage tends to become more common during drought years because suburban lawns and fairways may be the only areas that receive significant amounts of moisture. Of course, this also brings hogs into closer proximity to people, and the potential for human injury increases.

An increasing hog population has also led to an increase in hog-vehicle accidents. When struck, wild hogs, with their low center of gravity and potential for being large, often cause the vehicle to roll over, causing significant mechanical damage and injuries to the vehicle's occupants (Nunley 1999). Even the hogs' rooting and wallowing activities can lead to mechanical damage when farm equipment inadvertently falls into hidden holes (Nunley 1999).

In addition, hogs are potentially significant disease carriers. Some of these diseases can be transmitted to humans, but most only pose a risk to livestock and wildlife. Diseases such as swine brucellosis and pseudorabies are significant threats to the livestock industry. While their potential as a disease vector is often emphasized by those who are not fans of wild hogs, David Stallknecht of the Southeastern Cooperative Wildlife Disease Study said in his 2004 presentation for the symposium The Biology, Management and Control of Wild Pigs (Augusta, Georgia) diseases "can be used to support management decisions but may not be significant enough to drive them."

Quantifying the total effect that any species has on the environment is difficult at best, and potentially impossible. This is particularly true for a species as generalist in habitat and food preference as the wild hog. In their native lands, they are a natural component that makes positive contributions to those ecosystems. For example, Eurasian wild boar are thought to enhance the growth of pines in poor soils (Andrzejewski and Jezierski 1979), and their control is suspected to have decreased the rate of nutrient cycling and upset the stability of European forests (Grodzinski 1975).

It is important to remember the wild hog is an exotic species in North America. Our native habitats have not evolved along with it and wild hogs often have negative environmental impacts on plant communities, soils, aquatic systems, and native fauna through their movements, habitat use, and food habits. Hog rooting impacts plant communities through direct consumption of vegetation as well as indirect effects such as trampling and uprooting of uneaten plants. These impacts result in a reduction in plant cover on the forest floor (Tate 1984) and provide an opportunity for early succession plants, including exotic and invasive species, to invade the habitat (Springer 1977). While this may be perceived as a positive impact if an area is being managed for maximum floristic diversity, the reverse is true for lands being managed as natural areas. And, the potential exists for wild hogs to adversely impact endangered plant species through their rooting activities.

In California's oak-dominated ecosystems, Sweitzer and Van Vuren (2002) found wild hog rooting significantly reduced oak regeneration. Reduced seedling survival was linked to hog rooting, and acorn consumption is blamed for reduced acorn survival to potential germination as well as reduced forage available for native wildlife.

The presence of wild hogs in an area also impacts native animal species, either through direct predation or through competition for resources. As stated above, wild hogs are opportunistic predators on a variety of wildlife species including at least two threatened species, Jones' middle-toothed snail (*Mesodon jonesianus*) and the Jordan's red-cheeked salamander (*Plethodon jordani*) (Tate 1984). Hogs are reported to prey upon turtle eggs and hatchlings (Hanson and Karstad 1959). Although they do not frequently prey upon them, intensive hog rooting alters the appropriateness of leaf litter habitat for at least two vertebrate species, southern red-backed voles (*Clethrionomys gapperi*) and northern short-tailed shrews (*Blarina brevicauda*) (Singer, Swank, and Clebsch 1984).

Wild hogs compete with a number of species for available food sources. In Tennessee, hogs are thought to compete with deer (*Odocoileus virginianus*), bear (*Ursus americanus*), gray squirrels (*Sciurus carolinensis*), turkey (*Meleagris gallopavo*), raccoons (*Procyon lotor*), and opossums (*Dideplhis virginiana*) for acorns and other fruits and seeds. Henry (1969) proposed that, while wild hogs were only minor predators on bird nests, they competed with other native nest predators such as raccoons, opossums, and snakes for the available resource. The wild hog's ability to take advantage of an array of foods gives it a competitive advantage when compared to more specialized feeders.

While there have been no studies that have documented hogs having a significantly adverse direct impact on another species, Roemer, Donlan, and Courchamp (2002) reported how the presence of a large population of wild hogs on the California Channel Islands encouraged golden eagles (*Aquila chrysaetos*) to colonize the islands to take advantage of an abundant food supply, that is, piglets. Eagles soon became the top predator on the island and fed heavily on the resident island fox (*Urocyon littoralis*) population. With the reduction in foxes, island spotted skunk (*Spilogale gracilis amphiala*) populations increased significantly to fill the available niche. So, while the hogs were not preying upon any of the other three vertebrates in the study, their presence had a far-reaching ecological impact on all three and the entire island system by causing predation to replace competition as the dominant force shaping the island communities. Another study (Long, Sweitzer, and Ben-David 2001) hypothesizes that the growing hog population in California is partly responsible for the increase in mountain lions in the state and is therefore indirectly leading to an increase in human-mountain lion encounters.

Some of the potentially more devastating environmental impacts that can be attributed to hogs are directed against the abiotic components of the ecosystem. Hog rooting accelerates decomposition and loss of nutrients from the forest floor and upper soil horizons (Singer, Swank, and Clebsch 1984). Specifically, rooting reduced phosphorus (P), magnesium (Mg), and copper (Cu) levels available for plant uptake. Singer, Swank, and Clebsch (1984) found that soil erosion did not significantly increase because of hog rooting in their upland forest study site, but attributed this to the natural porosity of the soil and the decrease in soil bulk density caused by the rooting. However, water sampling in rooted watersheds indicated a significant increase in soil water NO_3 levels. Hog rooting and wallowing is suspected of causing siltation and contamination of Appalachian streams (Howe, Singer, and Ackerman 1981). Tate (1984) reported higher concentrations of fecal coliform bacteria in areas occupied by wild hogs.

Management of wild hog populations ranges throughout the spectrum from a desire to eradicate the species to trying to introduce the species into new areas. Because of its status as an exotic species and its adverse impacts to the ecosystem, most land managers are strongly entrenched on the side of eradication, although many admit that eradication of such a prolific species may be impossible because of budget constraints and public perceptions. Although eradication has been (or nearly been) achieved in some island habitats (Kessler 2002; Schuyler, Garcelon, and Escover 2002), controlling wild hog populations at a level where they are not severely impacting native habitats is the best that can be hoped for in areas where the species is firmly entrenched and such factors as immigration and human-aided introduction cannot be controlled.

Wild hog control efforts take many forms but usually involve a combination of techniques. Traps, controlled shooting from the ground or air, public hunts, capture with dogs, and poisoning have all been used. Attention has been paid to biological controls such as introducing hog diseases and using chemosterilants. Fortunately, the idea of introducing disease into the wild hog population has not been attempted because of the potential for the disease to pass into the domestic livestock industry. Work continues in chemosterilants, but with little reported practical success (Killian et al. 2003). The ultimate choice of techniques employed depends on the availability of money and personnel to conduct a control program, the expertise of the personnel, legal limitations, safety concerns, site conditions, and public sentiment.

Trapping is the most common control method and can be extremely successful, especially when used in conjunction with other techniques. Trapping is a labor-intensive and costly endeavor that requires considerable scouting and moving, prebaiting, and checking traps. Success is dependent on the expertise of the personnel doing the trapping (E.K. DeLozier, National Park Service, personal communication), but can result in large numbers of wild hogs being removed from the population in a short period of time. Once trapped, hogs can be euthanized quickly and as humanely as possible via a properly placed gunshot (Beaver 2001). Unfortunately, trapping alone will not eradicate wild hogs from an area, because some hogs, enough to breed and replenish the population, will become trap shy and must be taken by another method.

Controlled shooting, either from the ground or the air, is conducted by trained marksmen. Aerial shooting can result in very large numbers of hogs removed from the population if the hogs inhabit an area with little overhead cover. Ground shooting usually involves stalking hogs in their activity areas at night using night vision equipment and high-powered rifles equipped with scopes. Because of the potentially large numbers of hogs that can be taken in a relatively short period of time, controlled shooting is usually more cost effective than trapping. Safety is of the utmost importance, especially on public lands, when controlled shooting is used, and the potential for an accident can eliminate this technique from consideration in some areas.

In areas where public hunting is allowed, public wild hog hunts, often as an adjunct to deer hunts, are sometimes used as a population control measure. In most cases, this has not proven to be an effective control, because hunters do not remove as many individuals as is necessary either through lack of success or lack of interest in hunting hogs. Many areas that have tried to use public hunts to control hog populations have actually seen an increase in hog populations instead of a decrease.

Hunting with dogs usually involves contracting with a local hunter or dog handler, because few areas can afford to train and support a pack of dogs throughout the year. Dogs can be quite effective in locating, bringing to bay, and capturing wild hogs, although it involves great potential for danger for the dogs and their handlers. Once caught, the wild hog is usually dispatched using a firearm or through exsanguination. Hunting with dogs can raise the ire of animal rights/animal welfare groups, causing them to oppose the program.

Poisoning is used to control wild hog populations in Australia (Choquenot, Kay, and Lukins 1990) and New Zealand (Thomas and Young 1999). Yellow phosphorus (CSSP), sodium mono-fluoroacetate (1080), and warfarin have all been used with success, but the potential for nontarget animal poisoning prevents the use of poison in the United States.

Because of the wild hog's dual status as both a nuisance and a game animal, public perception of control programs is of utmost importance. Some may believe that nonlethal control measures, such as trapping and relocating, are more appropriate and may oppose other, lethal measures. Trapping and relocation have been used in the Great Smoky Mountains National Park with limited impact on the wild hog population. It is also possible that relocated animals simply return to the area from which they were removed or become a nuisance in their new range.

Wild hogs are here to stay in North America. Total eradication, at least with current technology, is not possible. Because of their generalist nature and great adaptability, wild hogs will continue to expand their range on the continent and invade new habitats. As the human population continues to grow, encounters with wild hogs, as well as the damage they cause, whether in the wilderness or in a subur-ban yard, will increase, and management of this exotic, invasive species will grow in importance.

The Fort Worth Nature Center and Refuge (FWNC&R) is a 1450+ hectare (3600+ acre) urban green space owned by the City of Fort Worth and managed by employees of the City's Parks and Community Services Department as a natural, native landscape. Resource management objectives are to conserve, maintain and/or restore plant and animal communities native to North Central Texas; this includes controlling the introduction and spread of exotic plants and animals with eradi-cation as the ultimate goal.

The park lies wholly within the political boundaries of Fort Worth on the northwest edge of the city and is bisected by a major river, the West Fork of the Trinity. The wetland and riparian zones of the West Fork and its tributaries provide many acres of prime habitat for feral hogs.

In July 1999, during a routine visit to a remote area on the north side of the park, FWNC&R staff found many areas of major soil disturbance consistent with the type of damage caused by feral hog rooting. Wallows began to appear along watercourses, and the soft, sandy soil in upland areas was plowed both randomly and in long, straight furrows. Soon thereafter, the staff made the first visual records of live hogs and started receiving visitor reports of hog sightings along trails in heavier traffic areas of the park. It is unknown whether the hogs moved into the FWNC&R via natural migration along the river or were purposely introduced by neighboring landowners for ille-gal hunting.

In keeping with the goal of controlling the introduction and spread of exotic animals, staff immediately began planning for removal of the hogs. An extensive literature search was undertaken to compile information on feral hog natural history, population assessment methods, and existing control options. In addition, staff consulted with local representatives of other agencies, including the U.S. Department of Agriculture's Wildlife Services Division, the U.S. Fish and Wildlife Service, and Texas Parks and Wildlife Department. Formulation of a control plan based on the information and experience gathered by others dealing with feral hog problems began in fall 2000, and a draft was ready to begin the approval process by spring 2001.

In formulating the control plan, the staff had first to make a decision about whether to remove the hogs via lethal or nonlethal means. A number of options were considered, including live trap-ping and euthanization by staff, exclusion with fencing, live trapping and relocation to another site, live trapping and euthanization by Fort Worth Police, hiring professional trappers, or conducting a regulated public hunt. As each option was weighed and the unique set of challenges faced by the FWNC&R staff considered, most proved unfeasible for implementation. For example, (1) the cost of fencing for total exclosure was (and still is) prohibitive; (2) staff were informed by the local USDA Wildlife Services agent that permission for relocation of live hogs would not be granted; (3)

providing access for police officers and professional trappers to remote and/or environmentally sensitive areas of the Refuge would tax the resources of a small, overburdened staff and create a different set of resource protection challenges; (4) funding for professional trappers was nonexistent; and (5) the risk to the public and potential liability of the city was considered too great to allow a public hunt. Live trapping and euthanization by staff was finally settled on as the only viable option under the circumstances.

Staff also made contact with a variety of external stakeholders during the process of developing the control plan to gather opinions and to generate an inclusive spirit of total disclosure. These stakeholders included the Friends of the Fort Worth Nature Center (a 501(c)3 support organization), Fort Worth Audubon Society, Humane Society of the United States (HSUS), People for the Ethical Treatment of Animals (PETA), North Texas Herpetological Society, Dallas/Fort Worth Herpetological Society, and neighbors and other nearby landowners. Input from these external stakeholders was then presented to internal stakeholders, departments within the City of Fort Worth including Police, Risk Management, Legal, and Animal Control, who then provided guidance on safety and liability issues.

Once the trap and euthanize option was chosen, the next steps were to decide trap type, euthanasia means, and carcass disposition method. Staff decided to begin the control program with modified box traps equipped with trip wire, spring-loaded, side swing doors. This type of trap is reasonably portable and small enough to be carried into remote areas with a standard off-road vehicle. Use of snares was considered and has not been totally ruled out as an option, particularly for trap-resistant individuals; however capture of nontarget animals and public perception of the method need to be considered.

Firearms were chosen over chemical means of euthanasia for several reasons. The AVMA Panel on Euthanasia recommends gunshot as a humane method for euthanizing hogs in the field (Beaver 2001). Chemicals are slower to act than firearms, leading to increased trap stress and longer, closer staff interaction with the angry hogs. Cost was also a factor, as the chemicals are expensive and require specialized training in their use. Euthanasia drugs are a controlled substance and are difficult to obtain, they require special secure storage, and a veterinarian must monitor their use. Staff was also concerned that nontarget animals that fed on drug euthanized-hog carcasses could be harmed.

The method chosen for carcass disposition led to the greatest amount of controversy during the formulation of the control plan. Staff chose to move the carcasses to discreet disposition sites and let natural nutrient recycling occur rather than attempt to utilize the meat. There were several reasons for this decision—ecologically, the hogs are robbing resources from the native flora and fauna that should be returned to the native ecosystem; from the risk management standpoint of safety and liability, hogs carry several diseases that are transmittable to humans; and practically, staff time and availability are very limited. In addition, this option was the most acceptable to animal rights/ animal welfare organizations that agreed in principle to the need for the removal of exotic animals to protect a native landscape, but objected to using those animals for the benefit of humans.

As the process of getting buy-in for a feral hog management plan from the City of Fort Worth governing body and FWNC&R stakeholders continued, three negative hog-human encounters occurred, two at the Nature Center and one at a neighboring ranch, that proved pivotal in steering sentiment in favor of implementing the plan. In April 2001, a woman was walking her dog on a leash down an old road in a relatively remote area of the park when she surprised a sow with numerous piglets. The sow tried to escape but a fence blocked its chosen route, so the sow turned and attacked the woman's dog. The woman managed to free her pet and, fortunately, the sow turned and ran away with her offspring instead of continuing the attack. The dog sustained severe injuries but did survive.

Shortly after this incident, a photographer who was a long-time supporter and frequent visitor to the Nature Center reported that he was hiking on a remote trail with his camera and tripod when he encountered a lone hog. The hog turned and charged the photographer, who used his tripod to defend himself against the hog. The animal gave up its attack quickly and departed, leaving the man shaken but uninjured.

A few months after the second incident, a neighboring rancher who had already begun removing hogs from his property, called to report one of his ranch hands had been attacked by a lone hog while on horseback in an area just across the Trinity River from the FWNC&R. His employee was not hurt, but the horse's legs were badly wounded. It is not known whether or not the horse survived or had to be euthanized. The rancher expressed his concern for the safety of visitors to the Nature Center and urged the staff to implement a control program.

These three negative encounters gave weight to the Nature Center staff's recommendation to implement the new management plan; permission to begin was granted soon afterwards. The staff began devising a public relations strategy to introduce the situation and proposed solution to the general public, but before the public relations plan could be implemented, the story hit the media via an internal leak within the City of Fort Worth. A Section B front-page story in the *Fort Worth Star-Telegram* led to featured news items on the local FOX, CBS, and ABC stations (Sidebar 12.1). This media exposure led to extensive public response—approximately 100 comments were received, with less than 6 percent negative. Most of those who made comments, in addition to supporting the control program, offered their assistance, which was politely declined.

With the go-ahead given, the staff began final preparations for removing hogs from the property. Members of the FWNC&R feral hog control team first obtained appropriate caliber firearms. Team members then began practicing and familiarizing themselves with the weapons and scouting for prime locations of hog activity to place traps. Trapping began on January 9, 2003, and the program started very well, with seven hogs taken from two traps after the first trap night. Trapping efforts slacked off considerably during the late spring, summer, and fall when the animals had access to plenty of wild foods, but resumed in intensity in December. By the close of 2003, thirty-three animals had been removed from the park, including at least ten boars and ten sows of reproductive size.

Trapping continued into 2004 and, as of August of that year, an additional forty-one animals had been removed. Anecdotally, the staff has noticed a significant decrease in hog activity since the program started. A series of exclosures have been erected in areas of high hog use for qualitative study of effects on the floral community and to document activity levels. Due to increasing urbanization around the park, which should eventually close most immigration routes, the staff now believes that total eradication is achievable and continues to work toward that goal.

SIDEBAR 12.1   Wild Hogs Threaten Texas Nature Center

Wayne Clark was exploring a remote part of the Fort Worth Nature Center when he saw deep holes in the ground—as if someone had been digging for buried treasure. Over the next few months, he learned that the nature retreat had become infested by a growing population of feral hogs that could put visitors and the environment at risk.

To eliminate the animals' threat, Clark, who supervises the center, and staff members will start a program as soon as next month to track and kill as many feral hogs as they can find.

"People have died from feral hog attacks," Clark said. "It's not extremely common, but the potential is there. You've got a wild animal that weighs up to 400 pounds, has sharp teeth, and isn't afraid to attack." At least one dog has been attacked and nearly killed by wild swine at the center.

At the same time, the feral hogs compete with other animals for available food and have damaged ecosystems by their rooting and wallowing, he said. "We think this program can work," Clark said. "We don't think we can eradicate the problem, but we can control it. We feel we can reduce the hogs' numbers, which reduces damage and risk to visitors."

This has drawn criticism from groups such as People for the Ethical Treatment of Animals, which say the feral hogs shouldn't be killed. "We regret that the Nature Center wants to remove the feral pigs from the area, but to do so using lethal methods is unconscionable when other methods exist," said Stephanie Boyles, a wildlife biologist with the north Virginia-based PETA. "We don't believe the feral pigs should be slaughtered in the name of conservation. They should use methods such as contraception and fencing to keep these animals away from areas of concern."

Others defend the decision to remove the wild swine from the Nature Center by any means necessary. "People must remember and know what the purpose of the Nature Center is," said Bill Meadows, a former city councilman and chairman of the Nature Center Endowment Fund. "The whole point is to preserve and nurture what is natural in this area—the indigenous plants and animals. When a species comes in and serves to significantly alter the mission, there's no question that you have to take a dramatic action. If that means you have to eliminate some of those animals, then that's what you have to do."

There are as many as 300 to 400 feral hogs at the nature center, a population that is rapidly growing, said Richard Zavala, director of the city's parks and community services department.

Over the past few years, city and center officials have researched ways to thin the feral hog population and have consulted with national and state experts at agencies ranging from the U.S. Department of Agriculture to the Texas Parks and Wildlife Department.

Feral hogs can't be transported from one area to another because of the threat of spreading livestock disease such as swine brucellosis and pseudorabies, according to the Texas Animal Health Commission.

"We will trap and euthanize them by shooting them on site," Zavala said. "It will be done in a controlled process, with the traps and euthanasia done by nature center staff in nonpublic areas during nonpublic times. The remains will be left out at the Nature Center to decompose. That's the way we do things at the Nature Center. It's a circle of life."

Feral hogs, which are a mix of domestic hogs and the more aggressive Russian boars, typically are dark and furry and have longer tusks than domestic hogs. They are unprotected, nongame animals. But they are prolific breeders, destroy the landscape, and can be a danger to people and other animals, said John Davis, one of two Texas Parks and Wildlife urban biologists who serve the Fort Worth/Dallas region.

"These hogs can begin breeding by the time they are 6 months old and in a 'good' year can bear 20 to 24 piglets. As of 1991, there were an estimated 1 million feral hogs —who have a lifespan of about 8 years—in Texas," Davis said. Trapping and shooting the hogs is a common way to control their population and remove them from areas they are destroying. Leaving their carcasses at the center is an ecologically sensitive way to rid the area of the animals, he said. "This has to be done," he said. "There are a lot of people who aren't going to be comfortable removing the animals, euthanizing a species found on the Nature Center. If it doesn't happen, there will be so much more damage done to the Nature Center and so many other animals will suffer. There's no other answer."

The easiest time to catch the hogs is during the cooler months, when they are more mobile, searching for food such as acorns. That's when Clark and the other nature center employees plan to start laying traps. The traps will be mostly steel boxes or metal fences with doors that allow hogs to enter, to find the bait left inside. After the hogs enter, they will not be able to escape, but others would be able to get inside, Clark said. He hopes to give the City Council a yearly update on the progress of the feral hog management program. "We feel like we're doing the best we can on this," he said.

Anna M. Tinsley, *Star-Telegram*, September, 2002

CHAPTER ACTIVITIES

1. Research your state's endangered species lists (both plants and animals). Identify any that have distributions within or close to an urban area. Arrange an interview with someone from your state's wildlife agency about the management of this species. Ask questions concerning the challenges of managing endangered species in or close to urban areas. Ask them how these challenges are different from management in rural landscapes or on conservation lands (state parks, nature preserves, etc.).

2. Contact your local animal shelter about the numbers of cats that are euthanized each year in your city. Conduct a survey to determine the density of cats in your city. Discuss the various ways you could conduct this survey.

3. Locate aerial photographs or conduct ground habitat surveys to evaluate if your city could support a population of large predators (e.g., coyotes, cougars, bears). Has there been any anecdotal or scientific evidence suggesting the presence of large predators in your city?

LITERATURE CITED

Ablett, P. 1981. Public reaction to control. In *Proceedings of UFAW Symposium*, 60–62. Royal Holloway College, University of London, September 23–24, 1980, London.

Adamec, R.E. 1976. The interaction of hunger and preying in the domestic cat (*Felis catus*): An adaptive hierarchy? *Behavioral Biology* 18(2):263–272.

Andrzejewski, R. and W. Jezierski. 1979. Management of a wild boar population and its effects on commercial land. *Acta Theriologica* 23(19):309–339.

Antos, M.J., J.A. Fitzsimons, G.C. Palmer, and J.G. White. 2006. Introduced birds in urban remnant vegetation: Does remnant size really matter? *Australian Ecology* 31(2):254–261.

Ash, S.J. 2001. Ecological and Sociological Considerations of Using the TTVAR (Trap, Test, Vaccinate, Alter, Return) Method to Control Free-Ranging Domestic Cat (*Felis catus*) Populations. Dissertation, Texas A&M University, College Station.

Baber, D.W. and B.E. Coblentz. 1986. Density, home range, habitat use and reproduction of feral pigs on Santa Catalina Island. *Journal of Mammalogy* 67:512–525.

Baber, D.W. and B.E. Coblentz. 1987. Diet, nutrition and conception in feral pigs on Santa Catalina Island. *Journal of Wildlife Management* 51:306–317.

Baroch, J. and M. Hafner. 2002. Biology and natural history of the nutria, with special reference to nutria in Louisiana. In Genesis Laboratories Inc., Nutria (*Myocaster coypus*) in Louisiana, Wellington, CO.

Baron, J. 1982. Effects of feral hogs on the vegetation of Horn Island, Mississippi. *American Midland Naturalist* 107:202–205.

Beach, R. 1993. Depredation problems involving feral hogs. In *Feral Swine: A Compendium for Resource Managers*, ed. C.W. Hanselka and J.F. Cadenhead, 67–75. Kerrville, TX: Texas Agricultural Extension Service.

Beaver, B.V. (Chair). 2001. 2000 Report of the AVMA panel on Euthanasia. *Journal of the American Veterinary Medical Association* 218:669–696.

Beyer, W.N., E.E. Connor, and S. Gerould. 1994. Estimates of soil ingestion by wildlife. *Journal of Wildlife Management* 58:375–382.

Brown, L.N. 1985. Elimination of a small feral swine population in an urbanizing section of central Florida. *Florida Scientist* 48:120–123.

Campbell, L. 1996. *Endangered and Threatened Animals of Texas: Their Life History and Management*. Austin, TX: University of Texas Press.

Castillo, D. and A.L. Clarke. 2003. Trap/neuter/release methods ineffective in controlling domestic cat "colonies" on public lands. *Natural Areas Journal* 23(3) 247–253.

Choquenot, D., B. Kay, and B. Lukins. 1990. An evaluation of warfarin for the control of feral pigs. *Journal of Wildlife Management* 54:353–359.

Churcher, P.B. and J.H. Lawton. 1987. Predation by domestic cats in an English village. *Journal of Zoology* 212:439–455.

Clout, M. and C. Gillies. 2003. The prey of domestic cats (*Felis catus*) in two suburbs of Auckland City, New Zealand. *Journal of Zoology* 259(3):309–315.

Coleman, J.S. and S.A. Temple. 1996. On the prowl. *Wisconsin Natural Resources* 20(6):4–8.

Cypher, B.L. and N. Frost. 1999. Condition of San Joaquin kit foxes in urban and exurban habitats. *Journal of Wildlife Management* 63(3)930–938.

Cypher, B.L. and K.A. Spencer. 1998. Competitive interactions between coyotes and San Joaquin kit foxes. *Journal of Mammalogy* 79:204–214.

Cypher, B.L. and G.D. Warrick. 1993. Use of human-derived food items by urban kit foxes. *Transactions of the Western Section of the Wildlife Society* 29:34–37.

Dards, J.L. 1978. Home ranges of feral cats in Portsmouth Dockyard. *Carnivore Genetic Newsletter* 3:242–255.

Dickman, C.R. 1996. *Overview of the Impacts of Feral Cats on Australian Native Fauna*. Sydney, Australia: Institute of Wildlife Research and Biological Science, University of Sydney.

Dickson, J.D. III. 1955. An ecological study of the Key deer. Technical Bulletin 3. Tallahassee, FL: Florida Game and Fresh Water Fish Commission,.

Dixon, J.R. 2000. *Amphibians and Reptiles of Texas*. College Station:Texas A&M Press.

Dunlap, B.G. and M.L. Thies. 2002. Giardia in beaver (*Castor canadensis*) and nutria (*Myocaster coypus*) from East Texas. *Journal of Parasitology* 88(6):1254–1258.

Eberhard, T. 1954. Food habits of Pennsylvania house cats. *Journal of Wildlife Management* 18(2):284–286.

Ehrlich, P.R., D.S. Dobkin, and D. Wheye. 1988. *The Birder's Handbook: A Field Guide to the Natural History of North American Birds*. New York: Simon and Schuster.

Eno, S. (1996). *House sparrows*. Bluebirds across Nebraska. http://www.bbne.org.

Errington, P.L. 1936. Notes on food habits of the southern Wisconsin house cats. *Journal of Mammalogy* 17(1)64–65.

Evans, J. 1970. *About Nutria and Their Control*. Resource Publication Number 86. Washington, DC: U.S. Department of the Interior, Sport Fisheries and Wildlife.

Fitzgerald, B.M. 1988. Diet of domestic cats and their impact on prey populations. In *The Domestic Cat: The Biology of Its Behavior*, ed. D.C. Turner and P. Bateson, 123–150. New York: Cambridge University Press.

Folk, M.L. 1991. Habitat of the Key Deer. Dissertation, Southern Illinois University, Carbondale.

Folk, M.L. and W.D. Klimstra. 1991. Urbanization and domestication of the Key deer (*Odocoileus virginianus clavium*). *Florida Field Naturalist* 19:19.

Frank, P.A., B.W. Stieglitz, J. Slack, and R.R. Lopez. 2003. The Key deer: Back from the brink. *Endangered Species Bulletin* 28:20–21.

Gallagher, D. 1991. Impact of the built environment on the natural environment. In *Monroe County Environmental Story,* ed. J. Gato, 226–229. Big Pine Key, FL: Monroe County Environmental Education Task Force.

Genovesi, P., M. Besa, and S. Toso. 1995. Ecology of a feral cat (*Felis catus*) population in an agricultural area of northern Italy. *Wildlife Biology* 1(4):233–237.

George, W.G. 1974. Domestic cats as predators and factor in winter shortages of raptor prey. *Wilson Bulletin* 86:384–396.

George, W.G. 1978. Domestic cats as density independent hunters and "surplus killers." *Carnivore Genetic Newsletter* 3:282–287.

Gipson P.S., B. Hlavachick, and T. Berger. 1998. Range expansion by wild hogs across the central United States. *Wildlife Society Bulletin* 26(2):279–286.

Gottschalk, J.S. 1970. United States list of endangered native fish and wildlife. Federal Register 35, 16047-16048.

Grodzinski, W. 1975. The role of large herbivore mammals in functioning of forest ecosystems: A general model. *Polish Ecological Studies* 1:10–15.

Guthery, F.S. 1997. A philosophy of habitat management for northern bobwhites. *Journal of Wildlife Management* 61:291–301.

Haag-Wackernagel, D. and H. Moch. 2004. Health hazards posed by feral pigeons. *Journal of Infection* 48:307–313.

Hanson, R.P. and L. Karstad. 1959. Feral swine in the Southeastern United States. *Journal of Wildlife Management* 23:64–74.

Hardin, J.W. 1974. Behavior, SocioBiology, and Reproductive Life History of the Florida Key Deer, *Odocoileus virginianus clavium.* Dissertation, Southern Illinois University, Carbondale.

Hardin, J.W., W.D. Klimstra, and N.J. Silvy. 1984. Florida Keys. In *WhiteTailed Deer: Ecology and Management,* ed. L.K. Halls, 381–390. Harrisburg, PA: Stackpole Books.

Hatfield, J.S., A.H. Price, D.D. Diamond, and C.D. True. 2004. Houston toad (*Bufo houstonensis*) in Bastrop County, Texas: Need for protecting multiple subpopulations. In *Species Conservation and Management: Case Studies,* ed. H.R. Akcakaya, M.A. Burgman, O. Kindvall, C.C. Wood, P. Sjogren-Gulve, J.S. Hatfield, and M.A. McCarthy. Oxford, UK: Oxford University Press.

Hawkins, C.C. 1998. Impact of a Subsidized Exotic Predator on Native Biota: Effect of House Cats (*Felis catus*) on California Birds and Rodents. Dissertation, Texas A&M University, College Station.

Hellgren, E.C. 1993. Biology of feral hogs (*Sus scrofa*) in Texas. In *Feral Swine: A Compendium for Resource Managers,* ed. C.W. Hanselka and J.F. Cadenhead, 50–58. Kerrville, TX: Texas Agricultural Extension Service.

Hellgren, E. 1999. Reproduction of feral swine. In *Proceedings of the Feral Swine Symposium*, 67–68. Texas Animal Health Commission, Fort Worth, June 2–3.

Henry, V.G. 1968. Length of estrous cycle and gestation in European wild hogs. *Journal of Wildlife Management* 32:406–408.

Henry, V.G. 1969. Predation on dummy nests of ground-nesting birds in the southern Appalachians. *Journal of Wildlife Management* 33:169–172.

Henry, V.G. and R.H. Conley. 1972. Fall foods of the European wild hogs in the southern Appalachians. *Journal of Wildlife Management* 36:854–860.

Hillis, D.M., A.M. Hillis, and R.F. Martin. 1984. Reproductive ecology and hybridization of the endangered Houston toad (*Bufo houstonensis*). *Journal of Herpetology* 18(1):56–72.

Hole, D.G., M.J. Whittingham, R.B. Bradbury, G.Q. Anderson, P.L. Lee, J.D. Wilson, and J.R. Krebs. 2002. Widespread local house-sparrow extinctions. *Nature* 418(6901):931–932.

Howe, T.D., F.J. Singer, and B.B. Ackerman. 1981. Forage relationships of European wild boar invading northern hardwood forests. *Journal of Wildlife Management* 45:748–754.

Howerth, E.W., A.J. Reeves, M.R. McElveen, and F.W. Austin. 1994. Survey for selected disease in nutria (*Myocaster coypus*) from Louisiana. *Journal of Wildlife Diseases* 30(3):450–453.

Hubbs, E.L. 1951. Food habits of feral house cats in the Sacramento Valley. *California Fish and Game* 37(2):177–189.

Ilse, L.M. and E.C. Hellgren. 1995. Resource partitioning by sympatric populations of collared peccaries and feral hogs in southern Texas. *Journal of Mammalogy* 76:784–799.

Jacobson, N.L. 1989. Breeding dynamics of the Houston toad. *Southwestern Naturalist* 34(3):374–380.

Janiga, M., B. Mankovska, M. Bobalova, and G. Durcova. 1990. Significance of concentrations of lead, cadmium, and iron in the plumage of the feral pigeon. *Archives of Environmental Contamination and Toxicology* 19(6):892–897.

Johnson, K. 1995. National pet alliance report on trap/alter/release programs. *Cat Fanciers Almanac,* July, 92–94.

Johnson, L.A. and A.L. Foote. 1997. Vertebrate herbivory in managed coastal wetlands: A manipulative experiment. *Aquatic Botany* 59(1–2):17–32.

Johnson, S.A. and G. Donaldson-Fortier. 2009. Florida's Introduced Birds: The Eurasian Collared-Dove (*Streptopelia decaocto*). Publication - WEC256, Department of Wildlife Ecology and Conservation, University of Florida/IFAS.

Johnston, R.F. and M. Janiga. 1995. *Feral Pigeons.* New York: Oxford University Press.

Jones, P. 1959. *The European Wild Boar in North Carolina.* Raleigh: North Carolina Wildlife Resources Commission.

Kamb, L. 2009. As U.S. Scouts sell off land, S.A. council buying it. *San Antonio Express-News*, January 31, 2009.

Kerby, G. and D.W. Macdonald. 1988. Cat society and the consequences of colony size. In *The Domestic Cat: The Biology of Its Behavior,* ed. D.C. Turner, and P. Bateson, 67–81. New York: Cambridge University Press.

Kessler, C.C. 2002. Eradication of feral goats and pigs and consequences for other biota on Sarigan Island, Commonwealth of the Northern Mariana Islands. In *Turning the Tide: The Eradication of Invasive Species,* ed. C.R. Veitch and M.N. Clout, 132–140. IUCN SSC Invasive Species Specialist Group. Gland, Switzerland and Cambridge, UK: International Union for Conservation of Nature.

Killian, G., L. Miller, J. Rhyan, T. Dees, D. Perry, and H. Doten. 2003. Evaluation of GNRH contraceptive vaccine in captive feral swine in Florida. In *Proceedings of the 10th Wildlife Damage Management Conference,* ed. K.A. Fagerstone and G.W. Witmer, 128–133, Bethesda, MD: The Wildlife Society.

Kinler, N.W., G. Linscombe, and P.R. Rarnsey. 1987. Nutria. In *Wild Furbearer Management and Conservation in North America*, ed. M. Novak, J.A. Baker, M.E. Obbard, and B. Malloch. Concord, Ontario: Ontario Fur Managers Federation.

Klimstra, W.D., J.W. Hardin, N.J. Silvy, B.N. Jacobson, and V.A. Terpening. 1974. *Key Deer Investigations Final Report: Dec 1967–Jun 1973.* Big Pine Key, FL: U.S. Fish and Wildlife Service.

Lepczyk, C.A., A.G. Mertig, and J. Liu. 2004. Landowners and cat predation across rural-to-urban landscapes. *Biological Conservation* 115:191–201.

Liberg, O. and M. Sandell. 1988. Spatial organization and reproductive tactics in the domestic cat and other felids. In *The Domestic Cat: The Biology of Its Behavior*, ed. D.C. Turner and P. Bateson, 83–98. New York: Cambridge University Press.

Littauer, G.A. 1993. Control techniques for feral hogs. In *Feral Swine: A Compendium for Resource Managers,* ed. C.W. Hanselka, and J.F. Cadenhead, 139–148. Kerrville, TX: Texas Agricultural Extension Service.

Long, E.L., R.A. Sweitzer, and M. Ben-David. 2001. Historic changes of mountain lion diets in relation to introduced feral pigs. *Abstracts of the Wildlife Society 8th Annual Conference.* Reno/Tahoe, NV.

Lopez, R.R. 2001. Population ecology of the Florida Key Deer. Dissertation, Texas A&M University, College Station.

Lopez, R.R., T.R. Peterson, and N.J. Silvy. 2001. Emphasizing numbers in the recovery of an endangered species: A lesson in underestimating public knowledge. In *Proceedings of the Second International Wildlife Management Congress,* ed. R. Field, R.J. Warren, H. Okarma, and P.R. Sievert, 353–356, Bethesda, MD: The Wildlife Society.

Lopez, R.R., N.J. Silvy, B.L. Pierce, P.A. Frank, M.T. Wilson, and K.M. Burke. 2003a. Population density of the endangered Florida Key deer. *Journal of Wildlife Management* 68(3):570–575.

Lopez, R.R., N.J. Silvy, R.N. Wilkins, P.A. Frank, M.J. Peterson, and N.M. Peterson. 2004. Habitat use patterns of Florida Key deer: Implications of urban development. *Journal of Wildlife Management* 68:900–909.

Lopez, R.R., M.E.P. Viera, N.J. Silvy, P.A. Frank, S.W. Whisenant, and D.A. Jones. (2003b). Survival, mortality, and life expectancy of Florida Key deer. *Journal of Wildlife Management* 67:34–45.

Loranger, S., G. Demers, G. Kennedy, E. Forget, and J. Zayed. 1994. The pigeon (*Columba livia*) as a monitor for manganese contamination from motor vehicles. *Archives of Environmental Contamination and Toxicology* 27(3):311–317.

Lucas, E.G. 1977. Feral hogs: Problems and control on national forest lands. In *Research and Management of Wild Hog Populations,* ed. G.W. Wood, 17–21. Georgetown, SC: Belle Baruch Forest Science Institute of Clemson University.

Maffei, M.D., W.D. Klimstra, and T.J. Wilmers. 1988. Cranial and mandibular characteristics of the Key deer (*Odocoileus virginianus clavium*). *Journal of Mammalogy* 69:403–407.

Mahlow, J.C. and M.R. Slater. 1996. Current issues in the control of stray and feral cats. *Journal of American Veterinary Medical Association* 209(12):2016–2020.

Maruska, E.J. 1986. Amphibians: Review of zoo breeding programmes. *International Zoo Yearbook* 24(1):56–65.

Mayer, J.J. and I.L. Brisbin, Jr. 1991. *Wild Pigs of the United States: Their History, Morphology, and Current Status.* Athens, GA: University of Georgia Press.

McLean, R. G., J. Mullenix, J. Kerschner, and J. Hamm. 1983. The house sparrow (*Passer domesticus*) as a sentinel for St. Louis encephalitis virus. *American Journal of Tropical Medicine and Hygiene* 35(5):1120–1129.

McMurray, F.B. and C.C. Sperry. 1941. Food of feral house cats in Oklahoma, a progress report. *Journal of Mammalogy* 22(2)185–190.

McShea, W.J., H.B. Underwood, and J.H. Rappole. 1997. Deer management and the concept of deer overabundance. In *The Science of Overabundance: Deer Ecology and Population Management,* ed. W.J. McShea, H.B. Underwood, and J.H. Rappole, 1–11. Washington, DC: Smithsonian Institution Press.

Meffe, G.K. and C.R. Carroll. 1997. *Principles of Conservation Biology.* Sunderland, MA: Sinauer Associates, Inc.

Miller, J.E. 1993. A national perspective on feral swine. In *Feral Swine: A Compendium for Resource Managers,* ed. C.W. Hanselka and J.F. Cadenhead, 9–16. Kerrville, TX: Texas Agricultural Extension Service.

Myers, P. 1993. The Wild Boar Conservation Association. In *Feral Swine: A Compendium for Resource Managers,* ed. C.W. Hanselka and J.F. Cadenhead, 130–131. Kerrville, TX: Texas Agricultural Extension Service.

Nam, D.-H. and D.-.P Lee. 2006. Monitoring for Pb and Cd pollution using feral pigeons in rural, urban and industrial environments of Korea. *Science of the Total Environment,* 357(1–3):288–295.

National Geographic. 1999. *Field Guide to the Birds of North America.* Washington, DC: National Geographic.

Natoli, E. 1985. Spacing patterns in a colony of urban stray cats (*Felis catus,* L.) in the historic centre of Rome. *Applied Animal Ethology* 14:289–304.

Nelson, M. 1999. Habitat conservation planning. *Endangered Species Bulletin* 24(6):12–13.

Nettles, V.F., C.F. Quist, R.R. Lopez, T.J. Wilmers, P. Frank, W. Roberts, S. Chitwood, and W.R. Davidson. 2002. Morbidity and mortality factors in Key deer, *Odocoileus virginianus clavium. Journal of Wildlife Diseases* 38:685–692.

Nowak, R.M. 1999. *Walker's Mammals of the World,* 6th edition. Baltimore, MD: Johns Hopkins University Press.

Nunley, G.L. 1999. The Cooperative Texas Wildlife Damage Management Program and feral swine damage management. In *Proceedings of the Feral Swine Symposium,* 27–30. Texas Animal Health Commission, Fort Worth, June 2–3.

Oppenheimer, E.C. 1980. *Felis catus*: Population densities in an urban area. *Carnivore Genetic Newsletter* 4:72–80.

Parmalee, P.W. 1953. Food habits of the feral house cat in east-central Texas. *Journal of Wildlife Management* 17(3):375–376.

Passanisi, W.C. and D.W. Macdonald. 1990. The fate of controlled feral cat colonies. UFAW Animal Welfare Research Report No. 4. Hertfordshire, UK: Universities Federation for Animal Welfare.

Patronek, G.J. 1998. Free-roaming and feral cats: Their impacts on wildlife and human beings. *Journal of American Veterinary Medical Association* 212(2):218–226.

Peterson, M.N., S.A. Allison, M.J. Peterson, T.R. Peterson, and R.R. Lopez. 2004. A tale of two species: Habitat conservation plans as bounded conflict. *Journal of Wildlife Management* 68(4):743–761.

Peterson, M.N., T.R. Peterson, M.J. Peterson, R.R. Lopez, and N.J. Silvy. 2002. Cultural conflict and the endangered Florida Key deer. *Journal of Wildlife Management* 66:947–968.

Pimentel, D., S. McNair, J. Janecka, J. Wightman, C. Simmonds, C. O'Connell, E. Wong, L. Russel, J. Zern, T. Aquino, and T. Tsomondo. 2002. Environmental and economic costs associated with non-indigenous species in the United States. In *Biological Invasions: Economic and Environmental Costs of Alien Plant, Animal and Microbe Species*, ed. D. Pimentel, 285–306, Boca Raton, FL: CRC Press.

Pritchard, P.C.H. 1995. Conservation of reptiles and amphibians. In *Conservation of Endangered Species in Captivity: An Interdisciplinary Approach,* ed. E.F. Gibbons, Jr., B.S. Durrant, and J.Demarest. 147–168, Albany, NY: SUNY Press.

Proctor May, R. 2005. Houston toad clings to survival in Bastrop. *The Austin Chronicle*, December 30, 1.

Quammen, D. 1998. Superdove on 46th Street. In *Wild Thoughts from Wild Places.* New York: Simon and Schuster.

Quinn, H.R. and G. Mengden. 1984. Reproduction and growth of *Bufo houstonensis* (Bufonidae). *Southwestern Naturalist* 29:189–195.

Remfry, J. 1996. Feral cats in the United Kingdom. *Journal of American Veterinary Medical Association* 208(4):520–523.

Rochlitz, I. 2007. *The Welfare of Cats.* New York: Springer.

Roemer, G.W., C.J. Donlan, and F. Courchamp. 2002. Golden eagles, feral pigs, and insular carnivores: How exotic species turn native predators into prey. *Proceedings of the National Academy of Sciences U.S.A.* 99:791–796.

Rollins, D. 1993. Statewide attitude survey on feral hogs in Texas. In *Feral Swine: A Compendium for Resource Managers,* ed. C.W. Hanselka and J.F. Cadenhead, 1–8. Kerrville, TX: Texas Agricultural Extension Service.

Rollins, D. and J.P. Carroll. 2001. Impacts of predation on quail. In *The Role of Predator Control as a Tool in Game Management.* Publication SP-113. College Station, TX: Texas Agricultural Extension Service.

Rosatte, R.C., M.J. Power, and C.D. MacInnes. 1991. Ecology of urban skunks, raccoons and foxes in Metropolitan Toronto. In *Proceedings Symposium: Wildlife Conservation in Metropolitan Environments,* ed. L.W. Adams, and D.L. Leedy, 31–38. Columbia, MD: National Institute for Urban Wildlife Publishing.

Rubin, E.S, W.M. Boyce, C.J. Stermer, and S.G. Torres. 2002. Bighorn sheep habitat use and selection near an urban environment. *Biological Conservation* 104(2):251–263.

Schuyler, P.T., D.K. Garcelon, and S. Escover. 2002. Eradication of feral pigs (*Sus scrofa*) on Santa Catalina Island, California, USA. In *Turning the Tide: The Eradication of Invasive Species,* ed. C.R. Veitch, and M.N. Clout, 274–286. IUCN SSC Invasive Species Specialist Group. Gland, Switzerland and Cambridge, UK: International Union for Conservation of Nature.

Sidra, M.S. and C.H. Walker. 1980. The metabolism of p,p-DDT by the feral pigeon (*Columba livia*). *Pesticide Biochemistry and Physiology* 14(1):62–71.

Singer, F.J., W.T. Swank, and E.E.C. Clebsch. 1984. Effects of wild pig rooting in a deciduous forest. *Journal of Wildlife Management* 48:464–473.

Soulé, M.E., D.T. Bolger, A.C. Alberts, J. Wright, M. Sorice, and S. Hill. 1988. Reconstructed dynamics of rapid extinctions of chapparal-requiring birds in urban habitat islands. *Conservation Biology* 2(1):75–92.

Springer, M.D. 1975. Food Habits of Wild Hogs on the Texas Gulf Coast. Thesis, Texas A&M University, College Station.

Springer, M.D. 1977. Ecological and economic aspects of wild hogs in Texas. In *Research and Management of Wild Hog Populations,* ed. G.W. Wood, 37–46. Georgetown, SC: Belle Baruch Forest Science Institute of Clemson University.

Stevens, R.L. 1996. *The Feral Hog in Oklahoma.* Ardmore, OK: Samuel Roberts Noble Foundation.

Steyer, G.D. and D.W. Llewellyn. 2000. Coastal Wetlands Planning, Protection and Restoration Act: A programmatic application of adaptive management. *Ecological Engineering* 15(3):385–395.

Swannack, T.M. 2007. Modeling Aspects of the Ecological and Evolutionary Dynamics of the Endangered Houston Toad. PhD Dissertation, Texas A&M University, College Station.

Sweeney, J.M. and J.R. Sweeney. 1982. Feral hog. In *Wild Mammals of North America: Biology, Management, and Economics,* ed. J.A. Chapman and G.A. Feldhammer, 1099–1113. Baltimore, MD: Johns Hopkins University Press.

Sweeney, J.M., J.R. Sweeney, and E.E. Provost. 1979. Reproductive biology of a feral hog population. *Journal of Wildlife Management* 43:555–559.

Sweitzer, R.A. and D.H. Van Vuren. 2002. Rooting and foraging effects of wild pigs on tree regeneration and acorn survival in California's oak woodland ecosystems. USDA Forest Service General Technical Report PSW-GTR-184. Washington, DC: U.S. Department of Agriculture.

Tate, J., ed. 1984. Techniques for controlling wild hogs in the Great Smoky Mountains National Park. In Proceedings of a workshop, November 29–30, Research/Resources Mgmt, Rpt, SRE-72. United States Department of the Interior, National Park Service, Southeast Regional Office, Atlanta, GA.

Taylor, R.B., E.C. Hellgren, T.M. Gabor, and L.M. Ilse. 1998. Reproduction of feral pigs in southern Texas. *Journal of Mammalogy* 79:1325–1331.

Thomas, M. and N. Young. 1999. Preliminary trial of a water-resistant bait for feral pig control. *Science for Conservation* 127:49–55.

Tinsley, A.M. 2002. City will trap, kill wild hogs. *Fort Worth Star-Telegram,* September 16, 2002, B1.

Toner, G.C. 1956. House cat predation on small animals. *Journal of Mammalogy* 37(1):119.

Towne, C.W. and E.N. Wentworth. 1950. *Pigs from Cave to Cornbelt.* Norman, OK: University of Oklahoma Press.

USFWS. 1998. *Recovery Plan for Upland Species of the San Joaquin Valley, California, Region 1.* Portland, OR: U.S. Fish and Wildlife Service.

USFWS. 2000. Recovery Plan for Bighorn Sheep in the Peninsular Ranges, California. Portland, OR: U.S. Fish and Wildlife Service.

Wahlenberg, W.G. 1946. *Longleaf Pine.* Washington, DC: Charles Lathrop Pack Foundation.

Wakely, P.C. 1954. *Planting the Southern Pines.* Agricultural Monograph 18. Washington, DC: U.S. Department of Agriculture Forest Service.

Whitehouse, D.B. 1999. Impacts of feral hogs on corporate timberlands in the south eastern U.S. In *Proceedings of the Feral Swine Symposium,* 108–110. Texas Animal Health Commission, Fort Worth, June 2–3.

Wildlife Society. 2001. Feral and free-ranging cats: position statement. *The Wildlifer* 306:57.

Wood, G.W. and R.H. Barrett. 1979. Status of wild pigs in the United States. *Wildlife Society Bulletin* 7:237–246.

Wood, G.W. and R.E. Brenneman. 1977. Research and management of feral hogs on Hobcaw Barony. In *Research and Management of Wild Hog Populations,* ed. G.W. Wood, 23–35. Georgetown, SC: Belle Baruch Forest Science Institute of Clemson University.

Wood, G.W. and D.N. Roark. 1980. Food habits of feral hogs in coastal South Carolina. *Journal of Wildlife Management* 44:506–511.

Woods, M., R.A. Macdonald, and S. Harris. 2003. Predation of wildlife by domestic cats, *Felis catus,* in Great Britain. *Mammal Review* 33(2):174–188.

Yumane, A., J. Emoto, and N. Ota. 1997. Factors affecting feeding order and social tolerance to kittens in the group-living feral cat (*Felis catus*). *Applied Animal Behavior Science* 52:119–127.

Zaunbrecher, K.I. and R.E. Smith. 1993. Neutering of feral cats as an alternative to eradication programs. *Journal of American Veterinary Medical Association* 203(3):449–452.

CHAPTER 13

Zoonoses and Management Considerations

This form of interspecies leap [zoonosis] is common, not rare; about 60% of all human infectious diseases currently known are shared between animals and humans.

—**David Quammen**

KEY CONCEPTS

1. Zoonoses—infectious diseases that can be transmitted between human and nonhuman vertebrate species.
2. Many zoonoses have the potential to be "weaponized," and are therefore a concern for governments around the globe.
3. Because of the increased potential for interface between humans and wildlife, urban habitats present a greater potential for the spread of zoonoses than rural habitats.
4. There are various methods for classifying zoonoses, including their infection agents and routes of infection.
5. There are five general types of infection agents: parasites, fungi, bacteria, viruses, and prions.

WHAT ARE ZOONOSES?

Zoonoses—infectious diseases that can be transmitted between human and nonhuman vertebrate species (World Health Organization website)—can be viewed as a subset of wildlife conflict, but one that deserves special attention. Zoonotic pathogens are of particular concern, or should be, for those who work in the wildlife management profession because they are more likely to come into close contact with infected animals. Other populations of concern include those whose age or health status may affect their immune system. Initial diagnosis of many zoonotic diseases can be challenging, because they tend to present with vague "flu-like symptoms." A doctor may not automatically consider zoonoses, so it may be up to the patient to suggest it.

Over 200 zoonoses have been described, and there are more than 140 known diseases or parasitic infections in the United States for which nonhuman animals may serve as a vector or reservoir (Bisseru 1967; Beran 1994). The impacts of zoonoses are not restricted to human morbidity and mortality. These pathogens can affect the well-being of rural and urban communities as well as national economies (Stephen et al. 2004). Globalization, and the ease with which humans are able to move from one country or continent to another, means we can no longer count on physical distance from outbreaks of zoonotic diseases to be a safeguard against infection. Even more disturbing is the potential for genetically modified, "weaponized" strains of zoonotic pathogens created for use as biological weapons against humans and domesticated livestock (Dudley 2004). A deliberate or accidental release of weaponized disease pathogens could have devastating effects on human populations, directly or through disruption of agricultural productivity, and severe indirect effects on wildlife and biodiversity in general. Various events over the past decade have demonstrated the critical need for zoonotic disease surveillance as a first line of defense and response against terrorist attacks involving bioweapon diseases (Dudley 2003).

As we've discussed throughout this book, the opportunities for people and wildlife to interact are much greater in an urban environment than in a rural landscape. The fact that the urban wildlife population densities are often much greater than elsewhere also increases the chance of intraspecies transmission of pathogens, thereby increasing the chance of interspecies transmission to humans.

The U.S. Centers for Disease Control and Prevention (CDC) keeps statistics on reported occurrence of nationally notifiable zoonotic diseases. A notifiable disease is "one for which regular, frequent, and timely information regarding individual cases is considered necessary for the prevention and control of the disease" (CDC website). Surveillance allows public health authorities to monitor the impact of notifiable conditions, measure disease trends, assess the effectiveness of control and prevention measures, identify populations or geographic areas at high risk, allocate resources appropriately, formulate prevention strategies, and develop public health policies. Diseases are added when new pathogens are identified and deleted as prevalence declines. Disease reporting is mandated by legislation or regulation at the state and local levels, but state reporting to the CDC is voluntary. As a result, reporting completeness is highly variable and related to the condition or disease being reported (Doyle, Glynn, and Groseclose 2002).

In 2008, the *Summary of Notifiable Diseases—United States, 2006* was released (U.S. Centers for Disease Control and Prevention 2008). The list includes diseases "for which wildlife species may serve as a vector or reservoir," such as anthrax, brucellosis, several forms of encephalitis, Lyme disease, malaria, plague, psittacosis, rabies, Rocky Mountain spotted fever, and tularemia. A sampling of some of the more common zoonotic diseases can be found in Table 13.1, along with the number of cases in 2006 (the most recent available summary) and 2001.

There are various methods for classifying zoonoses, including their infection agents (i.e., parasites, fungi, bacteria, viruses, and prions) and routes of infection (i.e., ingestion, inhalation, direct contact, or vector). We have organized this chapter using the former. Covering all zoonoses reported in the United States is beyond the scope of this book, but there are many publications devoted entirely to this subject. Our hope is that by discussing each of the five categories of zoonoses, and providing an example or two, we will motivate our readers to explore the subject in greater depth on their own.

PARASITIC DISEASES

Using a strict definition of the term parasite—an organism that grows, feeds, and is sheltered on or in a different organism while contributing nothing to the survival of its host (Merriam-Webster online dictionary)—all zoonotic pathogens could be classified as parasitic. Fungi, bacteria, viruses, and pria are discussed later in this chapter; here, we'll focus on protozoa, helminthes (trematodes, nematodes, and cestodes), and pentastomids (Krauss et al., eds. 2003). Some authors include arthropods but these organisms are more often than not the vectors for bacterial and viral zoonoses; we will discuss them in the context of the pathogen they transmit. The increasing popularity of consuming raw, undercooked, smoked, pickled or dried meat, fish, crustaceans, and molluscs has increased the prevalence of zoonoses caused by protozoa, trematodes, cestodes, and nematodes (Macpherson 2005).

Table 13.1 Reported Cases of Ten Notifiable Diseases for 2001 and 2006 in the United States for Which Wildlife Species May Serve as a Vector or Reservoir[a]

Disease	2001 Cases	2006 Cases
Brucellosis	136	121
Encephalitis[b,c]	216	4,357
Hansen disease (leprosy)	79	66
Hantavirus pulmonary syndrome	8	40
Lyme disease	17,029	19,931
Malaria	1,544	1,474
Plague	2	17
Psittacosis	25	21
Rabies (human)	1	3
Rocky Mountain spotted fever	695	2,288
Tularemia	129	95

Source: U.S. Center for Disease Control and Prevention, United States Department of Health and Human Services, Centers for Disease Control, Atlanta, GA, 2006.

[a] Humans, companion animals, and livestock also can serve as the vector or reservoir to the infectious agent. Wildlife are involved in an unknown proportion of these cases.

[b] Includes California serogroup viral.

[c] Eastern equine, Powassan (neuroinvasive), St. Louis, and West Nile (neuroinvasive and nonneuroinvasive).

Protozoa

Kingdom Protozoa includes single-celled, microscopic organisms such as amoebas, sporozoans, and paramecia. Two of the more well-known protozoic zoonoses are *Cryptosporidium parvum* and *Giardia lamblia*. Both have become more prevalent over the past two decades.

C. parvum, a sporozoan parasite, causes a debilitating diarrhea that is commonly referred to as cryptosporidiosis. Other nonspecific symptoms include weakness, fatigue, headache, and anorexia. Infections are readily transmissible between different host species, although some research suggests a degree of host specificity may occur within vertebrates. All stages of *Cryptosporidium* spp. development (asexual and sexual) occur within one host. The entire life cycle may be completed in two days. Infections may be short or last for several months (O'Donoghue 1995). *Cryptosporidium* was first identified in mice, but it has since been detected in a wide variety of wildlife hosts (Appelbee, Thompson, and Olson 2005).

Giardia spp. are flagellated protozoans that cause intestinal infections in vertebrate animals, including humans. It was one of the first protozoans to be described, by van Leeuwenhoek in 1681, and then in greater detail in 1859 by Lambl (for whom the species associated with humans is named). Symptoms of a giardia infection vary from asymptomatic to chronic diarrhea. Untreated, the illness lasts an average of six weeks, although symptoms seldom last less than a week. *Giardia* has a simple life cycle consisting of two phases—one in the host and one outside the host. The infectious cyst is resistant to environmental desiccation, the host's gastric acid, and even chlorination. After it has been ingested it settles in the small intestine. Transmission to another host occurs when cysts are passed in the feces and ingested by another host (Adam 1991). Beavers (*Castor canadensis*) have been, on various occasions, singled out as contributing to *G. lamblia* infection in humans, also referred to as "beaver fever." In 1984, for example, sixty-three residents of northeastern Pennsylvania contracted giardiasis. Beavers living in the watershed were implicated but the evidence to support this was inconclusive (Muller-Schwarze and Sun 2003). Beavers are but one host for giardia; others include everything from ducks (Anseriformes) and herons (Ardeidae) to muskrats (*Ondrata zibethicus*), coyotes (*Canis latrans*), and deer (*Odocoileus* spp.) (Erlandsen et al. 1990, 1991; Dunlap and Thies 2002; Appelbee, Thompson, and Olson 2005).

Helminthes

The helminthes—worm-like parasites—include platyhelminthes (trematodes and cestodes) and nematodes. The life cycle of these organisms includes three stages: egg, larvae (juvenile), and adult. In most cases, adult platyhelminthes and nematodes produce eggs that are passed in excretions or secretions of the host (Castro 1996).

Trematodes (Flukes)

Adult flukes are flatworms that are vaguely leaf-shaped, and they range from a few millimeters to seven to eight centimeters in length. Flukes have prominent oral and ventral suckers that help them to maintain position in the host. Trematodes are hermaphroditic, meaning each individual has both male and female reproductive organs, and they self- and cross-fertilize; blood flukes (schistosomes) are the bisexual exception. Eggs are passed in the feces, urine, or sputum of the host into an aquatic environment, where they hatch and release ciliated larvae. The larvae either penetrate or are eaten by an intermediate snail host. After several larval stages, a saclike sporocyst stage develops within the tissues of the snail. Cercariae develop asexually from the sporocyst and migrate out of the snail to the external environment (usually aquatic). The cercariae either penetrate the final host and transform directly into adults, penetrate a second intermediate host (e.g., crustaceans, fish, frogs) and continue development, or they encyst on vegetation or some other substrate. When the cyst is ingested, digestion releases an immature fluke that migrates to a specific organ site and develops into an adult worm (Castro 1996). Examples of zoonotic trematodes include liver flukes, including ones that cause dicrocoeliasis (seen worldwide in grazing livestock) and fascioliasis (once rare in humans but increasingly reported around the world; Krauss et al. 2003); intestinal flukes that cause anisakiasis (found in fish [Agnatha, Chondrichthyes, Osteichthyes] and marine mammals [Cetacea, Sirenia, Pinnipedia]; Chai, Murrell, and Lymbery 2005); and blood flukes, which cause schistosomiasis (through contact with specific types of snails [Gastropoda]; CDC website).

Cestodes (Tapeworms)

Cestodes, known commonly as tapeworms, have many of the same structural characteristics of flukes, but there are also distinct differences. Adult tapeworms are elongated, segmented flatworms that vary in length from two to three millimeters to ten meters. Like most flukes, they are hermaphroditic. The life cycle is rather complicated, including several larval stages and intermediate hosts before becoming adults in the final host (Castro 1996). Tapeworms inhabit the intestinal lumen of the host. Cestode infection is usually asymptomatic in humans, although some people experience abdominal discomfort, diarrhea, or loss of appetite. The fish tapeworm (*Diphyllobothrium latum*) can cause anemia because it absorbs B12, a nutrient needed by maturing red blood cells (CDC website).

A specific example of a zoonotic cestode can be found on the Michigan Department of Natural Resources website, which warns hunters about cysticercosis in cottontail rabbits (*Sylvilagus floridanus*) and snowshoe hares (*Lepus americanus*). The parasites are conspicuous when hunters are dressing the carcasses, appearing as "spots on the liver" that are misinterpreted as evidence the animal is infected with tularemia (see the section on Bacteria later in this chapter). Actually, the rabbits and hares are an intermediate host of two types of tapeworms, commonly found as adults in domesticated and wild canids (e.g., dogs [*Canis familiaris*], foxes [*Vulpes* spp., *Urocyon cinereoargenteus*], and coyotes). *Taenia pisiformis* are most commonly seen in cottontails, less frequently in snowshoe hares. *Multiceps serialis* is common in snowshoe hares, less so in cottontails.

Nematodes (Roundworms)

Many people learn about roundworms when they get a puppy and the veterinarian explains the need for deworming. Nematodes are the second most diverse group of animals; estimates place the number of different species at over 500,000. They live in every moist and aquatic habitat and can number in the billions per acre of topsoil. Only a small percent are dangerous to humans as parasites. Unlike platyhelminthes, nematodes are—as one might guess from the name—round or cylindrical rather than flattened. Nematodes are usually bisexual. Males are often smaller than females. Copulation between a female and a male nematode is necessary for fertilization, with the exception of the genus *Strongyloides,* in which parthenogenesis (the process in which an unfertilized egg develops into a new individual) occurs. The developmental process in nematodes involves egg, larval, and adult stages. Each of four larval stages is followed by a molt in which the cuticle is shed. The nematode formed at the fifth stage is the adult (Castro 1996).

One zoonotic nematode that is quite dangerous in humans, *Baylisascaris procyonis* or raccoon (*Procyon lotor*) roundworm, has become more of a problem as the number of urban raccoons has increased (see Chapter 1). It is a large intestinal roundworm. The adult worms produce eggs that are shed in the raccoons' feces. Humans and other animals become infected when they accidentally ingest the eggs. *B. procyonis* larvae have been identified in more than ninety species of wild and domesticated animals (Sorvillo et al. 2002). While the impact of this nematode on a healthy raccoon is minimal, in an intermediate or aberrant host, such as humans, the larvae undergo aggressive somatic migration, which can result in blindness, traumatic damage and inflammation in the brain, coma, and death. Sources of infection include any areas or objects contaminated with raccoon feces, including "latrines" at the base of trees and other horizontal structures, such as buildings; these areas become important long-term sources of infection. Infection is rare but increasing, particularly among infants; young children and developmentally disabled persons are also at risk because they are more likely to come into direct contact with raccoon latrines (Sleeman 2005).

To reduce the risk of *B. procyonis* infection, people should be advised to avoid contact with raccoons; this includes backyard feeding and other methods of encouraging raccoons to congregate near homes and parks. The eggs can survive for years and are resistant to common disinfectants. In cases of heavy contamination, removal of raccoon feces and several inches of soil may be necessary. The public should be educated regarding basic personal hygiene, with particular attention given to people who have close contact with raccoons as a result of their vocation or avocation (e.g., hunters, trappers, wildlife rehabilitators, animal control officers), but it is important to do this in a way that does not cause an overreaction. Anyone who suspects they have been infected should consult a health care provider immediately (Sleeman 2005).

MYCOTIC DISEASES

Mycotic or fungal zoonoses common in North America are usually receptive to treatment, but they can pose a serious problem in people who are immune-compromised. One of the best known mycotic zoonoses doesn't sound like a fungal infection at all. Ringworm is a dermatophyte. Some dermatophytes derive from human hosts (e.g., *Microsporum rubrum; M. audouini*) or soil (*M. gypseum*), but dermatophyte species of the genera *Microsporum* and *Trichophyton* are primarily zoonotic from domestic livestock, companion animals, and wild mammal sources. Other mycotic infections that originate from an animal source include *Cryptococcus neoformans* (a yeast), from inhalation of bird droppings (e.g., pigeons), aspergillosis, and histoplasmosis (Goldsmid and Leggat 2005).

Aspergillosis

Found worldwide, *Aspergillus* has been reported in almost all domestic mammals and birds and in many wild species. Aquatic and scavenging wild species have the highest incidence rate. There are many forms of aspergillosis. Allergic broncho-pulmonary aspergillosis (also called ABPA) is a condition in which the fungus causes wheezing and coughing but does not actually invade and destroy tissue. Invasive aspergillosis usually affects people with compromised immune systems. This fungus does invade and damage tissues, most commonly the lungs but other organs may be affected, and it can spread throughout the body (CDC website). Aspergillosis has been identified as a mortality factor in the American black duck (*Anas rubripes*), American robin (*Turdus migratorius*), bald eagle (*Haliaeetus leucocephalus*), Canada goose (*Branta canadensis*), canvasback (*Aythya valisineria*),common grackle (*Quiscalus quiscula*), common loon (*Gavia immer*), common merganser (*Mergus merganser*), eastern bluebird (*Sialia sialis*), evening grosbeak (*Coccothraustes vespertinus*),great horned owl (*Bubo virginianus*), herring gull (*Larus argentatus*), mallard (*Anas platyrhynchos*), mute swan (*Cygnus olor*), peregrine falcon (*Falco peregrinus*), purple martin (*Progne subis*), raven, redhead, red-tailed hawk, ring-billed gull (*Larus delawarensis*), ring-necked pheasant (*Phasianus colchicus*), rose-breasted grosbeak (*Pheucticus ludovicianus*), ruffed grouse (*Bonasa umbellus*), snowy owl (*Bubo scandiacus*), teal (*Anas* spp.), tundra swan (*Cygnus columbianus*), whistling swan (*Cygnus columbianus*), wild turkey (*Meleagris gallopavo*), and wood duck (*Aix sponsa*). It has also been found, but is a minor mortality factor in white-tailed deer (*Odocoileus virginianus*) (Michigan DNR website).

Histoplasmosis

Once referred to as "cave sickness," histoplasmosis is caused by the fungus *Histoplasma capsulatum*, which grows in soil and other surfaces contaminated with bat or bird droppings. When the contaminated area is disturbed the spores become airborne. Infection occurs through inhalation; the disease is not transmitted between infected individuals. There are three primary forms: acute pulmonary, chronic caviary, and disseminated. Antifungal medications are used to treat severe cases, while mild cases usually resolve without treatment. Most infected persons are asymptomatic. The acute respiratory disease is characterized by a general malaise, fever, chest pains, and a dry or nonproductive cough. Distinct patterns may be observed on a chest x-ray. The disease resembles tuberculosis and can worsen over months or years. Disseminated histoplasmosis affects the lungs and can be fatal if untreated. Infants, young children, the elderly (particular those with chronic lung disease), people with cancer, and those with AIDS or other forms of immunosuppression are at increased risk (CDC website).

Forms of the disease occur in skunks (*Mephitis* spp.), rats (*Rattus* spp.), opossums (*Didelphis virginianus*), foxes (*Vulpes* spp.), and other wildlife. Although bat guano has certainly been implicated as a source of *H. capsulatum*, Hoff and Bigler (1981) suggested the role of bats in human histoplasmosis is probably limited, given the high infection rates they observed in bats in areas where there are low rates of human infection. Emmons (1958) hypothesized that bats living near humans could be involved in sporadic rural and/or urban cases of human histoplasmosis, but the majority of bat-related cases of human histoplasmosis have been traced to caves and trees where exposure to fungi laden soil and guano is more intense. Birds have also been shown to play a role in histoplasmosis infection. Species that roost communally, including blackbirds (Icteridae), European starlings (*Sturnus vulgaris*), ring-billed gulls, and pigeons (*Columba livia*), appear to be a greater problem than those that do not congregate, because the fungus is associated with large quantities of bird droppings (Stickley and Weeks 1985).

BACTERIAL DISEASES

Microbial flora is a normal, everyday part of the daily life of humans and other animals. The human body commonly hosts about 10^{14} bacteria, which constitute the normal microbial flora. Viruses and parasites are not considered to be members of the normal microbial flora by most researchers. The normal microbial flora is associated with the skin and mucous membranes from shortly after birth until death, and is relatively stable. Specific genera can be found in different parts of the body during specific periods in an individual's life. Some of these microorganisms are advantageous to the host, some are detrimental, and some are neither. When the host organism's health becomes compromised, otherwise harmless normal microbial flora may cause disease (Davis 1996).

Bacterial pathogen infections range in severity from asymptomatic to rampant, and everything in-between. Bacteria can be organized into three major groups based on their capacity to cause disease: frank or primary pathogens, which cause disease in otherwise healthy individuals; opportunistic pathogens, which cause disease in an individual who has been injured or immunocompromised; and nonpathogens, which rarely or never cause disease. The latter categorization can be misleading, however, because bacteria are highly adaptable. Some bacteria previously categorized as nonpathogenic are now known to cause disease (e.g., *Serratia marcescens*, a common soil bacterium that causes pneumonia, urinary tract infections, and bacteremia in compromised individuals; Peterson 1996).

Virulence is related to the ability of the organism to cause disease despite the host's resistance mechanisms. Variables such as quantity of infecting bacteria, route of entry into the host, and specific and nonspecific host defense mechanisms all affect virulence. Pathogenesis is the step-by-step development of the disease. The pathogenic mechanisms of many bacterial diseases are not well understood. The relative importance of an infectious disease to the health of humans and nonhuman animals does not always coincide with our knowledge of its pathogenesis (Peterson 1996).

Educating the public about the basic rules of hygiene is key to controlling bacterial zoonoses such as plague, brucellosis, and tularemia. Even so, the incidence and impact of some bacterial zoonoses appears to be on the rise, and new pathogens may appear as well. Surveillance and control of these emerging bacterial zoonoses is essential if we hope to prevent human and nonhuman animal deaths and to avoid subsequent economic disorders (Blancou et al. 2005). Food-borne zoonoses (e.g., *Escherichia coli, Salmonella* spp.) account for the greatest number of emerging bacterial zoonoses, and this is due, in part, to the globalization of food resources (Thorns 2000). Other factors include human and nonhuman diet changes, increased densities of wildlife populations and livestock, human and nonhuman animal population displacement, increased contact with wildlife reservoirs, often through outdoor leisure activities, accelerated degradation of the natural environment, and global climate change (Blancou et al. 2005). We'll examine more closely two bacterial zoonoses: plague and tularemia. Additionally, in Chapter 14 we've provided information on Lyme disease as it relates to urban deer.

Plague

Plague is a flea-borne, rodent-associated zoonosis caused by the gram-negative bacterium *Yersinia pestis*. The disease can be fatal in humans if antimicrobial treatment is delayed or unavailable. Following a two- to six-day incubation period, infection begins at the flea bite site. "Flu-like symptoms" include fever, headache, chills, fatigue, muscle pain, and upset stomach. Bubonic plague causes the lymph nodes to enlarge—the disease actually gets its name from the "buboes" that are formed. The rarer but more dangerous pneumonic plague, caused by the same bacterium, may be transmitted through airborne droplets.

Although plague is treatable, the disease is notorious for its role in three massive pandemics, including the Black Death, which killed nearly one-third of Europe's population, and an identified outbreak can still create fear and even hysteria (Gage and Kosoy 2005). Because of its virulence and transmissibility, *Y. pestis* is considered an important potential agent of biological terrorism that requires special countermeasures to protect the public's health (Russo and Johnson 2008). Factors governing the occurrence of plague are not well understood, but the disease does seem to fluctuate in response to climactic factors (CDC website).

Y. pestis is transmitted between adult fleas and certain rodent (Rodentia) hosts and, in some cases, lagomorphs (rabbits, hares [*Lepus* spp.], and pikas [*Ochotona* spp.]). Plague infection also has been identified among the Artiodactyla (even-toed ungulates), Carnivora, Hyracoidea (hyraxes), Insectivora, Marsupialia, and Primates, suggesting that virtually any mammal can become infected. Birds, reptiles, and amphibians are generally thought to be resistant to *Y. pestis* infection. Animals that prey on rodents and lagomorphs may play an indirect role in the spread of plague (Gage and Kosoy 2005). Rodent population control using exclusion and habitat management must be proactive. Public education should encourage people to be aware that rodent die-offs may signal an outbreak. Prairie dogs (*Cynomys* spp.) and ground squirrels (Marmotini) are highly susceptible to the disease, and may serve as a bellwether to warn communities of an outbreak. Black-tailed prairie dogs (*Cynomys ludovicianus*) experience almost 100 percent mortality during outbreaks; other sciurid species, including other prairie dog species, are similarly susceptible (Biggins and Kosoy 2001).

Tularemia

Tularemia, also known as "rabbit fever," is caused by the bacterium *Francisella tularensis*. It is usually found in rodents, rabbits and hares, and has been reported in all U.S. states with the exception of Hawaii. There are several modes of infection, including the bite of an infected insect (e.g., ticks, deerflies), by handling sick or dead animals, by consuming contaminated food or water, or by inhaling airborne bacterial. On average, 200 human cases of tularemia are reported in the United States annually, with most in the south-central and western states (CDC website).

The incubation period for tularemia is typically three to five days, but can range from one to fourteen days. Symptoms depend on how the individual was exposed; they include skin ulcers, swollen and painful lymph glands, inflamed eyes, sore throat, mouth sores, diarrhea and/or pneumonia. If the bacteria are inhaled, symptoms are "flu-like." Diagnosis is not straightforward; *F. tularensis* is difficult to culture, and handling the bacterium poses a significant risk to laboratory personnel (Ellis et al. 2002). Symptoms are similar in companion animals and livestock. Tularemia can be fatal if the individual is not treated with appropriate antibiotics (CDC website).

Insecticides and repellants can help minimize exposure in areas where tularemia is transmitted by ticks or insects. Anyone who handles potentially infected animals, particularly rabbits, should wear impervious gloves and practice proper hygiene. Meat should be thoroughly cooked prior to consumption (Hadidian 2007). Tularemia is a potential agent of biological terrorism, particularly in an airborne form (Dennis et al. 2001).

VIRAL DISEASES

Viruses are small, subcellular agents that are incapable of multiplying outside a host cell. The assembled virus (virion) includes only one type of nucleic acid (RNA or DNA) and, in the simplest viruses, a protective protein coat. The nucleic acid contains the genetic information required to program the host cell for viral replication. The protein coat protects the nucleic acid and permits attachment of the virion to the membrane of the host cell. Viruses are the only microorganisms with such an absolute and extreme dependence on the host cell. Some viruses establish a "silent" infection of cells, but their multiplication usually causes cell damage or death.

Still, since viruses depend on host survival for their own survival, they tend to establish mild infections; death of the host is more of a deviation or accident than the norm. Some notable exceptions include HIV, ebola virus, hantavirus, and rabiesvirus (Baron, Fons, and Albrecht 1996). Viruses can infect all forms of life, and may have played a role in the natural selection of animal species. Epidemiologic studies indicate viral infections in developed countries are the most common cause of acute disease that does not require hospitalization, while in developing countries viral diseases tend to result in high mortality and permanent disability. Viruses use all known portals of entry, mechanisms of transmission, target organs, and sites of excretion (Baron, Fons, and Albrecht 1996).

Nearly all recent emergent disease episodes have involved zoonotic or species-jumping infectious agents. One of the factors that allows viruses to cross species is their potential to adapt rapidly to a changing environment by nature of their high genetic variability. The emergence or reemergence of diseases has been influenced by the interaction of factors as diverse as environmental and ecological changes, social factors, decline of health care, human demographics, and behavior (Ludwig et al. 2003).

Immunization, one of the more effective responses to viral infection, involves administering a virus preparation that stimulates the body's immune system to produce its own specific immunity. Viral vaccines now available for use include the following types: (1) attenuated live viruses, (2) killed viruses, and (3) recombinant produced antigens. Other management options include eliminating nonhuman reservoirs, eliminating the vector, and improving sanitation (Goldenthal, Midthun, and Zoon 1996).

Rabies

Rabies (*Lyssavirus* spp.) may well be the most famous, and infamous, viral zoonosis. This bullet-shaped virus causes acute infection of the central nervous system. Five general stages are recognized in humans: incubation, prodrome (early, nonspecific symptoms), acute neurologic period, coma, and death. The incubation period is widely variable, ranging from fewer than ten days to longer than two years, but on average one to three months (Rupprecht 1996).

In the continental United States four wildlife species are recognized as rabies reservoir or vector species: raccoons, bats (various species), skunks, and foxes. In 2006, bats became the second most reported species with rabies (CDC website). As a result of the success of vaccination programs, rabies has been reported more frequently in wildlife in the United States than in domesticated populations, accounting for 90 percent of all reported cases.

Humans can be exposed to rabies when the virus is introduced by a bite wound or when the virus contaminates existing cuts in the skin or mucous membranes. Pre- and postexposure vaccinations are available and effective; in the case of postexposure vaccination, immediate treatment is critical. Postexposure vaccination was once a frightening proposition; currently, treatment consists of five relatively painless vaccinations given in the upper arm over a one-month period. Once clinical signs appear, the virus is almost always fatal. Any bite from a known rabies vector species should be assumed to have the potential for transmitting rabies.

The persistence of variants of rabies virus in wild bats and canids is a continuing challenge to medical, veterinary, and wildlife management professionals. Oral rabies vaccination (ORV) targeting specific canid species has become an accepted addition to conventional rabies control strategies. ORV has been used in Europe and Canada with some success toward eliminating rabies in red foxes. Since 1995, coordinated ORV was implemented among eastern states in the United States to prevent the spread of raccoon rabies and to contain and eliminate variants of rabies virus in the gray fox and coyote in Texas (Slate et al. 2005).

Hantavirus Pulmonary Syndrome (HPS)

Hantavirus pulmonary syndrome (HPS) is a rare but deadly disease in humans. Hantaviruses were identified in the United States in the 1980s, but were not associated with human disease. The first U.S. outbreak of HPS was diagnosed in 1993, in the Four Corners region of the American Southwest. Since then, it has been identified throughout the United States. In the Old World, hantaviruses manifest in the form of hemorrhagic fever with renal syndrome (HFRS; Hadidian 2007). Hantaviruses belong to the Bunyaviridae family, a group of spherical, enveloped particles 90 to 100 nm in diameter. Except for hantaviruses, biologic transmission of Bunyaviridae is by a tick, mosquito (Culicidae), midge (Nematocera), or sand fly (Tabanidae and Ceratopogonidae) vector. Arthropods are infected for life, and vertebrates (wild or domesticated) usually are needed to maintain the cycle (Shope 1996).

New World hantavirus hosts include deer mice (*Peromyscus* spp.), cotton rats (*Sigmodon* spp.), white-footed mice (*Peromyscus leucopus*), and rice rats (*Oryzomys* spp.). Humans become infected by inhaling aerosolized particles of contaminated rodent feces, urine, or saliva. Rodent bites may be another route of transmission. Once again, initial symptoms are described as "flu-like"—headache, fever, muscle pain, and a dry cough. The disease rapidly progresses to acute pulmonary insufficiency and the patient may die within days of onset (Hadidian 2007).

The primary strategy for preventing hantavirus infection is rodent control in and around the home. Trapping and poisoning are temporary solutions; long term, habitat modification is the goal, so as to prevent rodent access to buildings and food sources. Appropriate clothing, including rubber gloves and respiratory protection, should be worn by anyone who must work in a rodent-infested location (Shope 1996; Hadidian 2007).

West Nile Virus (WNV)

West Nile virus (WNV) is a flavivirus originally found in Africa, West Asia, and the Middle East. It can infect humans, birds, mosquitoes, horses (*Equus ferus*), and some other mammals. Other flaviviruses cause encephalitis (St. Louis encephalitis, Japanese encephalitis, Powassan, and tick-borne encephalitis viruses), febrile illness with rash (dengue virus), hemorrhagic fever (Kyasanur Forest disease virus and sometimes dengue virus), and hemorrhagic fever with hepatitis (yellow fever virus). WNV emerged as a zoonotic concern in the United States in the summer of 1999, when New York's Bronx Zoo reported the deaths of several exotic birds. Dead American crows (*Corvus brachyrhynchos*) and blue jays (*Cyanocitta cristata*) were also found in the area. Before the 1999 outbreak subsided, sixty-two people had been diagnosed with a severe infection and seven died (Hadidian 2007).

The virus is carried by the *Culex* mosquito and was documented as having "overwintered," along with its mosquito host; by the summer, widespread transmission was observed (CDC website).

The most severe form of the disease is sometimes referred to as "neuroinvasive disease" because it affects the infected individual's nervous system. Specific types of neuroinvasive disease include West Nile encephalitis, West Nile meningitis, and/or West Nile meningoencephalitis. Encephalitis is an inflammation of the brain, meningitis is an inflammation of the membrane around the brain and the spinal cord, and meningoencephalitis is an inflammation of the brain and the membrane surrounding it. Another illness related to the virus is West Nile Fever, characterized by—you guessed it—flu-like symptoms. Some cases also present with a rash. Although the illness can be as short as a few days, full recovery can take much longer (CDC website).

Human outbreaks of WNV are not always associated with the severe neuroinvasive disease. Some infected individuals never exhibit any clinical symptoms. Several studies show about 1 out of every 150 infected individuals has clinical signs of encephalitis and/or meningitis.

Prevention is strongly correlated to control of mosquito populations and breeding habitats. Additionally, individuals can decrease the risk of exposure by wearing long-sleeved shirts, long pants, using an insect repellant with 45 percent DEET on any exposed skin, and limiting outdoor activity at dusk, when *Culex* mosquitoes are most active.

A study funded by the National Institutes of Allergies and Infectious Diseases (NIAID) and conducted by scientists at the Consortium for Conservation Medicine (CCM) at the Wildlife Trust New York, the New York State Department of Health, and the Smithsonian Migratory Bird Center has shown that American robins (*Turdus migratorius*) are the primary host of WNV in the northeastern United States. The authors found that mosquitoes carrying WNV seem to have a preference for robins, even in the presence of other, more abundant species such as house sparrows. They were also able to demonstrate that because American robins produce high levels of virus in their blood when infected, the majority of West Nile virus-infected mosquitoes at the study sites became infected from feeding on robins. WNV infects a variety of bird species, with high mortality rates in some (e.g., American crows). Previously, researchers hypothesized that crows, jays, or the abundant but much maligned house sparrows (*Passer domesticus*) were the key amplifying hosts (Kilpatrick et al. 2006).

PRION DISEASES

Also known as transmissible spongiform encephalopathies (TSEs), prion diseases are a family of rare progressive neurodegenerative disorders that affect both human and nonhuman animals. They have long incubation periods, characteristic spongiform (sponge-like) changes associated with neuronal loss, and a failure to induce inflammatory response. TSEs are believed to be caused by prions. A prion is an abnormal, infectious protein particle similar to a virus but lacking a nucleic acid agent. Prions are able to induce abnormal folding of normal cellular prion proteins in the brain, leading to brain damage and the characteristic signs and symptoms of the disease. Prion diseases are usually rapidly progressive and always fatal (CDC website).

Prions are highly resistant to most organic and inorganic chemicals, ultraviolet light, and cobalt-60 radiation. They are heat stable, nonantigenic, and have no virus-like structure. Infection can be induced by medical therapy (e.g., implants of contaminated electrodes, injection of contaminated human pituitary gland-derived growth hormone; Gibbs and Asher 1996). Nonhuman prion diseases include bovine spongiform encephalopathy (BSE), scrapie, chronic wasting diseases (CWD), transmissible mink encephalopathy, feline spongiform encephalopathy, and ungulate spongiform encephalopathy (CDC website).

Bovine Spongiform Encephalopathy

Bovine spongiform encephalopathy (BSE) is a progressive neurological disorder of cattle. The nature of the transmissible agent is not well understood; it is generally accepted that the agent is a modified form of a normal prion protein. Research suggests the first probable infections of BSE in cows occurred during the 1970s, and an additional two cases were identified in 1986. BSE may have originated as a result of feeding cattle BSE-infected meat-and-bone meal. Strong epidemiologic and laboratory evidence exists for a causal association between a new human prion disease called variant Creutzfeldt-Jakob disease (vCJD), first reported from the United Kingdom in 1996, and the BSE outbreak in cattle. The time between the most likely period of initial extended exposure of the population to potentially BSE-contaminated food (1984 to 1986) and the onset of initial variant CJD cases (1994 to 1996) is consistent with known incubation periods for the human forms of prion disease (CDC website). BSE does not appear to have crossed to wild ruminants in North America (e.g., bison, deer), but Walawski et al. (2005) have reported a characteristic of the bovine prion protein (PRNP) gene polymorphism within the European bison (*Bison bonasus*) population, as well as in two endemic cattle breeds (Polish Red and Polish Whitebacked).

Scrapie

Scrapie is a fatal, degenerative prion disease that affects the central nervous system of sheep and goats. The agent responsible for scrapie is not well understood. Hamsters (Cricetinae), mice, rats, voles (Arvicolinae), gerbils (Gerbillinae), mink (*Neovison* spp. and *Mustela lutreola*), cattle (*Bos primigenius*), and some species of monkeys (Simiiformes) have been infected with scrapie via inoculation in the laboratory. There is, however, no evidence to date that scrapie poses a threat to human health, and no epidemiologic evidence that scrapie is transmitted to humans through contact on the farm, at slaughter plants, or butcher shops (USDA-APHIS website).

Chronic Wasting Disease

Chronic wasting disease (CWD), a prion disease that affects mule deer (*Odocoileus hemionus*), white-tailed deer (*Odocoileus virginianus*), elk (*Cervus canadensis*), and moose (*Alces alces*), is not technically a zoonosis because, to date, no strong evidence of CWD transmission to humans has been reported. Still, it remains a concern for hunters and wildlife managers. CWD was first identified as a fatal wasting syndrome in captive mule deer in Colorado in the late 1960s, and in the wild in 1981. It was identified as a spongiform encephalopathy in 1978. By the mid-1990s, CWD had been diagnosed among free-ranging deer and elk in a contiguous area in northeastern Colorado and southeastern Wyoming, where the disease is now endemic. Subsequently, CWD has been found outside of this zone, including Illinois, New York, West Virginia, and Wisconsin. The geographic range of diseased animals currently includes eleven U.S. states and two Canadian provinces and is likely to continue to grow (CDC website).

Transmissible mink encephalopathy results from feeding scrapie-infected animal renderings (Gibbs and Asher 1996). Feline spongiform encephalopathy (FSE) has been identified in domesticated cats (*Felis catus*) and captive wild cats, including cheetahs (*Acinonyx jubatus*), pumas (*Puma concolor*), ocelots (*Leopardus pardalis*), tigers (*Panthera tigris*), lions (*Panthera leo*), and Asian golden cats (*Pardofelis temminckii*). We were unable to find any reports of FSE in wild North American felids.

SPECIES PROFILE: AMERICAN ROBIN (*TURDUS MIGRATORIUS LINNAEUS*)

A robin redbreast in a cage puts all of heaven in a rage.

—**William Blake, Poet**

The American robin (Figure 13.1) is as much a harbinger of spring as April showers and May flowers, which is odd because the bird actually stays put during the winter in much of its breeding range, despite being a migratory species. They do, however, spend less time in backyards and congregate in large flocks during the cold months, so we don't notice them as much as when the daffodils are nodding in the breeze (Cornell Lab of Ornithology website; Sallabanks 2000).

Technically, it isn't a robin at all … it's a thrush. But the birds' red breast reminded early European settlers of a favorite bird from home, the European robin (*Erithacus rubecula*) so even though the two species have only the most superficial resemblance (*E. rubecula* looks more like a bluebird than a thrush) they hung the moniker on this native American and it stuck. We continued

Figure 13.1   American robin. (Courtesy Lee Karney/USFWS)

with the trend, and there are now four North American thrushes posing as robins—the American robin, the rufous-backed robin (*Turdus rufopalliatus*), the white-throated robin (*Turdus assimilis*), and the clay-colored robin (*Turdus grayi*). Only *T. migratorius* can be found across the continental United States, having expanded its range with the help of *Homo sapiens,* as they too, expanded into the Great Plains, planting trees and installing irrigation, creating nesting sites and moist grasslands for foraging (Cornell Lab of Ornithology website; Ehrlich, Dobkin, and Wheye 1988; National Geographic 1999).

American robins (let's just call them robins from here on out) are a well-known sight in the land of lawns as they run, stop, cock their heads, and then pull earthworms from the ground like a magician with a rabbit in a hat. They do eat other things as well, including a variety of invertebrates; fruit, including berries, is a dietary mainstay during the winter months (Cornell Lab of Ornithology website). Nestling robins are fed insects. Both the male and female are active parents, with the male caring for the fledgling first brood of the season while the female incubates a second clutch of robin's egg blue eggs.

A long-term study by Walcott (1974) found that robins and the gray catbird (*Dumetella carolinensis*) were the only species of birds that remained in the study area despite disturbance as development went from 5 percent of plots covered by houses to 85 percent. Unlike several other urban bird species, including European starlings, pigeons, and house finches (*Carpodacus mexicanus*), robins have not adapted to the presence of humans to the extent that it uses man-made structures for nesting (Morneau et al. 1995). This may be one of the reasons robins are not considered to be a human-wildlife conflict concern. They go about their business shoulder to wing with their human neighbors while rarely even acknowledging our presence. The exception is when parent birds feel their nest and young are being threatened. In such cases, they will aggressively defend their own.

Because robins forage mainly on lawns, they are vulnerable to pesticide poisoning. The use of DDT for Dutch elm disease in the 1950s and its effect on robins helped to create a public outcry over the potential for the *Silent Spring* described by Rachel Carson. DDT has since been banned in the United States, but robins remain an important indicator of chemical pollution (Ehrlich, Dobkin, and Wheye 1988). More recently, robins have been found to be the primary host of West Nile virus in the northeastern United States (Kilpatrick et al. 2006).

CHAPTER ACTIVITIES

1. Investigate which zoonoses are of concern in your area by visiting the state health department website.
2. Chose one of the zoonoses described in this chapter; when was the last outbreak (if ever) in your state?
3. Which zoonoses not described in this chapter are a concern in your state and/or region?
4. Visit the USGS National Wildlife Health Center Web site (www.nwhc.usgs.gov). What are the currently listed "hot topics"? Are there any zoonoses listed? If so, are any of them a concern for your state and/or region?
5. Check out the Wildlife Disease Association Web site (www.wildlifedisease.org) to learn about jobs in this field—this career path is not just for veterinarians!

LITERATURE CITED

Adam, R. D. 1991. The biology of *Giardia* spp. *Microbiological Reviews* 55(4):706–732.

Appelbee, A.J., R.C.A. Thompson, and M.E. Olson. 2005. *Giardia* and cryptosporidium in mammalian wildlife: Current status and future needs. *Trends in Parasitology*, 21(8):370–376.

Baron, S., M. Fons, and T. Albrecht. 1996. *Medical Microbiology: General Concepts,* 4th edition. Galveston, TX: University of Texas Medical Branch.

Beran, G.W., ed. 1994. *Handbook of Zoonoses,* Vol. 1, 2nd edition. Boca Raton, FL: CRC Press.

Biggins, D.E. and M.Y. Kosoy. 2001. Disruptions of ecosystems in western North America due to invasion by plague. *Journal of the Idaho Academy of Science,* 37:62–65.

Bisseru, B. 1967. *Diseases of Man Acquired from His Pets.* Philadelphia, PA: Lippincott.

Blancou, J., B.B. Chomel, A. Belotto, and F.X. Meslin. 2005. Emerging or re-emerging bacterial zoonoses: Factor of emergence, surveillance and control. *Veterinary Research,* 36:507–522.

Castro, G.A. 1996. Helminths: Structure, classification, growth and development. In *Medical Microbiology: General Concepts,* 4th edition, ed. S. Baron. Galveston, TX: University of Texas Medical Branch.

Chai, J.-Y., K.D. Murrell, and A.J. Lymbery. 2005. Fish-borne zoonoses: Status and issues. *International Journal of Parasitology* 35(11–12):1233–1254.

Davis, C.P. 1996. Normal flora. In *Medical Microbiology: General Concepts,* 4th edition, ed. S. Baron. Galveston, TX: University of Texas Medical Branch.

Dennis, D.T., T.V. Inglesby, D.A. Handerson, J.G. Bartlett, M.S. Ascher, E. Eitzen, A.D. Fine, A.M. Friedlander, J. Hauer, M. Layton, S.R. Lillibridge, J.E. McDade, M.T. Osterholm, T. O'Toole, G. Parker, T.M. Perl, P.K. Russell, and K. Tonat. 2001. Tularemia as a biological weapon: Medical and public health management. *Journal of the American Medical Association* 285(21):2763–2773.

Doyle T.J., M.K. Glynn, and L.S. Groseclose. 2002. Completeness of notifiable infectious disease reporting in the United States: An analytical literature review. *American Journal of Epidemiology* 155:866–874.

Dudley, J.P. 2003. New challenges for public health care: Biological and chemical weapons awareness, surveillance, and response. *Biological Research for Nursing* 4:244–250.

Dudley, J.P. 2004. Global zoonotic disease surveillance: An emerging public health and biosecurity imperative. *BioScience,* 54(11):982–983.

Dunlap, B.G. and M.L. Thies. 2002. Giardia in beaver (*Castor canadensis*) and nutria (*Myocaster coypus*) from East Texas. *Journal of Parasitology* 88(6):1254–1258.

Ehrlich, P.R., D.S. Dobkin, and D. Wheye. 1988. *The Birder's Handbook: A Field Guide to the Natural History of North American Birds.* New York: Simon and Schuster.

Ellis, J., P.C.F. Oyston, M. Green, and R.W. Titball. 2002. Tularemia. *Clinical Microbiology Reviews* 15(4):631–646.

Emmons, C.W. 1958. Association of bats with histoplasmosis. *Public Health Report* 73:590–595.

Erlandsen, S.L., L.A. Sherlock, W.J. Bemrick, H. Ghobrial, and W. Jakubowski. 1990. Prevalence of *Giardia* spp. in beaver and muskrat populations in northeastern states and Minnesota: Detection of intestinal trophozoites at necropsy provides greater sensitivity than detection of cysts in fecal samples. *Applied Environmental Biology,* 56(1):31–36.

Erlandsen, S.L., A.R. Weisbrod, L.W. Knudson, R. Olereich, W.E. Dodge, W. Jakubowski, and W.J. Bemrick. 1991. Giardiasis in wild and captive bird populations: High prevalence in herons and budgerigars. *International Journal of Environmental Health Research* 1(3):132–143.

Gage, K.L., and M.Y. Kosoy. 2005. Natural history of plague: Perspectives from more than a century of research. *Annual Review of Entomology* 50:505–528.

Gibbs, C.J. and D.M. Asher. 1996. Subacute spongiform unconventional virus encephalopathies. In *Medical Microbiology: General Concepts*, 4th edition, ed. S. Baron. Galveston, TX: University of Texas Medical Branch.

Goldenthal, K.L., K. Midthun, and K.C. Zoon. 1996. Control of viral infections and diseases. In *Medical Microbiology: General Concepts,* 4th edition, ed. S. Baron. Galveston, TX: University of Texas Medical Branch.

Goldsmid, J.M. and P.A. Leggat. 2005. *Primer of Tropical Medicine.* Brisbane: The Australasian College of Tropical Medicine, Brisbane, Queensland, Australia.

Hadidian, J. 2007. *Wild Neighbors: The Humane Approach to Living with Wildlife.* Washington, DC: The Humane Society of the United States.

Hoff, G.L. and W.J. Bigler. 1981. The role of bats in the propagation and spread of histoplasmosis: A review. *Journal of Wildlife Diseases* 17(2):191–196.

Kilpatrick, A.M., P. Daszakl, M.J. Jones, P.P. Marra, and L.D. Kramer. 2006. Host heterogeneity dominates West Nile virus transmission. *Proceedings of the Royal Society B* 273(1599):2327–2333.

Krauss, H., A. Weber, M. Appel, B. Enders, H.D. Isenberg, H.G. Schiefer, W. Slenczka, A. von Graevenitz, and H. Zahner, eds. 2003. *Zoonoses: Infectious Diseases, Transmissible from Animals to Humans*, 3rd edition. Washington, DC: ASM Press.

Ludwig, B., F.B. Kraus, R. Allwinn, H.W. Doerr, and W. Preiser. 2003. Viral zoonoses: A threat under control? *Intervirology* 46:71–78.

McPherson, C.N.L. 2005. Human behavior and the epidemiology of parasitic zoonoses. *International Journal of Parasitology* 35(11–12):1319–1331.

Morneau, F., C. Lepine, R. Decarie, M.-A. Villard, and J.-C. DesGranges. 1995. Reproduction of American robin (*Turdus migratorius*) in a suburban environment. *Landscape and Urban Planning* 32:55–62.

Müller-Schwarze, D. and L. Sun. 2003. *The Beaver: Natural History of a Wetlands Engineer.* Ithaca, NY: Cornell University Press.

National Geographic. 1999. *Field Guide to the Birds of North America.* Washington, DC: National Geographic.

O'Donoghue, P.J. 1995. Cryptosporidium and cryptosporidiosis in man and animals. *International Journal for Parasitology* 25(2):139–195.

Peterson, J.W. 1996. Bacterial pathogenesis. In *Medical Microbiology: General Concepts*, 4th edition, ed. S. Baron. Galveston, TX: University of Texas Medical Branch.

Rupprecht, C.E. 1996. Rhabdoviruses: Rabies virus. In *Medical Microbiology: General Concepts*, 4th edition, ed. S. Baron. Galveston, TX: University of Texas Medical Branch.

Russo, T.A. and J.R. Johnson. 2008. Diseases caused by gram-negative enteric bacilli. In *Harrison's Principles of Internal Medicine*, 17th edition, ed. A.S. Fauci, E. Braunwald, D.L. Kasper, S.L. Hauser, D.L. Longo, J.L. Jameson, and J. Loscalzo. New York: McGraw-Hill Professional.

Sallabanks, R. 2000. The American robin. *Auk* 117(1):274–276.

Shope, R.E. 1996. Bunyaviruses. In *Medical Microbiology: General Concepts,* 4th edition, ed. S. Baron. Galveston, TX: University of Texas Medical Branch.

Slate, D., C.E. Rupprecht, J.A. Rooney, D. Donovan, D.H. Lein, and R.B. Chipman. 2005. Status of oral rabies vaccination in wild carnivores in the United States. *Virus Research* 111(1):68–76.

Sleeman, J. 2005. Wildlife zoonoses for the veterinary practitioner. *Journal of Exotic Pet Medicine* 15(1):25–32.

Sorvillo, F., L.R. Ash, O.G.W. Berlin, J. Yatabe, C. Degiorgio, and S.A. Morse. 2002. *Baylisascaris procyonis*: An emerging helminthic zoonosis. *Emerging Infectious Diseases* 8(4):355–359.

Stephen, C., H. Artsob, W.R. Bowie, M. Drebot, E. Fraser, T. Leighton, M. Morshed, C. Ong, and D. Patrick. 2004. Perspectives on emerging zoonotic disease research and capacity building in Canada. *Canadian Journal of Infectious Disease & Medical Microbiology* 15(6):339–344.

Stickley, A.R., Jr. and R.J. Weeks. 1985. Histoplasmosis and its impact on blackbird/startling roost management. *Proceedings of the 2nd Eastern Wildlife Damage Control Conference* 2:163–171.

Thorns, C.J. 2000. Bacterial food-borne zoonoses. *Revue Scientifique Et Technique* 19:226–239.

U.S. Centers for Disease Control and Prevention. 2008. *Summary of Notifiable Diseases– United States, 2006.* Atlanta, GA: United States Department of Health and Human Services, Centers for Disease Control.

Walawski, K., G. Grzybowski, U. Czarnik, T. Zabolewicz, M. Szeremeta, A. Kownacka, and M. Leszyczynska. 2005. The polymorphism of a prion protein gene in European bison (*Bison bonasus*, L. 1758) and Polish Red and Polish Whiteback cattle included in the genetic resources conservation programme. Animal Science Papers and Reports. Polish Academy of Sciences.

Walcott, C.F. 1974. Changes in bird life in Cambridge, Massachusetts from 1860 to 1964. *The Auk* 91:151–160.

CHAPTER 14

Distribution, Abundance, and Management Considerations of Resident Canada Geese and Urban White-Tailed Deer

This is the image of the anthropocentric man: he seeks not unity with nature but conquest.

—Ian L. McHarg (1969)

A thing is right when it tends to preserve the integrity, stability, and beauty of the biotic community. It is wrong when it tends to do otherwise.

—Aldo Leopold (1976)

KEY CONCEPTS

1. There are seven factors that contributed to geese and deer abundance in urban America.
2. A national assessment of human-white-tailed deer and -Canada geese conflicts revealed how pervasive these problems are in urban America.
3. White-tailed deer and Canada geese are found in nearly every state in the continental United States, but geese and deer abundance (density) is different within states.
4. There are serious ecological problems associated with overabundant geese and deer in urban communities.
5. There are serious health and safety issues associated with overabundant geese and deer in urban communities.
6. The spread of Lyme disease to humans is the result of complex interrelationships between white-tailed deer, white-footed mice, gypsy moths, oak trees, ticks, and a bacterium.
7. The four different approaches to geese and deer management in urban communities include avoiding the problem, getting at the root cause, attacking the symptoms, and doing nothing.

INTRODUCTION

Two animals of special significance in urban and suburban America are resident giant Canada goose (*Branta* sp.) and white-tailed deer (*Odocoileus virginianus*). These two animals (Figure 14.1 and Figure 14.2) share a common legacy in terms of: (1) those factors that contributed to their prominence in urban America, (2) distribution in the continental United States, (3) the human response

Figure 14.1 An urban deer herd on the Cornell University campus and the close relationship that commonly develops between urban deer and humans. (Courtesy Paul Curtis and Lindsay Thomas, QDMA, respectively)

to their presence, (4) ecological impacts on their habitats, (5) human health and safety issues, and (6) feasible and acceptable management strategies.

Hundreds of professional and popular articles have been written on geese and deer problems in urban America. It has been our practice throughout this book to find and identify synthesis documents that compile the body of literature related to a specific topic or to go through the task of assimilating all of the most recent relevant literature on a topic under a single chapter. For this case study, we found an entire issue of the *Wildlife Society Bulletin* (1997, Volume 25, No. 2) devoted to urban deer management. The magnitude of the issues related to Canada goose management was compiled by the U.S. Fish and Wildlife Service in a draft environmental impact statement cited below. These two sources contained many citations of the field studies conducted by the leading researchers on resident Canada geese and urban white-tailed deer management.

Figure 14.2   A large flock of urban Canada geese overwintering in Lubbock, Texas. (Courtesy K. Haukos)

FACTORS THAT CONTRIBUTED TO GEESE AND
DEER ABUNDANCE IN URBAN AMERICA

Geese were nearly eliminated in most parts of the United States by unrestricted hunting, harvesting of eggs, and draining of wetland habitat (Smith, Craven, and Curtis 1999). In 1999, the U.S. Fish and Wildlife Service (FWS) estimated the population of resident Canada geese in North American at 3.5 million (FWS 2002). There used to be only 100,000 deer in the United States in the early 1900s, reduced by commercial hunting, but now they number over 30 million. The restoration of geese and deer populations in North America has led to one of the most challenging problems facing wildlife managers today and in the future—geese and deer over-abundance (Warren 1997; FWS 2002). Both geese and deer are superbly adapted to exploit the resources in urban areas, and appear to be in exponential growth patterns caused by basically the same factors including:

1. Low abundance of natural predators. Large predators are the first species eliminated during urban sprawl.
2. Healthy breeding habitat conditions that include sheltered nesting sites for geese and patchy and defined edge habitat for deer.
3. Tolerance of urban disturbances including human presence and their activities, for example, golfing or picnicking.
4. Longer life spans in the city when compared to the country.
5. Lack of hunting in urban areas, and regulatory protection at the state and/or federal levels.
6. High production and survival rates in offspring.
7. Abundant alternative food resources in the form of ornamental shrubs, garden plants, succulent grasses, small plants, and supplemental feed.

EXTENT OF THE PROBLEM: A NATIONAL ASSESSMENT OF HUMAN-
WHITE-TAILED DEER AND CANADA GEESE CONFLICTS

In order to determine the extent of human-white-tailed deer and human-Canada geese conflicts nationally, an online survey of wildlife biologists was conducted in the fall of 2008. Respondents came from the TWS listserv, and big game, waterfowl, and urban wildlife biologists from each state's department of natural resources (DNR). There were 204 respondents to the survey representing forty-five states and the District of Columbia. There were multiple respondents for some states. Nonrespondent states were Hawaii, Mississippi, New Jersey, Oklahoma, and Wyoming. Other sources of information in the published literature and online were used to answer some of the questions for some of the nonresponding states.

All respondents ($n = 111$) who identified their job titles were engaged in some aspect of natural resource management in their states. They represented state (69 percent) and federal (9 percent) agencies, academia (8 percent), private industry (4 percent), extension (4 percent), or other (6 percent). The majority of other job titles were "wildlife" related indicating that respondents were the best audience to respond to survey questions.

Objectives and Questions Included in the National Assessment

The national assessment was designed around four objectives:

1. To determine the extent of human-wildlife conflicts with urban white-tailed deer and resident Canada geese at the state and community levels.
2. To determine historical trends of change in human conflicts with these two species.
3. To identify who is responsible for the management of white-tailed deer and Canada geese in urban areas.
4. To identify the most important issues that need to be considered in the management of these two species in urban areas.

The online survey consisted of the same series of six questions regarding human-white-tailed deer and human-Canada geese conflicts in urban areas. The first question determined if these types of human-wildlife conflicts existed in their state. If the answer was "yes," respondents were then asked to list up to ten cities in their state that are experiencing these types of conflicts and if the cities listed have implemented an urban wildlife management program for one or the other of the two species. Respondents were then asked whether there had been any change (e.g., increased, decreased, stabilized, do not know) in the number of citizen and community reports about white-tailed deer or Canada geese problems over the past ten years. Next, they were asked to identify who was in charge (community and state agency levels of analysis) of the urban wildlife management programs for these two species. Respondents ranked (most to least important) ten issues related to urban white-tailed deer and Canada goose management. The issues included: (1) state and federal game laws, (2) public safety, (3) damage to public property, (4) health of the animal, (5) method of population reduction, (6) public opinion, (7) public education, (8) public input in management decisions, (9) funding, and (10) media involvement.

Results of the National Assessment

Most (80 percent) of the 204 respondents reported that some or many communities in their state were experiencing human-white-tailed deer conflicts, and 84 percent of 140 respondents reported human-Canada geese conflicts. It would not be overstating the case to say that every state within the national distribution of white-tailed deer and Canada geese have human-wildlife conflicts with these game animals in urban areas. Nearly half (43 percent) of the respondents could not or would not identify specific community locations of human-white-tailed deer or human-Canada geese conflicts in their states. There were several reasons for this omission, for example, they did not know, did not want to answer the question, or the problem was so pervasive throughout the state that listing only ten communities would not do justice to the question. Examples of the latter explanation would

be most of the northeastern and eastern seaboard states and Ohio. However, twenty-one states identified 190 communities that had human-white-tailed deer conflicts and twenty states identified 307
communities with human-Canada geese conflicts. Washington, DC was included in both lists. Ohio
alone identified thirty-eight and forty-nine communities with human-white-tailed deer and human-
Canada geese conflicts, respectively. One Ohio respondent said that any urban center in the state
has human-white-tailed deer and human-Canada geese conflicts. This national assessment only
scratched the surface of the magnitude, that is, the universality, of human-white-tailed deer and
human-Canada geese conflicts in urban America.

Compared to ten years ago, there was majority agreement among respondents that citizen and
community reports concerning white-tailed deer (71 percent) and Canada geese (63 percent) conflicts had increased. However, respondents also reported that 25 percent of the listed communities
had no management plans for either urban white-tailed deer or resident Canada geese, or they did
not know if management plans existed (24 and 41 percent, respectively). Even though there are many
state-specific management plans for urban white-tailed deer and resident Canada geese posted on
the Internet, this raised the question of management responsibility—who is in charge? The majority
(>85 percent) of respondents identified the state game agency as the management authority at the
state level for both white-tailed deer and Canada geese. The community-level authority was the city
council for white-tailed deer, but USDA APHIS Wildlife Services was the second choice over city
council for resident Canada geese. Homeowners' associations, city mayor, and community residents
were considered players in the management process by only a few (<20 percent) of the respondents.

The top three issues of importance in urban white-tailed deer management were public safety,
property damage, and public education. Media involvement was of least importance. The top three
issues of importance in resident Canada goose management were state and federal laws, property
damage, and public education. Again, media involvement was the least important issue to be considered in the management process.

Management Implications of the National Assessment

There were several management implications that emerged from the results of the national
assessment. The first order of business in developing effective urban wildlife management programs
is to assess various ramifications of the problem(s) that need to be addressed. This first step can
be complicated by the absence of a population of individuals who can act as reputable and willing
informants for their state. The objectives of the national assessment were compromised by difficulty in getting the information required. For example, nearly 50 percent of the respondents who
started the online survey quit the effort after the first question. Frankly, this was surprising since all
survey questions could have been answered quickly by one or more wildlife biologists in any state.
For example, every state DNR has a big game, deer, urban, or waterfowl biologist. All who could
be identified were contacted personally and invited to participate in the study. Were they obligated
to participate? No, because participation in any survey is voluntary. Yes, because they represent
the only link that the outside world has to obtain information about any wildlife species under the
regulatory authority of their particular state and federal agencies. In addition, professional courtesy, dialogue, and reciprocity should be expected and honored even within the wildlife profession.
Based on the results of the national assessment, wildlife professionals in nearly every state face the
challenges of urban white-tailed deer and resident Canada goose management. Thus, from an urban
wildlife management standpoint, managers would benefit from maintaining a liberal communications policy with individuals and groups outside their agency and state.

There are national ramifications in the management of urban white-tailed deer and resident
Canada geese. These two species will continue to require constant attention, diverting agency personnel and monetary resources away from their traditional management agendas. Forewarned is
forearmed, meaning that academia needs to begin a vigorous process of training urban wildlife

biologists to meet the emerging and growing challenges of wildlife management in urban areas. In addition, state DNRs will need to begin providing employment opportunities for individuals specifically trained to manage wildlife in urban areas.

Finally, it is probably more common to find management plans for white-tailed deer and Canada geese as game animals rather than as urban species of management concern. Readers are directed to an Internet search for urban management plans for these two species in their own states (e.g., search for urban goose or white-tailed deer management). Later in this chapter the common considerations for developing management plans are listed and discussed.

DISTRIBUTION OF RESIDENT CANADA GEESE AND WHITE-TAILED DEER IN THE CONTINENTAL UNITED STATES

Both white-tailed deer and Canada geese are found in nearly every state (partially or completely) in the continental United States. There is a remarkable similarity in the abundance of resident Canada geese and urban white-tailed deer in some states (Figure 14.3 and Figure 14.4). This similarity is recognized in the states where they occur, and in some cases, the numbers of geese and deer that occur within specific states. The highest numbers (ranked from 1 to 5) of resident Canada geese were in Michigan, Virginia, Pennsylvania, Minnesota, and New York. The major factor that influenced the distribution of geese was the migratory patterns that include the Atlantic, Mississippi, Central, and Pacific flyways. However, resident Canada geese do not migrate, and in fact, probably recruit additional nonmigrants among migrating flocks by acting as living decoys.

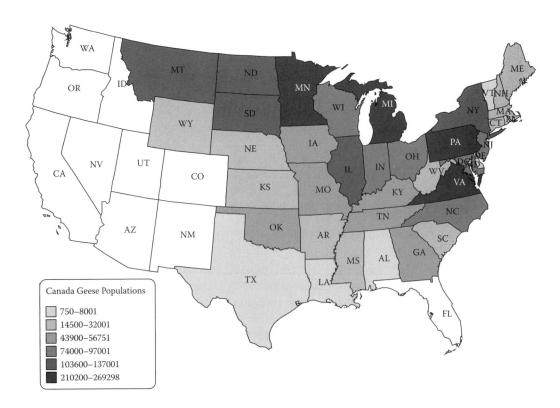

Canada Geese Populations

- 750–8001
- 14500–32001
- 43900–56751
- 74000–97001
- 103600–137001
- 210200–269298

Figure 14.3 Distribution and number of resident Canada geese in the Continental United States. There are no resident Canada geese data for nonshaded states, for example, Florida. (USFWS 2002)

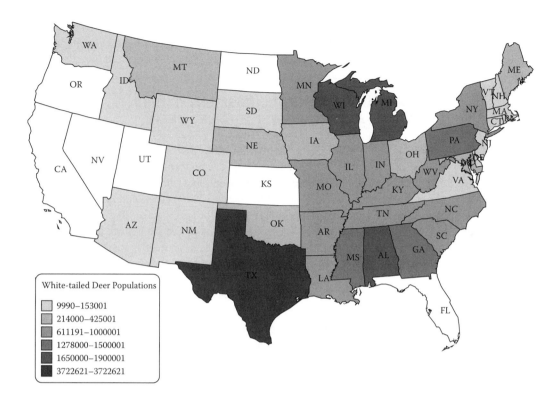

Figure 14.4 Distribution and number of white-tailed deer in the Continental United States in 1999. White-tailed deer are not present in nonshaded states or census data are not available. (Courtesy Quality Deer Management Association)

Furthermore, it is unlikely that offspring of resident Canada geese heed the environmental signals to head north or south.

The highest numbers (ranked from 1 to 5) of white-tailed deer were in Texas, Michigan, Alabama, Wisconsin, and Mississippi. In general, the highest numbers of white-tailed deer were found in Texas and Minnesota and in states adjacent to and east of the Mississippi River (Figure 14.4). Even though white-tailed deer are present in Oregon, North Dakota, and Kansas, there is no information available on the size of the deer herds in these states. Some states do not count their deer.

The methods of determining the number of geese and deer within each state vary in terms of agency census techniques and methods of generalizing actual count data to the entire state. Resident Canada geese and white-tailed deer are not run through turnstiles that give actual counts. The published data on geese and deer numbers within each state are, at best, estimates. In addition, the estimates may be based on a census of the geese and deer that live outside city borders. However, the process of urban sprawl has connected the rural populations of geese and deer with people.

This led us to determine whether some of the measurable variables that denote the process of urbanization (e.g., number of golf courses, human population size, and percent of population living in urban areas) could be better predictors of geese and deer numbers than farmland acreage, and square miles of land and water area.

We ranked each state in terms of the number of resident Canada geese ($N = 38$), white-tailed deer ($N = 41$), and golf courses, farmland acres, people per square mile, and the amount of land and water per square mile ($N = 43$). A Spearman rank correlation revealed that the number of geese, deer, and people were correlated ($r = 0.34$ to 0.59, $P = 0.05$) with the number of golf courses and square miles of surface water. The number of geese and deer were correlated ($r = 0.42$ and 0.50, respectively, $P = 0.05$) with the number of people. There may not be any relationship between correlation and cause, but apparently, people, geese, and deer have a lot in common when it comes to the type of landscapes they prefer.

THE HUMAN RESPONSE TO RESIDENT CANADA GEESE AND URBAN WHITE-TAILED DEER

The human response to the presence of resident Canada geese or white-tailed deer is measured in terms of a range of values they attribute to each species in terms of aesthetic, recreational, ecological, educational, utilitarian, and potential danger. Both species have aesthetic value when viewed as part of the inherent natural beauty of the landscape. The recreational value is based on the enjoyment derived from viewing a pair of geese gliding effortlessly across a city lake followed by eight to ten goslings or a doe attending twin or triplet fawns in the front yard. The ecological value is realized by determining the relationship of geese and deer to other species, and the urban environment, community, and ecosystem. Geese and deer have educational value in what can be learned about their life history, methods of resource utilization in urban environments, and how to manage their numbers within the carrying capacity of their ecosystem.

If geese and deer can be considered useful to humans in some tangible way, for example, food, then they have utilitarian value. White-tailed deer are revered as the state mammal in eleven states and appear on the Vermont state flag and seal. Both geese and deer represent a potential for danger in terms of negative impacts on human health or economics, particularly when their population size increases beyond a level considered tolerable by urban residents. In general, human values are fickle given a host of variables that influence human attitudes, activities, knowledge, and expectations concerning Canada geese and white-tailed deer in urban communities.

The majority of residents in urban communities are three to four generations removed from rural life and all direct functional ties with the natural world. As such, the value systems of urban residents are based largely on a sanitized, synthetic, unrealistic portrayal of wildlife in the media, literature, and commercial aspects of their urban communities. Urban residents have a profound ignorance of the natural world and how it works. Marchinton (1997) coined the term "urbanism" to define this condition as a way of looking at life by urban residents. This way of looking at life has led to almost insurmountable problems in the effective management of urban white-tailed deer populations, as evidenced in Case Study 14.1, A Tale of Two Cities.

ECOLOGICAL IMPACTS OF RESIDENT CANADA GEESE AND URBAN WHITE-TAILED DEER

The key to understanding the ecological impacts of resident Canada geese and urban white-tailed deer is to consider effects on habitat when a species approaches or exceeds the carrying capacity. The lesson on population dynamics in Chapter 6 demonstrated why no population of organisms can be allowed to increase exponentially indefinitely without causing serious damage to other species within the biotic community and itself. There is ample evidence suggesting that resident Canada geese and urban white-tailed deer are approaching the carrying capacity of the urban ecosystems they occupy.

For example, Canada geese are "eating-excreting" machines. A large flock of geese can compact soil and/or denude an area of vegetation, which leads to erosion. On average, an adult Canada goose defecates twenty-eight times per day, which equates to 0.1 pound of feces per day. Imagine the fecal mess that can be produced by a flock of 200 Canada geese in one park every day! The ecological consequence of the fertilizer load in aquatic ecosystems is nutrient enrichment (eutrophication) of city lakes and ponds. The enormous amount of goose feces deposited in city lakes and ponds turns them into a thick green soup (caused by green feces and excessive algal growth), destroys much of the plant and animal diversity that once existed, and alters the water chemistry (e.g., O_2 concentration). Only those organisms that have the widest ranges of tolerance to the altered abiotic conditions can survive. Agricultural and natural resource impacts include losses to grain crops, overgrazing of pastures, and degrading water quality. On land, the high nitrogen content of the feces leads to overfertilization of plants and, if concentrated in one area, leaves dead spots on grassy areas (Smith et al. 1999).

The ecological impact of overabundant white-tailed deer in urban areas either: (1) affects the distribution or abundance of many other species, (2) can affect community structure by strongly modifying patterns of relative abundance among competing species, or (3) affects community structure by influencing the abundance of species at multiple trophic levels (Waller and Alverson 1997:218). The heavy and constant browsing of urban white-tailed deer can destroy entire vegetative communities (e.g., forest understory) upon which other species depend for their survival. In addition, browsing deer promote the selection of vegetative communities with less plant and animal biodiversity. These profound impacts on the ecology of urban ecosystems led Waller and Alverson (1997) to designate urban white-tailed deer as a "keystone" species.

HEALTH AND SAFETY ISSUES RELATED TO RESIDENT CANADA GEESE AND URBAN WHITE-TAILED DEER

There is a large body of literature that documents the human health and safety problems caused by resident Canada geese and urban white-tailed deer. Both species transmit diseases to other animals and humans. For example, geese can transmit coccidiosis, avian influenza, schistosomes, chlamydiosis, salmonella, and avian cholera to other birds, and cholera to cattle (Smith et al. 1999).

Evidence of conflicts between geese and people include health and safety issues, or damage to property, agriculture, and natural resources. Common problem areas include public parks, airports, public beaches and swimming facilities, water-treatment reservoirs, corporate business areas, golf courses, schools, college campuses, private lawns, athletic fields, amusement parks, cemeteries, hospitals, residential subdivisions, and along or between highways (U.S. Fish and Wildlife Service 2002).

Property damage usually involves landscaping and walkways, most commonly on golf courses, parks, and waterfront property. In parks and other open areas near water, large goose flocks create local problems with their droppings and feather litter. Surveys have found that while most landowners like seeing some geese on their property, eventually, increasing numbers of geese and the associated accumulation of goose droppings on lawns cause many landowners to view geese as a nuisance, which results in a reduction of both the aesthetic value and recreational use of these areas (U.S. Fish and Wildlife Service 2002).

Negative impacts on human health and safety occur in several ways. At airports, large numbers of geese can create a serious threat to aviation. Resident Canada geese have been involved in a large number of aircraft strikes, resulting in dangerous landing and take-off conditions, costly repairs, and loss of human life. One of the most recent (January 14, 2009) and widely publicized events was the crash of US Airways Flight 1549 which had to make an emergency landing in the Hudson River because resident Canada geese had been sucked into both engines during takeoff. Fortunately, there was no loss of human life as a result of this human-Canada goose encounter. Many airports have active resident Canada goose and other bird and animal control programs (U.S. Fish and Wildlife Service 2002).

Excessive goose droppings are a disease concern (e.g., human influenza from fecal *Escherichia coli* bacteria) for many people. Public beaches in several states have been closed by local health departments due to excessive fecal coliform levels that in some cases have been traced back to geese and other waterfowl. Additionally, during nesting and brood-rearing, aggressive geese have bitten and chased people and injuries have occurred due to people falling or being struck by wings (U.S. Fish and Wildlife Service 2002).

White-Tailed Deer and Lyme Disease

Urban white-tailed deer carry one of the stages in the Lyme disease life cycle which can be transmitted to humans (Figure 14.5). Lyme disease exists because of a complex set of interactions between oak trees, white-footed mice, white-tailed deer, gypsy moths, ticks, and a bacterium. Lyme disease is transmitted to humans by the bite of deer ticks (*Ixodes* sp.) carrying the bacterium *Borrelia burgdorferi*. For this disease to exist in an area at least three closely interrelated elements must be present: the Lyme disease bacteria, ticks that can transmit them, and mammals (such as mice and deer) to act as hosts for the ticks during various stages in their life cycle.

The life cycle of an *Ixodes* tick requires two years to complete. Adult ticks feed and mate on large animals, especially deer, in the fall and early spring. Female ticks then drop off the host animals to lay their eggs on the ground. By summer, eggs hatch into larvae. The larvae feed on mice and other small mammals and birds in the summer and early fall, then go dormant until the next spring when they molt into nymphs. Nymphs feed on small rodents and other small mammals and birds in the late spring and summer and molt into adults in the fall, completing the two-year life cycle.

Larvae and nymphs typically become infected with Lyme disease bacteria when they feed on infected small animals, particularly the white-footed mouse. The bacteria remain in the tick as it changes from larva to nymph or from nymph to adult. Infected nymphs and adult ticks then bite and transmit Lyme disease bacteria to other animals and humans in the course of their normal feeding behavior (http://www.cdc.gov/).

So how do gypsy moths (*Lymantria dispar*), mice (*Peromyscus* sp.), deer, and oak (*Acer* sp.) trees become players in the spread of Lyme disease to humans? Ostfeld et al. (1999) documented a web of biotic interrelationships involving episodic acorn production by oak trees, how white-footed mice and white-tailed deer respond to acorn abundance, and how the former two events affected

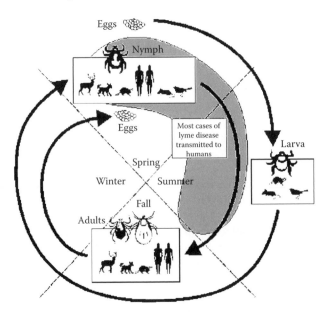

Figure 14.5 Life cycle of Lyme disease ticks. (Data from Centers for Disease Control, diagram by Linda Causey)

the population dynamics of tick parasites and defoliating insects (Figure 14.6). Gypsy moths are well known for the destructive effect they can have by defoliating entire forests. Bumper crops of acorns increase mouse populations, which are voracious consumers of the pupa stage of the gypsy moth life cycle. Predation by mice provides control of gypsy moth populations. When mice and deer are drawn to the acorn crop it concentrates these two prime tick vectors in a relatively small area, increasing the chance for transmission of Lyme disease.

In contrast, an unhealthy forest ravaged by gypsy moths does not produce bumper acorn crops. Mice and deer must go elsewhere for food or starve, which also increases the chance of Lyme disease transmission. To some extent, the choice is between a healthy forest (which people find aesthetically and economically pleasing) or healthy people (Wright 2004:407).

Lyme disease is the leading vector-borne infectious illness in the United States, with about 14,000 cases reported annually, though the disease is greatly under-reported. The number of reported cases of Lyme disease in the United States has been steadily increasing during the past fifteen years (e.g., 1993 to 2007, Figure 14.7) Ten states account for nearly three-quarters (72 percent) and the highest incidences of reported cases of Lyme disease (Table 14.1). Nationally, there is a much higher incidence of Lyme disease in the northeastern Atlantic seaboard states, and in Minnesota and Wisconsin than in the rest of the United States (Figure 14.8).

It seemed reasonable to assume that the incidence of Lyme disease would be a density-dependent value correlated with population densities of people and white-tailed deer in each state. In fact, the rank of the number of cases of Lyme disease in each state was highly correlated ($r = 0.74$, $P = 0.05$) with the rank of population density followed by the rank of white-tailed deer densities ($r = 0.33$, $P = 0.05$). As mentioned earlier in this chapter, the methods of estimating the number of deer in each state is less than a perfect census unlike the methods used to determine the actual number of people in each state. The deer census methods may be inaccurate enough to offset the expectation of a higher correlation between deer densities and incidence of Lyme disease in each state. There may also be a problem with under-reporting the actual cases of Lyme disease in each state. However, readers can take note of the occurrence of Lyme disease in their state of residence and be aware how it is transmitted and understand the role that white-tailed deer play in the disease life cycle.

In 2008, the Highway Loss Data Institute (HLDI), an affiliate of the Insurance Institute for Highway Safety (IIHS, http://www.iihs.org/), found that car collisions with deer and other animals (e.g., cattle, horses, dogs, and bear) spike in November and fatal crashes are up 50 percent since the year 2000. Fatal crashes have steadily increased since 1993 (Figure 14.9). One study found that white-tailed deer were involved in 1.5 million car collisions per year, killing between 100 and 200 motorists annually, and causing in excess of $2 billion in damage (Rondeau and Conrad 2003). It is important to point out that the majority of people killed were not killed by contact with the animal. The IHDI study found that 60 percent of the people killed while riding in vehicles were not using safety belts, and 65 percent of those killed on motorcycles were not wearing helmets.

FEASIBLE AND ACCEPTABLE MANAGEMENT STRATEGIES
FOR OVERABUNDANT RESIDENT CANADA GEESE

There is a point when the number of resident Canada geese and urban white-tailed deer has negative impacts on some aspect of urban residents' quality of life. In recent years, numbers of both species have undergone dramatic population growth and have increased to levels that are increasingly coming into conflict with people and causing personal and public property damage. In many communities, increasing numbers of locally breeding Canada geese and white-tailed deer have resulted in an example of the conflict and disagreement that can occur among various publics when wildlife becomes locally overabundant and exceeds the tolerance level of some people and communities. Overall, complaints related to personal and public property damage, agricultural damage,

Mast Year

Fall/Winter

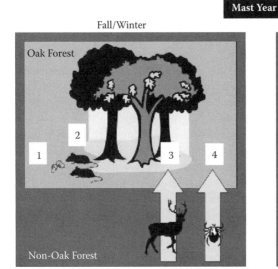

Spring/Summer

1) Many acorns are produced. 2) Mice in oak forest eat and store acorns and survive winter well. 3) Deer immigrate into oak forest and 4) Transport their burdens of adult ticks into oak forest. In spring/summer following a mast year. 5) Mouse density in oak forest is high. 6) Many mice emigrate from oak to non-oak forest carrying both larval and nymphal ticks. 7) Density of larval ticks in oak forest is high, mice are infested. 8) Gypsy moth survive poorly due to high mouse populations, resulting in low egg mass densities by the end of summer.

Following Year

Fall/Winter

Spring/Summer

Figure 14.6 Dynamic interrelationships between oak trees, white-footed mice, and white-tailed deer that promote the spread of Lyme disease during a mast year and the following year. (After Ostfeld et al. 1999, diagram rendered by Linda Causey)

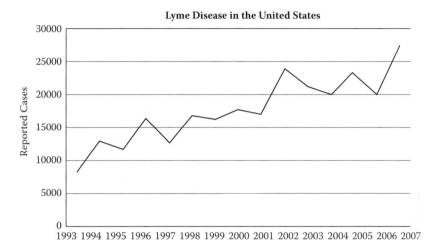

Figure 14.7   Reported cases of Lyme disease in the United States from 1993 to 2007. (Data from http://www. cdc.gov.)

Table 14.1   Top Ten U.S. States with Highest Annual Incidences of Lyme Disease in 2007

State	No. of Cases Reported	Annual Incidence[a]
Connecticut	3,058	87.3
Delaware	714	82.7
New Hampshire	896	68.1
Massachusetts	2,988	46.3
Maryland	2,576	45.8
Maine	529	40.2
New Jersey	3,134	36.1
Wisconsin	1,814	32.4
Pennsylvania	3,994	32.1
Vermont	138	22.2

Source: http://www.cdc.gov.
[a] Per 100,000 population.

public safety concerns, and other public conflicts have increased as resident Canada goose and white-tailed deer populations increased (Warren 1997; USFWS 2002). McShea, Underwood, and Rappole (1997) and Warren (1997) present comprehensive overviews of the problems associated with deer overabundance, and various programs that have been employed to manage these over-abundant deer herds (as cited by Warren 2000).

Interestingly, the first activity in resident Canada geese and urban white-tailed deer management is to assess urban residents' attitudes, activities, expectations, and knowledge concerning the species in question, that is, the human dimensions of wildlife management. Any urban residential neighborhood will have a diverse set of stakeholder groups that embrace compatible or contrary wildlife management agendas (see Case Study 14.1). The key to effective urban wildlife management is to bring all the various stakeholder groups together so everyone has an opportunity to voice their position on the management problem. The wildlife manager takes on the role of a facilitator to identify the various stakeholder groups and to open the lines of communication between them during a town hall meeting. The stakeholder groups need to be organized under some banner of recognition (e.g., Deer or Goose Action Committee) that identifies them as a group ready and willing

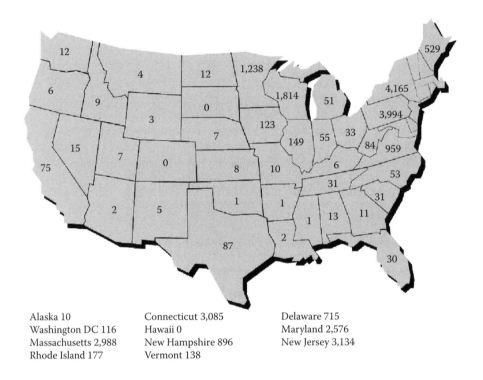

Alaska 10   Connecticut 3,085   Delaware 715
Washington DC 116   Hawaii 0   Maryland 2,576
Massachusetts 2,988   New Hampshire 896   New Jersey 3,134
Rhode Island 177   Vermont 138

Figure 14.8 Number of reported cases of Lyme disease in the United States in 2007. (Data from http://www. cdc.gov.)

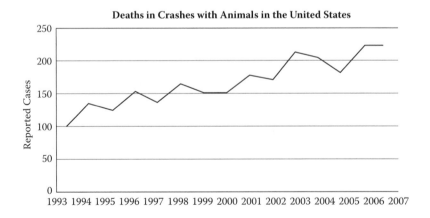

Figure 14.9 An increasing trend in the number of fatal vehicle-animal crashes in the United States from 1993 to 2007. (Data from http://www.iihs.org/.)

to address the urban wildlife management problem that affects them all. To learn more about the stakeholder process, the reader is referred to the premiere journal titled: *Human Dimensions of Wildlife* which is the only international journal that focuses on research concerning the association between people and wildlife management (http://www.tandf.co.uk/journals/titles/10871209.asp).

The town meeting is a relatively consistent phenomenon in terms of participants and the points of view expressed by them (Kirkpatrick and Turner 1997). Participants will include: (1) those who want to save the geese or deer, (2) those who object to hunting in general, (3) those who object to management of any kind, (4) those who hate the geese or deer because of the damage they have done or the potential that they can spread diseases, (5) city and county officials who want to be reelected, (6) at least one representative of the state game agency, (7) some shotgun hunters, (8) some bow hunters, (9) a representative from either an animal rights or animal-welfare organization, and (10) the media. Expect discussions between the various stakeholder groups to be spirited, contentious, and sometimes combative. Also expect the town hall meeting events to be extended in the form of a media feeding frenzy for several weeks, and that dialogue will supersede sound science in the end.

The second consideration in resident Canada geese and urban white-tailed deer management is the availability of background data that defines the nature of the goose or deer overabundance problem. The types of information that are needed include:

1. The actual numbers in the neighborhood (a difficult task at best).
2. Each species' rate of annual increase and the projected growth in population size with and without management intervention.
3. Problems they are causing with property, human health and safety, agriculture, and natural resources.
4. Legal ramifications, such as deed restrictions, city ordinances, state hunting laws, and federal migratory bird acts.
5. Factors (listed at the beginning of this chapter) that contribute to each species' overabundance.
6. Identified or suspected ecological, economic, sociological, and political consequences (Coffey and Johnston 1997).

The integrated pest management (IPM) strategy proposed by Coffey and Johnston (1997) was designed to address white-tailed deer overabundance, but applies equally to resident Canada geese overabundance. IPM is a problem-solving approach that begins by acquiring the information listed above, that is, management plans cannot be developed in an intellectual vacuum! This information allows the development of a management plan based on a specific problem. Problem definition facilitates the statement of clear, precise IPM goals and objectives. Goals identify the desired end points in the management process and objectives provide the blueprint for action in achieving the goals. The final component in IPM is the development and implementation of a monitoring program. The monitoring program provides the baseline data needed by managers to measure one action against another; progress, success, or failure; and to evaluate future management actions.

IPM addresses the criteria that define overabundance. Coffey and Johnston (1997) identified a set of consequences that managers can use to demonstrate that populations have or will exceed the environmental or cultural carrying capacity of the neighborhood. These consequences include (1) unacceptable damage; (2) changes in ecological processes; (3) destruction of native plants, agricultural crops, and landscapes; (4) unacceptable health or safety conditions; (5) displacement or loss of native species; or (6) prevention of the attainment of established goals and objectives.

The identification of the specific problem associated with resident Canada geese and urban white-tailed deer management guided by the goals and objectives leads to the selection of a specific management strategy. Urban residents, however, will opt for strategies that do not harm humans or animals (e.g., geese or white-tailed deer), do not cost a lot of money, do not impact negatively on their quality of life, but solves the problem permanently. Needless to say, there is no management strategy that can satisfy these criteria. So a management compromise needs to be designed that represents the best management practice (BMP).

The management strategies that can be applied to resident Canada geese and urban white-tailed deer are identical in terms of lethal and nonlethal alternatives (Coffey and Johnston 1997; Conover 2002). In general, management strategies should be focused on the reduction and control of geese and deer populations and to reduce related damage. There are four BMP options: (1) avoid the problem altogether, (2) get at the root cause of the problem, (3) attack the symptoms, or (4) do nothing.

Avoiding the Problem

There are ways to avoid the problems of wildlife overabundance in urban communities altogether, that is, a proactive management strategy. This involves urban development that integrates rather than excludes nature. It involves community designs that preserve the natural habitat, are sustainable in the use of natural resources, and reconnect human society with the natural world they live in. There are several rubrics that identify an alternative form of urban community development that includes sustainable communities, smart growth, design with nature, and conservation design for subdivisions.

Sustainable communities improve the quality of human life by living within the carrying capacity of supporting ecosystems. Sustainability can be achieved by not exceeding the carrying capacity of natural resources and ecosystems; reducing the impact that human activities have on the environment (e.g., rates at which renewable and nonrenewable resources are used); integrating long-term economic, social, and environmental goals; and preserving biological, cultural, and economic diversity (Farrell and Hart 1998). When serious consideration is given to sustainability issues in community development decisions, the conditions that promote wildlife overabundance are considered immediately, and factored into the decision-making process. For example, the issues of sustainability and overabundance present a paradox of terms. In sustainable communities, living within the carrying capacity of the ecosystem precludes allowing any species to exceed the carrying capacity.

In communities across the nation, there is a growing concern that current development patterns— dominated by what some call "sprawl"—are no longer in the long-term interest of our cities, existing suburbs, small towns, rural communities, or wilderness areas. Supporters of smart growth communities question the economic costs of abandoning the city infrastructure only to rebuild it further out. Factors promoting the smart growth movement are demographic shifts, a strong environmental ethic, increased fiscal concerns, and more nuanced views of growth. The result is both a new demand and a new opportunity for smart growth (http://www.smartgrowth.org/).

Design with nature and conservation design are similar concepts. Each promotes a form of urban development around an open space framework that includes meadows, fields, and woodlands that would otherwise be cleared, graded, and converted into households and streets. The open space that is conserved in this way can be laid out so that it will ultimately coalesce to create an interconnected network of protected lands (Arendt 1996; McHarg 1969). The "design with nature" and "conservation design" approaches to urban development integrate the existing habitat into the urban development process. The "business as usual approach" produces habitats that are fragmented, unnatural, and conducive to invasions of suburban adapter species such as resident Canada geese and white-tailed deer.

Getting at the Root Cause

Getting at the root causes of animal overabundance in urban areas requires an analysis of factors that promote and prevent the presence of geese and deer in a "typical" urban community. Many aspects of urban sprawl provide a landscape and resources that invite the presence of resident Canada geese and white-tailed deer. The most effective management strategy to get at the root causes of the overabundance of geese and deer are to "clean up the neighborhood," which, in general, means to alter the existing habitat so it is less inviting to geese and deer (Humane Society of the United States 1997). Examples of "cleaning up the neighborhood" would include:

1. Removing food sources altogether or change the landscape vegetation to plants that are less palatable to geese and deer.
2. Surrounding lakeshores with tall grass and/or dense hedges to restrict goose access to the water and residential lawns.
3. Removing accumulated nesting materials or nests that do not contain eggs.
4. Planting landscapes and agricultural crops that are not preferred deer food.

Attack the Symptoms

The most prevalent management strategies to reduce or control resident Canada geese and white-tailed deer overabundance in urban communities do little more than attack the symptoms, that is, a reactive management strategy. The cliché "winning the battle, but losing the war" is used appropriately in this case. Management strategies that attack the symptoms include those that (1) clean up the mess—feces and automobile or plane wrecks, (2) cull the flock or herd of some members, (3) move the problem elsewhere—translocation, (4) use high or electric fences, (5) employ behavioral modification—aversion techniques, (6) introduce pseudo-natural predators—dogs, and (7) use fertility control—egg addling or contraceptives (Smith et al. 1999; Conover 2002). These strategies are usually applied using the triage approach, that is, they are applied only in the most problematic areas. Further, management strategies that cause the least harm to the animal are the most expensive on a per animal basis.

More often than not, resident Canada geese or white-tailed deer become so overabundant in an urban community that the management strategies listed above are unsuccessful in reducing animal numbers or the damage they cause. Therefore, more aggressive techniques that significantly reduce the geese or deer populations may have to be used. For example, large numbers of geese can be herded into a pen when they are in molt (cannot fly), and taken to a local slaughtering house where the meat can be processed for the food for the needy program (Smith et al. 1999). Likewise, urban white-tailed deer can be harvested by sharpshooters (sometimes off-duty police), and the meat used similarly. During a three-year-long deer control program in a coastal Georgia residential community, Butfiloski et al. (1997) donated nearly twenty metric tons of edible venison to needy families and organizations.

Do Nothing

There are groups of stakeholders that advocate a "live and let live (or die)" or "let nature take its course" philosophy of management regardless of population size. These philosophies cannot or do not consider the resultant environmental impacts (habitat degradation) when a species exceeds the carrying capacity of its environment. The height of the browse line is an indicator of the degree to which overabundant deer populations are capable of deriving sufficient food from the urban habitat. Starvation is a slow and uncomfortable way to die. Emaciated animals (particularly young ones) are a pitiful sight. A do nothing approach to resident Canada geese and urban white-tailed deer management will probably bring little satisfaction to anyone as a credible approach to the problem.

SUMMARY

There is an abundant body of information on the issues pertaining to overabundant geese and deer in urban communities. This chapter attempted to reduce the complexities of the issues by highlighting the principal components involved in the management of these two species. Both geese and deer can be cast in the same light regarding their capabilities of exponential growth, environmental damage, and danger to human health and safety in urban environments. It seems that the only difference between them is feather or fur. Nevertheless, it is important for the urban resident to be exposed to the fundamental ecological, economic, cultural, and political issues that need to be considered with any species that exceeds the carrying capacity of its environment. Resident Canada geese and urban white-tailed deer are charismatic personifications of this problem which is why we chose to focus on these two species in this chapter.

CASE STUDY 14.1: A TALE OF TWO CITIES

The national assessment documented that white-tailed deer are a growing urban wildlife management problem in many metropolitan areas throughout the United States. As urban sprawl increases, the natural habitat required by many wildlife species disappears, but white-tailed deer are able to adapt to urban environments and related human lifestyles. When humans and wildlife live close to one another, an increase in human-wildlife conflicts occurs (Organ and Ellingwood 2000). Most human reactions to wildlife are based on previous knowledge and experience with the species involved, and what the animal was doing during the human-wildlife encounter. Many urban residents enjoy watching white-tailed deer browsing in parks and yards until the deer population becomes a nuisance by eating expensive plants, trees, or gardens (Stout and Knuth 1995).

Previous studies have evaluated the public's interests, attitudes, perceptions, and opinions about overabundant white-tailed deer populations, and acceptable management options (Curtis and Hauber 1997; Chase, Schusler, and Decker 2000; Fulton et al. 2004; Koval and Mertig 2004). This case study examined the degree to which urban residents' deer management preferences were grounded in an adequate knowledge base, an understanding and acceptance of available and appropriate management strategies, personal involvement in deer management, and selected demographic characteristics. The case study focused on community as a possible determinant of differences in response. It seemed reasonable to expect that residents from two different communities, each with an overabundant deer herd, would also have similar attitudes, activities, expectations, and knowledge concerning deer management. If this proved to be the case, then a tailored set of educational materials could be developed that provide the baseline information on how to address deer management in urban communities.

Lindsey and Adams (2006) used reference tracing, electronic databases, and Internet searches to review published literature on the transfer of wildlife information to the public. They found a high demand for wildlife information from the public and a need for effective information-transfer strategies by state and federal wildlife management agencies. All too often, citizens are not adequately informed about the efficacy of using different management techniques on the wildlife in their community. Communication between stakeholder groups (see Chapter 11) is a key factor in reducing conflict within the community. The primary goal of communication is to educate all interested citizens so they can make informed choices about where they stand on an issue; not just information that will persuade them in a certain direction. For example, citizens changed their attitudes about management via contraception after being given relevant information on the appropriateness, effectiveness, and humaneness of contraception (Lauber and Knuth 2004).

The Two Cities

The two Texas communities used in this case study were Lakeway, located northwest of Austin, Texas, and Hollywood Park, located in the north central area of San Antonio, Texas (Figure 14.10). San Antonio is the second and Austin is the fourth largest city in Texas (U.S. Census Bureau 2006). Both communities have similar geographic, ecological, and socioeconomic characteristics, which include:

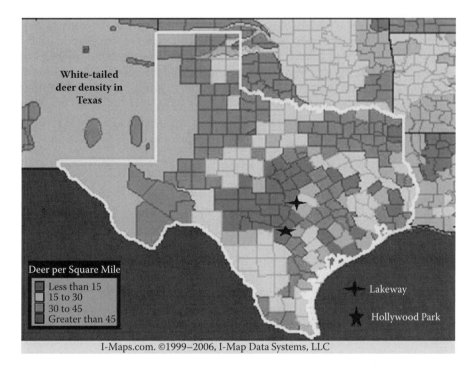

I-Maps.com. ©1999–2006, I-Map Data Systems, LLC

Figure 14.10   Locations of Hollywood Park and Lakeway, Texas. (Courtesy Quality Deer Management Association

1. Being located in counties that have more than forty-five deer per square mile.
2. High density (actual values not available) of urban white-tailed deer populations.
3. Large expanses of open green space and lot sizes.
4. The required habitat for increased white-tailed deer population growth, for example, lack of natural predators, highly controlled hunting, increased survival of offspring, and abundant food resources.
5. Residents who are predominantly white, with college-level education, income of $100,000 or above, who are older (average age = 57). Hollywood Park residents have lived longer in Texas (forty-one years) and at their current residence (fourteen years) than Lakeway residents, thirty years and eight years, respectively.

Since the early 1990s, the numbers of deer-vehicle collisions, defined browse lines, and human-deer conflicts have increased in both communities. Each community has a Deer Control Committee that decides, in cooperation with Texas Parks and Wildlife (TPWD), management strategies to reduce deer numbers. The success of the urban deer management plan in Lakeway and Hollywood Park is determined by changes in the number of deer-vehicle collisions, human-deer encounters, deer found injured or dead, and fence-related accidents. Strong differences of opinion between stakeholder groups in Hollywood Park has led to a more contentious urban deer management environment than is the case in Lakeway.

Both communities have implemented a city ordinance that prohibits feeding deer, restricts public access to deer control areas, and assigns penalties for damage or destruction of deer control equipment. Violators are charged with a Class C misdemeanor punishable by a fine of $1 to $500 if enforced by the local police department. These laws and regulations have created a conflict between residents and deer managers. For example, some residents have cut down expensive drop nets, walked their dogs around the trapping sites, and continue to feed the deer even though it is prohibited. Their behavior could be due to a lack of definitive information on, or experience with, the consequences of not controlling urban deer herds.

Deer managers in Lakeway and Hollywood Park thin urban deer herds using one of two options known as Trap/Tranquilize, Transport, and Transplant (TTT) and Trap, Transport, and Process (TTP). To trap deer, managers use drop nets with remote-controlled, silent release systems to prevent scaring the deer when the net is dropped. Automatic feeders are used to bait the trap sites. All trapping sessions are well documented, and a police escort is always present. Deer management activities typically occur from October through March. In order to use either of these two methods, the city must obtain a permit from TPWD. Lakeway is currently using the TTP method and Hollywood Park uses the TTT (tranquilizing) method to control the urban deer in their communities (Kevin Schwausch, personal communication).

Deer may be transported to and released at lease-hunting operations, but only after meeting some TPWD guidelines. For example, before deer are trapped or tranquilized, TPWD must: (1) inspect the release sites for sustainable habitat, (2) receive and approve a Wildlife Management Plan (WMP), and (3) receive Site Information Forms for every release location. The state requires that 10 percent of transported deer be tested for Chronic Wasting Disease (CWD) before they are relocated. CWD is a transmissible neurological disease that produces small lesions on the brains of white-tailed deer, mule deer, and Rocky Mountain elk. The symptoms of this disease include loss of body condition, behavioral abnormalities, and eventually death. It is not known exactly whether the disease is spread in feces, urine, or saliva. Human infection is a concern, but there have been no documented cases (Williams and Miller 2002). Deer that cannot be relocated are taken to the processing plant, and the meat is donated to charitable organizations.

The current deer management programs in Lakeway and Hollywood Park seem to be effectively controlling the deer populations. For the 2007 to 2008 trapping season, Lakeway removed 95 white-tailed deer, which included 47 bucks and 48 does (Robert Latham, personal communication). Hollywood Park removed a total of 117 deer, of which 114 were relocated to ranches, and 3 were sent to the processing plant and tested for CWD (Will Mangum, personal communication). In the past, Lakeway and Hollywood Park have removed an excess of 250 deer a year in order to sustain a healthy deer population and ecosystem within their cities.

The Citizen Survey

A citizen survey was conducted to determine what a sample of Lakeway and Hollywood Park residents know about the ecology and management of urban white-tailed deer populations with the intent to develop educational materials to inform citizens about the complexities of urban wildlife management. The specific objectives of the citizen survey were to:

1. Categorize the levels of public attitudes, activities, expectations, and knowledge about urban white-tailed deer population ecology and management techniques.
2. Identify the type of educational materials the citizens needed to provide current and accurate information about urban deer ecology and management.

Questionnaire Development

An Internet survey was designed using the guidelines of Schonlau, Fricker, and Elliott (2002) and Dillman (2007). Postcard invitations were sent to a random sample of Lakeway ($n = 616$) and Hollywood Park ($n = 704$) residents. Following a brief explanation of the purpose of the survey, residents were asked a series of questions about the deer herd in their community, which included (1) perceptions of numbers, (2) personal interactions, (3) personal economic losses, (4) acceptable and actual management methods and personal involvement, (5) management responsibility—who is in charge, (6) expected management results, (7) knowledge about urban deer, and (8) best methods of disseminating information and educational materials to urban residents. Response rates were 50 percent for Hollywood Park and 33 percent for Lakeway.

The Response

On average, residents of both communities reported seeing only eight deer per day and twenty-three per week, and nearly half said the number of deer was just about right compared to those who thought there were too many (28 percent), not enough (23 percent), or had no opinion (5 percent). Even though community respondents reported a total economic loss of between $20 and $30,000 due to deer-vehicle collisions and replacement of landscape shrubs, the average loss was $114 or less in both cases. A study by West and Parkhurst (2002) found that respondents who experienced severe damage and economic losses were more likely to consider and support dramatic reductions in deer herds. Since Hollywood Park and Lakeway respondents had minimal economic losses, it was expected that they would support little or no deer management in their communities. In fact, over half (51 percent) opposed or strongly opposed efforts to reduce the size of the deer herd in their communities compared to 40 percent who favored reduction programs.

For some questions, community of residence was a significant determinant of acceptable and actual management methods and personal involvement. For example, more Hollywood Park than Lakeway residents advocated the removal of deer, allowed access to their property to trap deer (no Lakeway respondents allowed this), protested deer control measures, and attended meetings regarding deer management in their community. Residents from both communities were polarized (nearly a 1:1 split) on whether to save or reduce the size of the deer herd. Yet more Hollywood Park than Lakeway resident disapproved of doing nothing at all, allowing nature to take its course, and any method of killing the deer (e.g., regulated hunting with firearms or archery by anyone). The preferred management methods were trap and transfer and fertility control by both communities. Residents also thought that trap and transfer was more expensive than fertility control methods. Rudolph, Porter, and Underwood (2000) concluded that immunocontraception was time consuming and costly (ranging from $802 to $1100 per treated female). The TTT and TTP deer management techniques used in Hollywood Park and Lakeway cost between $110 and $200 per deer.

There were few differences in community responses regarding what they considered to be evidence of a successful urban deer management program. For both communities the top five signs, in order, would be improved health (left undefined) of the deer herd, reduced number of deer-vehicle collisions, reduced loss of ornamental landscapes, lowered fertility rates in the deer herd, and survival of more deer per acre. Less than 14 percent felt that it was impossible to successfully manage urban deer herds.

The primary method of deer damage control used by the residents themselves was to use landscape plants that deer will not eat (75 percent). One-third or less used repellents, high fences, protected edible plants, or allowed deer traps on their property. Nearly one in five did nothing. Related to this was what residents knew about existing deer management activities in their communities. More Hollywood Park (80 percent) than Lakeway (51 percent) respondents knew about the TTT and TTP programs in their communities. Conversely, more Lakeway (26 percent) than Hollywood Park (14 percent) respondents were not aware of any urban deer management activities in their community. This difference might be explained by the longer tenure of Hollywood Park compared to Lakeway residents as discussed above.

When asked what type of stakeholder (a check all that apply question) they would represent at a "Deer Action" committee meeting, nearly half would represent someone who wanted to reduce the size of the deer herd, but 60 percent would represent someone who wanted to save the deer. This was consistent with a previous question that demonstrated how the residents of both communities had totally opposite opinions on whether to regulate the size of their urban deer herds at all.

Residents were asked to identify who makes the decisions about deer management in their community. The majority (84 to 86 percent) of Hollywood Park and Lakeway residents stated that the City Council decided how deer were going to be managed in their community (which is correct). The city mayor and county judge also play a role in the decision-making process. TPWD is not responsible for making decisions about how deer are managed in communities. However, many Hollywood Park (33 percent) and Lakeway (21 percent) residents believed the TPWD was involved. How the TPWD becomes involved in urban deer management is discussed later.

Residents were given a list of ten issues that might be considered when selecting a deer management option in their communities. Except for two cases, community was not a determinant of response. In rank order of most to least important, the issues were: (1) method is humane, (2) deer do not suffer, (3) health of deer, (4) allows deer to coexist with humans, (5) safety around people, (6) never fatal to deer, (7) cost to taxpayers, (8) similar to way nature would balance the herd, (9) easy to use, and (10) works quickly. The management message from residents of both cities seems to be that if anything has to be done to control the deer herd, the deer should not be aware of it. Strict adherence to the resident's responses to this question would confound any wildlife biologist's attempts to facilitate a needed, timely, and cost-effective method of regulating the size of the deer herds in these two communities. Herein lies the fundamental difference when managing wildlife in urban when compared to rural areas, and reinforces the critical need for training in the human dimensions of wildlife management (see Chapter 9) before applying the discipline in urban areas.

Residents were asked to identify how TPWD was involved in urban deer management. The primary role of TPWD is to provide technical guidance. First, a TPWD biologist will assess the size of deer herds and advise the city council on how to deal with their urban deer population. TPWD staff also visit communities to help form deer action committees, issue permits for the community to legally manage the deer herd, and enforce the game laws associated with urban deer management.

The majority of Hollywood Park and Lakeway respondents knew that the TPWD did not provide funding to implement deer management in their communities. Few Hollywood Park (6 percent) and Lakeway respondents (2 percent) knew that TPWD was not involved in managing deer in urban communities. About half of Lakeway and Hollywood Park respondents were unsure how TPWD was involved in urban deer management. The role of TPWD in urban deer management is explained in three informational papers/brochures: *Living with Overabundant White-Tailed Deer in Texas*, *Local Deer Control Methods*, and *Deer Management within Suburban Areas*, available on the TPWD web page (http://www.tpwd.state.tx.us/).

Educational Program Development

A series of sixteen questions was used to determine what Hollywood Park and Lakeway residents know about some aspects of urban deer ecology and management (see Table 14.2). Residents' score ranged from five to a perfect score of sixteen. The average score for residents of both communities was ten correct answers. Over half of the residents of both communities responded incorrectly to statements 4, 5, 9, 12, and 16 (Table 14.2). The majority were unaware of how urban deer feed, breed, and die (statements 4, 5, and 9), and the consequences and effectiveness of two management options (statements 12 and 16). Obviously, a sixteen-point examination does not provide all of the information required for developing the appropriate educational programs on urban deer ecology and management. However, what emerged from the examination was the need to develop educational programs that make the public aware of the need to: (1) control local deer populations, (2) inform citizens about why certain management alternatives are recommended, and (3) successfully implement long-term solutions to problems caused by urban deer (see Coluccy et. al. 2001 on resident Canada geese).

Today, Lakeway deals with minimal controversy from residents regarding urban deer management practices, which was not true in the 1990s when deer management first began. In comparison,

Table 14.2   Score Card on a Series of Statements Concerning Urban Deer Ecology and Management by 256 Residents of Hollywood Park and Lakeway, Texas, in 2008

Statements	Answer	% Correct
1. The size of urban habitats prevents deer herds from growing too large.	Fiction	60
2. Urban deer can carry diseases that affect humans.	Fact	59
3. Urban deer can destroy habitat used by other animals.	Fact	52
4. In order to survive, urban deer rely heavily on supplemental food sources.	Fact	36
5. Urban deer can begin to produce fawns when they are six months old.	Fact	17
6. Twins are a common result from urban deer reproduction.	Fact	52
7. More deer per acre means the deer will be physically larger in size.	Fiction	86
8. The most common cause of death in urban deer herds is predators.	Fiction	75
9. The least common cause of death in urban deer herds is disease and starvation.	Fiction	49
10. Deer are an endangered species in Texas.	Fiction	91
11. Deer live in urban areas because they have adapted to living near people.	Fact	55
12. People can help urban deer the most by letting nature take its course.	Fiction	41
13. Deer live in urban areas because human development has pushed them out of their natural habitat.	Fact	81
14. People have done more harm than good for urban deer.	Fact	60
15. People can help urban deer the most by managing them.	Fact	69
16. Fertility control techniques for managing urban deer are cost effective and easy to implement.	Fiction	10

Hollywood Park continues to struggle with protesting residents who want to coexist with the deer. Many urban residents do not understand that if deer managers suddenly stopped management activities, deer populations would soon exceed the biological carrying capacity. At this point, residents would experience more economic losses and increased deer-human encounters, and begin to perceive the deer as a nuisance animal. It is important for managers to understand that when dealing with the public, a variety of issues will arise from different stakeholder groups. These issues must be addressed before management programs are implemented in order to reduce the potential conflict between managers and residents. Public education on urban deer ecology and management is an extremely important management implication derived from the citizen survey.

Jessica Alderson and Clark E. Adams

CHAPTER ACTIVITY

Identify one community within your state or in another state that has either a goose or deer (white-tailed or mule) overabundance problem. Use the case study analysis outline below to investigate the historical, economic, ecological, and societal ramifications of the problem in one of these communities. An example of this type of analysis is given in Case Study 14.1.

Case Study Outline

I. Introduction
   a. Give a historical overview of the factors that led to goose or deer overabundance in the community.
   b. Provide a separate discussion of the political, cultural, economic, and ecological ramifications of the goose or deer overabundance in the community.
   c. Identify the main stakeholders that represented the public sector, private sector, academic institutions, and the public in the management plan.

II. Statement of the problem
    a. Briefly define the agendas of the various stakeholder groups.
    b. Identify how stakeholders' agendas may have conflicted, why they occurred, and how they were resolved.
III. Facilitating a resolution to the problem
    a. Explain how the problem of goose or deer overabundance in the community was resolved in terms of who became the responsible parties and the management plan.
    b. The final outcome: does the problem of goose or deer overabundance in the community need constant monitoring or has it been resolved with a single solution? Explain your response.

LITERATURE CITED

Arendt, R. 1996. *Conservation Design for Subdivisions: A Practical Guide to Creating Open Space Networks.* Washington, DC: Island Press.

Butfiloski, J.W., D.I. Hall, D.M. Hoffman, and D.L. Forster. 1997. White-tailed deer management in a coastal Georgia residential community. *Wildlife Society Bulletin* 25:491–495.

Chase, L.C., T.M. Schusler, and D.J. Decker. 2000. Innovations in stakeholder involvement: What's the next step? *Wildlife Society Bulletin* 28:208–217.

Coffey, M.A. and G. H. Johnston. 1997. A planning process for managing white-tailed deer in protected areas: Integrated pest management. *Wildlife Society Bulletin* 25:433–439.

Coluccy, J.M., R.D. Drobney, D.A. Graber, S.L. Sheriff, and D.J. Witter. 2001. Attitudes of central Missouri residents toward local giant Canada geese and management alternatives. *Wildlife Society Bulletin* 29:116–123.

Conover, M.R. 2002. *Resolving Human-Wildlife Conflicts: The Science of Wildlife Damage Management.* Boca Raton, FL: Lewis Publishers.

Curtis, P.D. and J.R. Hauber. 1997. Public involvement in deer management decisions: Consensus versus consent. *Wildlife Society Bulletin* 25:399–403.

Dillman, D.A. 2007. *Mail and Internet Surveys: A Tailored Design Method.* Hoboken, NJ: John Wiley & Sons.

Farrell, A. and M. Hart. 1998. What does sustainability really mean? The search for useful indicators. *Environment* 40(4–9):26–31.

Fulton, D.C., K. Skerl, E.M. Shank, and D.W. Lime. 2004. Beliefs and attitudes toward lethal management of deer in Cuyahoga Valley National Park. *Wildlife Society Bulletin* 32:1166–1176.

Humane Society of the United States. 1997. *Wild Neighbors: The Humane Approach to Living with Wildlife.* Golden, CO: Fulcrum Publishing.

Kirkpatrick, J.F. and J.W. Turner, Jr. 1997. Urban deer contraception: The seven stages of grief. *Wildlife Society Bulletin* 25:514–519.

Koval, M.H. and A.G. Mertig. 2004. Attitudes of the Michigan public and wildlife agency personnel toward lethal wildlife management. *Wildlife Society Bulletin* 32:232–243.

Lauber, T.B. and B.A. Knuth. 2004. Effects of information on attitudes toward suburban deer management. *Wildlife Society Bulletin* 32:322–331.

Leopold, A. 1976. *A Sand County Almanac.* New York: Ballantine Books.

Lindsey, K.J. and C.E. Adams. 2006. Public demand for information and assistance at the human-wildlife interface. *Human Dimensions of Wildlife* 11:267–283.

Marchinton, R.L. 1997. Obstacles to future deer management. *Quality Whitetails* 4(1):21–23.

McHarg, I.L. 1969. *Design with Nature.* New York: Wiley.

McShea, W.J., H.B. Underwood, and J.H. Rappole. 1997. *The Science of Overabundance: Deer Ecology and Population Management.* Washington, DC: Smithsonian Institution Press.

Organ, J.F. and M.R. Ellingwood. 2000. Wildlife stakeholder acceptance capacity for black bears, beavers, and other beasts in the east. *Human Dimensions of Wildlife* 5:63–75.

Ostfeld, R.S., F. Keesing, C.G. Jones, C.D. Canham, and G.M. Lovett. 1999. Integrative ecology and dynamics of species in oak forests. *Integrative Biology* 1:178–186.

Rondeau, D. and J.M. Conrad. 2003. Managing urban deer. *American Journal of Agricultural Economics* 85:266–281.

Rudolph, B.A., W.F. Porter, and H.B. Underwood. 2000. Evaluating immunocontraception for managing suburban white-tailed deer in Irondequoit, New York. *Journal of Wildlife Management* 64:463–473.

Schonlau, M., R.D. Fricker, and M.N. Elliott. 2002. *Conducting Research Surveys via E-Mail and the Web.* Santa Monica, CA: Rand Corporation.

Smith, A.E., S.R. Craven, and P.D. Curtis. 1999. Managing Canada Geese in Urban Environments. Publication 16. Ithaca, NY: Jack Berryman Institute and Cornell University Cooperative Extension.

Stout, R.J. and B.A. Knuth. 1995. Effects of a Suburban Deer Management Communication Program, with Emphasis on Attitudes and Opinions of Suburban Residents. Human Dimensions Resolution Unit Publication 95–1, Cornell University, Ithaca, NY.

U.S. Census Bureau. 2006. American Fact Finder. http://www.census.gov/ (retrieved October 1, 2007).

U.S. Fish and Wildlife Service. 2002. Draft Environmental Impact Statement: Resident Canada Goose Management. FWS/AMBS-DMBM/006380. Washington, DC: U.S. Fish and Wildlife Service.

Waller, D.M. and W.S. Alverson. 1997. The white-tailed deer: A keystone herbivore. *Wildlife Society Bulletin* 25(2):217–226.

Warren, R.J. 1997. The challenge of deer overabundance in the 21st century. *Wildlife Society Bulletin* 25(2):213–214.

Warren, R.J. 2000. Overview of fertility control in urban deer management. In Proceedings of the 2000 Annual Conference of the Society for Theriogenology, 237–246. Nashville, TN: Society of Theriogenology.

West, B.C. and J.A. Parkhurst. 2002. Interactions between deer damage, deer density, and stakeholder attitudes in Virginia. *Wildlife Society Bulletin* 30:139–147.

Williams, E.S. and M.W. Miller. 2002. Chronic wasting disease in deer and elk in North America. *Scientific and Technical Review* 21(2):305–316.

Wright, R.T. 2004. *Environmental Science: Toward a Sustainable Future*, 9th edition. Upper Saddle River, NJ: Pearson Education.

Index

"f" indicates material in figures. "t" indicates material in tables.